Ecologies of Community in Performance

Karen Savage · Dominic Symonds

Ecologies of Community in Performance

Rewilding the Arts

palgrave
macmillan

Karen Savage
University of Lincoln
Lincoln, UK

Dominic Symonds
University of Lincoln
Lincoln, UK

ISBN 978-3-031-94342-3 ISBN 978-3-031-94343-0 (eBook)
https://doi.org/10.1007/978-3-031-94343-0

Cover credit: Drawing inspired by a pop-up screening at Ndiébène Gandiol, Senegal, during Festival StLouis'DOCS 2023" (Courtesy Estrella Sendra)

This Palgrave Macmillan imprint is published by the registered company Springer Nature Switzerland AG
The registered company address is: Gewerbestrasse 11, 6330 Cham, Switzerland

Acknowledgements They say it takes a whole village to raise a child, and perhaps something similar can be said for bringing a research project to fruition. While it may be our names on the cover of this book, there have been immeasurable contributions from the support and inspiration of our communities at home, at work, and at play.

Most of all, we'd like to thank our family and friends: our parents Mick and Ann Savage, and Jenny and Graham Symonds, for their unequivocal love and support; our children Verité, Evi, and Max, who have patiently put up with our obsession with ecologies and communities for the past five years; our godfather the late Michael Beard, whose passion for Cornwall and St Ives crystallized the thoughts around chapter five; and our dear friends in different countries, especially Sam North and Sharon Liszewski, who have always been ready to engage in conversation, hear our ideas, and offer invaluable insights.

We've also leant on colleagues and friends from our various research circles, to whom we are greatly indebted: Stephanie Hemelryk Donald, Kaya Davies Hayon, and John Erni, whose organization of the "Justice, Arts, and Migration" network, the "There's No Place Like Home" project, and the "Included-Outs" symposium (Lincoln and Hong Kong, 2018–2020), was formative in the writing of chapter seven; colleagues and friends in the New York-based Music Theatre Forum, including Joanna Dee Das, Ryan Donovan, Liza Gennaro, Raymond Knapp, Jeff Magee, Doug Reside, David Savran, Jessica Sternfeld, Stacy Wolf, Tamsen Wolff, and Liz Wollman, whose generosity in responding to various chapter drafts has been invaluable and perceptive; delegates at the Climate Reality Leadership Project event (Rome, 2024), whose companionship assisted in the development of chapter six; friends and acquaintances such as Rach and Wes, Colin Hopkirk, and Joe Pitts, who may not even realize the contribution that even the smallest of conversations has made; and colleagues and students at the University of Lincoln, whose community dynamics we engage with every day.

We'd like to thank the anonymous peer reviewers of our early proposals and draft chapters, whose responses were hugely encouraging, and whose feedback has helped us to craft the structure of this book. Finally, we'd like to thank our editors, production colleagues, and publishers at Palgrave Macmillan and Springer—especially Eileen Srebernik and Uma Vinesh—who have had faith in our project from the very start, who have waited patiently while we delayed its delivery, and who have been so very helpful during the final stages of production.

Competing Interests The authors have no competing interests to declare that are relevant to the content of this manuscript.

Contents

About the Authors

Karen Savage is Professor of Interdisciplinary and Collaborative Arts, and Head of Arts, Culture, and Heritage at the University of Lincoln, UK. She is a Principal Fellow of the Higher Education Academy, and a Climate Reality Project Leader. Previous Monographs include *Postdigital Performances of Care: Technology and Pandemic* (2023), co-authored with Liam Jarvis, and *Economies of Collaboration in Performance: More than the Sum of the Parts* (2018), co-authored with Dominic Symonds. Other publications include 'TechNO-fixes?: Performances within Ecological Emergencies", *International Journal of Performance and Digital Media* (2022), and *Avatars, Activism and Postdigital Performance: Precarious Intermedial Identities* (2021), both with Liam Jarvis.

Dominic Symonds is Professor of Musical Theatre, and co-Director of Research and Knowledge Exchange at the University of Lincoln. He was formerly co-editor of *Studies in Musical Theatre*, and was co-founder of the international conference series *Song, Stage and Screen* in 2006, and the "Justice, Arts, and Migration" network in 2019. He has published books including *We'll Have Manhattan: The Early Work of Rodgers and Hart* (2015), *Broadway Rhythm: Imaging the City in Song* (2017), and *Economies of Collaboration in Performance: More than the Sum of the Parts* (2018), with Karen Savage.

List of Figures

CHAPTER 1

Plant a Seed

It's hard to think of a more endearing environmental message than the one wrapped up in Dr Seuss's colourful picture book, *The Lorax* (1971). It tells the parable of the now-aging Once-ler, who settled in the valley in which the story is set many years ago to start a business. When he arrives, he is enraptured by the valley's beauty: Brown Barbaloots frisking about in the shade of the trees, Humming Fish splashing in the river, Swomee Swans flying overhead, and most of all the "bright-colored tufts of the Truffula Trees" waving in the wind. Judging that the tufts will be the perfect material to manufacture his new multi-purpose "Thneed", he establishes a factory, chops down the trees, and industrializes the valley. But so successful is the Thneed, and so blind is the Once-ler to the devastating effects his factory has on the countryside and its creatures, that he deforests the whole region, polluting the environment, driving away the wildlife, and leaving the landscape a blighted mess.

There's the message. It's one of a bottom-up *trophic cascade*: take away the trees, and the entire ecosystem disappears.

Framed as a regretful anecdote told to a young boy by the old Onceler from his garret at the far end of town, the parable concludes with a final, hopeful plea. The Once-ler has saved a single Truffula seed, which he gives to the boy, asking him to regenerate the land: "Plant a new Truffula", he begs; "Treat it with care. Give it clean water. And feed it

K. Savage and D. Symonds, *Ecologies of Community in Performance*, https://doi.org/10.1007/978-3-031-94343-0_1

fresh air. Grow a forest. Protect it from axes that hack. Then the Lorax and all of his friends may come back".[1]

The Lorax is a funny little environmental activist, who pops out of the stump of the first tree the Once-ler fells. He announces himself as the guardian of the trees, repeatedly haranguing the Once-ler as the factories are built, pleading on behalf of the wildlife for their habitat to be preserved, and finally—once everything has been destroyed—taking his leave with "a very sad, sad backward glance" (This is neat; he grabs the seat of his onesie, lifts off, and propels himself through a tiny hole in the smog). The only reminder he leaves is a small stone shrine bearing the word "Unless", which the Once-ler explains to the young boy: "Unless someone like you cares a whole awful lot, nothing is going to get better. It's not".

Thus Dr Seuss turns the parable from a morality tale into a call for action, inciting his young readers—kids like us—to become environmentally aware. In the words of one of those now grown-up kids, "It captures the basics of why mismanagement and overexploitation of ecosystems are a bad idea in a way that children can understand".[2]

The Root of the Matter

We'll be turning to the planting of trees throughout this book, much in the same way that we focused on bees and pollination in our previous book, as a way of exploring *Economies of Collaboration*. Like our use of the bees in that analogy, this summoning of the trees is not just a gimmick—in fact, it's fundamental to the argument we are going to make throughout the book: that a great deal can be learned about the activities of (human) communities by recognizing comparable dynamics operating in natural ecosystems.

We're not alone in making this link, as will become evident; the literature on communities has blossomed in recent years, and hundreds of researchers and commentators have invoked various comparisons to ecology. The trend is more than just a vogue, though, since the working dynamics of natural systems and the way these are understood seem to be uncanny analogues for the ways in which human communities also work. As many previous commentators have remarked: we can expand the bounds of knowledge by applying the thinking of ecologists to the study of human communities.

Indeed, in some ways, the clarity of thinking that has developed in the field of ecology enables us to avoid some of the confused and confusing narratives about communities. Because the notion of community means different things in different circumstances and to different people, and the ways in which communities are viewed, treated, and described are often complex and contradictory. As such, when we turn to various case studies in our later chapters, we will introduce a variety of differing communities and differing perceptions of what "community" means: not only communities of belonging or communities working together, but also communities under pressure, those beset by problems, and some thrown together; not only communities that are consciously active, or those that announce themselves with slogans, but also communities that are represented by others—sometimes in support, sometimes in attack, and sometimes in artworks. There are many different communities, and there are many different ways in which the notion of community is "performed", to use a term from our home area of performance studies. Somehow, the concept of an ecology seems to handle those complexities and contradictions; and the added objectivity of observing the topic from outside of our home area—obliquely and, to an extent, playfully—enables us to cut through the confusion. In this book, we will ask whether the conceptualization of an ecology can indeed provide clarity, and perhaps some of the solutions, that we think it offers in exploring the idea of community.

But there's another reason why our study is an important contribution to the literature. This is because we will go further and deeper into the language and dynamics of ecological thinking than previous commentators about community have done, and in so doing, we will offer original and significant insights. That's a bold claim, though one we stand by. In exposing some of the existing shortcomings of what's often termed an "ecological perspective",[3] we hope to advance, mature, and improve the application of ecological thought to the way in which we understand communities.

Because of our background, we will be particularly interested in focusing on communities in the arts, though again, that is a complex terrain. Sometimes these will be communities represented in artworks themselves—as is the case with *The Lorax*, in which the community of Barbaloots and Swomee Swans is depicted by Seuss's colourful brush; sometimes they will be communities of artists and practitioners creating their work; sometimes audiences and consumers, brought together for an event; and sometimes, lay communities engaging with the arts as part of a leisure pursuit.

Of course, no single study can be exhaustive, so in all of our multi-faceted pursuits we can only venture so far. There will be many stones left unturned, many crops left unharvested, and many layers of the under-story left undisturbed in our sometimes whimsical interventions into these communities, their ecologies, and their practices. But what is at stake, we believe, is significant, in a global context in which communities, the environment, and the arts are all at threat. Even as we write, humanity has just emerged from the worst pandemic for 100 years; a US President impeached for inciting insurrection has been re-elected for a second term; globally destabilizing conflicts are taking place in both Europe and the Middle East; countless communities have been displaced by food shortages, climate disasters, and conflict; sea-levels rise; glaciers melt; rain forests succumb to deforestation and fire; carbon levels escalate. And the arts—our most eloquent means of commenting on all of these things—slips inexorably down the list of priorities for governments around the world.

That may seem like a statement from the perspective of the privileged: how can the arts claim the global humanitarian significance of climate change or migration?

As the conscience of the world dimly awakes to our impending climate disaster; as humanity limpingly acquiesces to its responsibilities towards international aid; and as global politics wrestles with the paradigm shift of populism, we will argue that a commitment to "rewilding" the arts is not just desirable, but fundamental. That's why we have called this chapter "Plant a Seed". We—global communities—are the guardians of those precious seeds, the seeds of art: we carry them in our hands, in our voices, in our movements, and in our reflections of the world. But we need to cultivate the soil in which to plant them, nurture the habitats in which they'll survive, and steward the ecosystems of the present in order for a sustainable future to arise. Our case studies will evidence how the first faltering footsteps of rewilding have been made by communities as they propagate their art.

Our methodological approach is simple: to apply the thinking from one terrain (ecology) onto the landscape of another (communities in the arts). We will do this first by setting out the theoretical ideas in which our study is embedded—about ecologies, about communities, and about performance; and then by focusing on several case studies, through which we will explore differing ecologies of community. Each of our case studies will consider a different facet of community from the perspective of a specific ecological concept.

We will explore communities emerging with reference to *ecological succession*; communities of activism through the prism of *charismatic species*; communities of preservation in terms of *seedbanks*; incarcerated communities with reference to *mycorrhizae*; and thrown-together communities with a focus on *disturbance*. Finally, we will pick up on environmental initiatives that have replanted landscapes, reintroduced species, and restored ecosystems to transform our world. We'll make a similar call to action in relation to culture, proposing a *rewilding of the arts*.

As these topics attest, our application of ecological thinking engages with some nuanced ideas and terminology, in order to reap as much of a harvest from our short study as we can. But before we engage with these ideas, we need first to interrogate the way in which previous applications of ecology to community thinking have, in our view, fallen short. Through a look at the conceptual ways in which ecological terminology has been taken for granted, we will expose the need for a more systematic approach to the application of these metaphors.

Community/Ecology

Not only is it significant to our study that ecological terminology has currency in relation to communities, but also that the idea of community is used in the field of ecology. To give just one example, the Quaking Aspen is one of many foundation species that "structure a community by creating locally stable conditions for other species".[4] The tree itself, its herbaceous feeders, the understory of plant life, and the insects and pollinators relying on its habitat comprise the community as ecologist R.J. Putman understands the term: "an interactive assemblage of species occurring together in a particular geographical area, a set of species whose ecological function and dynamics are in some way interdependent".[5]

Indeed, this term is not just used fleetingly. There is a whole area of study that is called "Community Ecology", and a great deal of literature that has explored the concept. Perhaps there's no surprise that there is some contiguity between ecological and anthropological conceptualizations of community; in fact, as Peder Anker points out, the original use of the term in ecology emerged from Danish botanist Johannes Warming's research into human communities.[6] Community ecology in its essence, then, is already rooted in an analogous relationship to our understanding of human communities.

But if that is the case, what is to be gained in us reverse engineering this analogy? After all, there are plenty of ways in which ecological communities differ from human communities, and there are significant ways in which the analogy becomes stretched. For a start, human communities do not consist of different species, as ecological communities do (plants, animals, birds, insects, fungi). More to the point, the relations between humans within communities are considerably more *socialized* than those between flora and fauna in communities, even if we take into account research that reveals social, economic, and even political relations in nature.[7] Simplistically speaking, the glue that binds together communities in nature is a product of natural and life-sustaining urges (growth and decay; feeding and defecation; procreation and predation); on the other hand, the "most characteristic manifestations of collective life" in human communities, if we are to take the lead as many do from the work of Émile Durkheim, are "religious beliefs, morality, and the precepts of law": man-made impositions.[8] Now, it may be the case that Durkheim's assertion needs modernizing, but the principle remains: human communities behave with a much more pronounced societal logic than communities in nature—a "rational" as opposed to "natural" will, as Nancy L. Rosenblum puts it.[9] How then can we make productive use of natural ecological communities to understand the rational dynamics of human networks?

Our contention is that the reverse engineering of this analogy enables us to bring a different way of thinking to the focus of our work (human communities in the arts), and it is precisely because our point of reference (communities in ecology) has an oblique perspective that such an alternative way of thinking can be enabled. The differences require us to make use of them metaphorically, and through metaphor, for insightful findings to emerge. This is a similar approach to that of our previous study, *Economies of Collaboration in Performance*, in which we turned to economics to explore the structures, processes, and flows of collaborative practice.

To an extent, this reverse engineering has been done before: the term "ecology" has frequently been used in discussions of (human) community, though in a rather different approach to our own: R.D. McKenzie used the term in relation to human settlements as long ago as 1924[10]; and Robert Ezra Park of the Chicago School famously appropriated the idea to understand urban social environments during the 1930s.[11] In expanding that conceptualization, the field of Human Ecology since the 1950s has sought to understand "inter-relationships between the human species and its environment",[12] or "our place in nature", as Frederick R. Steiner puts it.[13] This field has accumulated reams of scholarship, and a recognition of humanity not just sharing the resources of the planet with the natural world, but intrinsically being a part of one shared ecosystem.

This is a fairly literal application of the term "ecology", though, and the exploration of "community" within that application has therefore also remained somewhat literal. Steiner's dedicated discussion of "community", chapter three of his book *Human Ecology* (2002), for instance, focuses on how human communities become "imprinted on the land" in the design, planning, and construction of our cultural habitats (villages and towns). Thus, "theories of community that address our surroundings (our environments) as well as our interactions (our ecologies)" tend to tease out functionalities within that straightforward conceptualization of Human Ecology—"our place in nature".[14] That's not to undermine the importance of Human Ecology: most recently, its popular expression has emphasized humanity's often negative impact on the planet through the concept of the Anthropocene, and the need for a global response to environmental concerns—something which has nothing less than existential significance, and something which, of course, features in the pages of this book.

Our methodological application is slightly different, however; indeed, in some ways we are working against the underlying assumptions of Human Ecology in our methodology, though with the intentions of more fully understanding "our place in nature" as a result. For our purposes, it is appropriate to maintain something of a distance between one area (Community Ecology) and another (Human Community) in order to triangulate a useful understanding of a third (Human Ecology). It will be in our metaphorical use of concepts from Community Ecology, applied to Human Community, that we will enable a profitable discussion of the whole.

One of the most useful conceptual models of human interaction in analogically ecological terms (in other words, as distinct from the surrounding environmental habitat) comes from Mark Warren. He bases his discussion of democracy on an "ecology of associations"[15] (associations being subtly different from "communities", but nevertheless, bearing significant similarities).[16] For Warren, the term "ecology" is significant, since it alerts us to the complex dynamics that interweave in society between individuals, groups (or "associations"), and the state. In particular, as he and many theorists of community and association attest, "the liberal-democratic constitutional government depends upon social mores, political culture, and habits of collective action cultivated by horizontal relations of association".[17] Horizontal is Warren's formulation—and here he distinguishes from more usual "vertical" hierarchies—but the idea is one that might ecologically be viewed in terms of a multi-dimensional network, a web, or a system. These are all phrases used in contemporary parlance to review the complexity of human and collaborative communities, and also terms used in the sphere of Community Ecology to understand the complexity of biotic communities.

Warren is not the first to employ ecological vocabulary in a different area—in his case, social politics. The linguist Einer Haugen wrote about the "ecology of language" in the early 1970s,[18] while the idea of a "news ecology" has become common in journalism since about 2000.[19] This in turn led to Henry Jenkins coining the term "media ecosystem",[20] by which time the analogy was already used in business,[21] digital technologies,[22] social care,[23] and other domains. Before long, scholars were acknowledging that "there seems to be hardly any area that cannot be considered the object of an ecology and thus open to an ecological reformulation".[24]

An Ecological Approach to the Arts: Case Studies to Date

There has also been a use of the analogy in the arts, and specifically, relating to arts policy: Arts Council England published a report in 2014 titled *This England: How Arts Council England uses its investment to shape a national cultural ecology*[25]; its 2011 report *Supporting Growth in the Arts Economy* conceptualized the idea of an "arts ecology" throughout[26]; and its 2016 *Analysis of Theatre in England* refers to the "national arts ecology" and, in frequent mentions, a "theatre ecology".[27] In his foreword to the ACE 2011 report, Minister for Culture, Communications, and Creative Industries Ed Vaizey wrote "I firmly believe that the lines will continue to blur between what we consider the arts ecology and the creative economy".[28] Meanwhile, in 2015, the UK's Arts and Humanities Research Council picked up on the same idea in *The Ecology of Culture*, whose author John Holden argues that "culture is often discussed as an economy, but it is better to see it as an ecology, because this viewpoint offers a richer and more complete understanding of the subject".[29]

Holden's ecological approach to the UK cultural sector "concentrates on relationships and patterns within the overall system, showing how careers develop, ideas transfer, money flows, and product and content move, to and fro, around and between the funded, homemade and commercial subsectors".[30] In other words, this is a very similar analogical application of the terminology to that seen in Warren's work, this time viewing the cultural sector as a network, web, or system. The aim of Holden's approach was "to examine the many ways in which each part of the overall cultural world [in the UK] co-operates and/or competes with, influences, and affects the others".[31] And he extends the metaphor, calling up common vocabulary from ecology to give it currency in the landscape of the arts: he gestures towards a "flowering" of cultural events, the "deep roots" of cultural moments, a "regenerative life cycle" of creativity, and a "taxonomy of culture"[32]; he refers to ecological terms such as "regeneration, symbiosis, fragility, positive and negative feedback loops, and mutual dependence",[33] and later, "species depletion; invasive species; pathological growth [; …] the existence of top predators; balanced populations of species; increasing complexity and fecundity".[34]

Holden is not alone in making use of these concepts, indicating that such vocabulary has been well tested; and "the concept of ecologies is so important", writes Michael Carolan, because of its ability "to push against the idea of the binary opposition of "capitalist/alternative" economic forms".[35] The spatial conceptualization of an ecological network, web, or system, if nothing else, works as an alternative to what can otherwise be perceived in linear or hierarchical—and therefore somewhat delimiting—formulations. Even so, one significant observation in Holden's work is that there is a "lack of agreement about what 'the ecology of culture' means", and that, therefore, the "whole field of cultural ecology is poorly understood".[36] The analogy may be common-place, but its metaphorical usage is yet to be consistently applied. To be frank, Holden only goes so far in reconciling this confusion: for all his use of ecological vocabulary, and for all his interviews with industry players on "the ecology of culture", his report does not fully interrogate the terminology it is appropriating—and this is a consistent weakness throughout literature that relies on a metaphorical application of ecological terms.

In Holden—as elsewhere—ecological language is transferred to cultural systems rather simplistically, relying on generic understandings of terms like "deep roots", "regeneration", and, for that matter, "ecology". The vocabulary is ported into use in this new context without any measure of its specificity or of the ecological features and processes it describes. As such, in Holden's report and elsewhere, the ecological dynamics that are noted in culture remain somewhat nebulous: he refers to "relationships and patterns within the overall system"; "the complex interdependencies that shape the demand for and production of arts and cultural offerings"; "its wholeness and interconnectedness".[37] None of these are phrases that give more than a generalized impression of their systems.

In such reports, we'd suggest, a more developed use of ecological language could be useful in extending the application of ecological metaphors to culture. Were Holden to call on terms such as, for example, "biome", or even to pick up on one of his respondents' uses of the viticultural term "terroir", one of the observations he makes towards the end of the report about recognizing regional ecological diversity could develop greater substance.[38] This is a pervasive shortcoming, not just in Holden's ecological approach, but more widely: the metaphors are seldom mined for detail; some terms are over-invoked but under-explored; and other vocabulary that could provide nuance gets discarded in favour of a generalized application.

Indeed, Holden's ecological metaphor weakens as he turns either to more straightforward economic concepts like "supply-chain links", or alternative analogies with alchemy and cooking.[39] Although the ecological is something he returns to repeatedly, his most promising applications of the metaphor are not pursued: for example, he writes that "thinking of the natural environment as a machine to be managed has resulted in many examples of species depletion", though quite how we are to understand "species depletion" in the context of culture is left to the reader's interpretation.[40] Nor do sections in his report on "emergence", "evolution", and "convergence" attempt to apply the ecological terms in a meaningful way, weakening the potential of a promising approach in favour of somewhat trite observations: "Natural ecologies are fecund, messy, highly competitive places, and so is the contemporary cultural world", Holden writes.[41] In short, like many other examples, Holden's project is, in principle, very compelling; but his application of ecological thinking only scratches the surface.

We've dwelt on this example not to dismiss Holden's work, but to emphasize that there is further to go in putting the analogy to use.

Indeed, it is because we agree with the value of an ecological approach to the arts that we seek to develop the methodology further in this book. Of course, we are well aware of the potential hubris of our critique, which sets us up as hostages to fortune in applying our own analysis through metaphor: for all the virtues of a metaphorical approach, metaphor is after all only an imprecise analogue—and we have written about this elsewhere, acknowledging its inherent inexactitude.[42] Nevertheless, we feel that the ecological approach can only benefit from a more rigorous consideration of its ecological terms, and in exploring the area as such, we hope to further the debate and introduce more functional reasoning to the metaphorical analysis.

To be fair, the cross-disciplinary currency of the term "ecology" makes a casual scholarly use sufficient in certain cases, especially when particular qualities of the metaphor are defined as the justification for its use. Thus when Amos Hawley refers to Human Ecology as "the perspective of collective life as an adaptive process consisting of an interaction of environment, population, and organization",[43] or when Paul Makeham and his colleagues note that "Ecologies are complex systems" in order to discuss the arts in terms of "the overall performance of [its] system",[44] we can recognize the value of a surface-level engagement with the terminology. Other scholars have used ecological vocabulary in similar ways to remark about the ecosystems of arts funding, independent theatre, theatre training, disability theatre, and many other areas[45]; and the reverse has also been brought into play, with ecologists adopting metaphors from the arts with similar generalized thinking.[46] Studies in psychology,[47] mental health,[48] education,[49] and media studies[50] have lightly appropriated ecological terms like "interdependence", "adaptation", "succession", "evolution", "hybridization", and "extinction", to illuminate their respective fields, and the metaphors of "life-cycle", "consumption", and "waste", are common-place in all manner of applications. But how successfully do they work? Our aim is to move beyond these simple metaphorical usages to interrogate what such thinking and vocabulary can bring to an understanding of the arts.

Digging up the Grassroots

Before we do this—and by way of example—it is worth exploring one further instance of the underdeveloped thinking that is common in ecological approaches to culture: this time relating to the term "grass-roots".

It's a term that has been widely used in relation to localized public participation in politics, activism, and community matters for well over a century. Grassroots organizations are "locally-based, volunteer-run, nonprofit, common-interest groups", according to one study.[51] The term "implies local-level, small-scale efforts that are driven by groups who are directly impacted by the problems or conditions they seek to change", writes another.[52] They "are characterized by more democratic and hier-archical forms of governance and accountability [than other types of organisation], the predominance of volunteers as opposed to paid staff, and a local focus", suggests a third.[53]

Definitions are broadly in agreement, then, though in spite of the term clearly having its origins in an ecological metaphor, none of the authors quoted above addresses its provenance or adequately probes its nuances. The closest these interrogations of the term get is to argue that "con-ceptions of the grassroots gain their power in part by functioning as a deeply moralizing spatial metaphor"; in Sarah Dempsey's reading, this spatial metaphor relates to a local orientation rather than a proximity to the grass or the ground, and though she does mention "a sense of root-edness, or deep connection with a local place", (clarified as "deeply held cultural beliefs tied to place"), it is the "romanticization of the local scale" rather than the rootedness that is ultimately explored.[54]

Thus many of the associative connotations of this metaphor seem to have slipped entirely from use. Indeed, in the literature we have encoun-tered, one of the only reports to recognize the term's etymology comes from a social diversity website: "When an organization or community is looking to make a change from the ground up, they will start at the grass-roots", writes Susanne Ricee; "In simpler terms, grassroots is the place where things begin and grow".[55]

To be sure, despite being distanced from its ecological origins, the term when appropriated by organizations themselves is frequently linked with an environmental agenda. Such is the case with the community gardening initiatives explored by Marianne E. Krasny[56] and Keith Tidball[57] in invoking their concept of *civic ecology*. Their case studies reflect the wealth of "grassroots greening practices"[58] that take place at a local level across the globe; as they put it, "thousands of individual practices that restore nature and community in broken places around the world".[59] The connection at the heart of these initiatives between place and people is key, leading to a "renewal of the residents, the community, and the landscape".[60]

A similar study is that of John C. Morris and his colleagues, who dedicate their book *The Case for Grassroots Collaboration* (2013) to "the tens of thousands of citizens who give freely of their time and money to make their communities cleaner, better places".[61] The work of the grassroots communities on which Morris et al. focus is directed principally towards restoring the ecosystems of watercourses, an approach that directly and indirectly improves the wider environment, thus producing an invaluable ecological service. Not only that, but the proactive engagement of local participants within these activities fosters a powerful sense of community pride, which in turn generates attitudes of care, commitment, and wellbeing to both people and places.

As this pair of projects attests, there is therefore at least a casual relationship between notions of the grassroots and ecological concerns, and that can also be witnessed elsewhere. The Fair Earth Foundation's "Grassroots database", for instance, is a list of around 100 community action groups that identify with the "grassroots" terminology. Many of these are eco-focused organizations intent on protecting environments like trees, oceans, or forests; species like turtles, baboons, and orang utans; or regions like Snowdonia, the Amazon, or Indonesia. As we can see, many of the uses of "grassroots" are indeed explicitly linked to ecological activities. Nevertheless, there is by no means an exclusive link between grassroots organizations and environmentalism. Other groups in the Fair Earth Foundation's list target support for health conditions such as dementia, brain tumours, or motor neurone disease; some engage with social concerns relating to homelessness, child poverty, victims of rape, and refugees; and several offer therapeutic participation in cultural activities such as tennis, surfing, or theatre. Elsewhere, Kunreuther's analysis of grassroots associations lists a number of types of organization, none of which are related to environmentalism: "a chapter of Alcoholics Anonymous (AA), an after-school program organized by parents, a storefront church, a neighborhood soup kitchen, a choral society, a block association, and many, many more".[62] Among those many, many more are numerous culture- and arts-based groups whose practices operate in what Holden considers the "homemade sector".

It's interesting, though, that even in his ecological approach, Holden himself resists associating these "homemade" activities with the arguably more common term, "grassroots". Elsewhere, scholars and report authors do rely on that term. In their approach for example, researchers Angus McCabe, Jenny Philmore, and Hilary Ramsden pick up on the same justification for using an ecological perspective that we have encountered before: "to understand grassroots arts practices, and the sector, in ways that extend usefully beyond a definition of evaluation simply in economic terms".[63] For them, the ecological approach enables them to explore a sector of the arts landscape that is commonly undervalued precisely because it does not fit neatly into a conventional economic framework. "The lack of a dedicated literature on grassroots arts groups means their contributions remain hidden, and under-researched", writes McCabe; "our research shows that the time has come to take amateur arts seriously".[64] In this, "grassroots" is not precisely a synonym for "amateur", since it "denotes a more activist, broader remit, also including 'community' arts".[65] Thus the adoption of the term "grassroots" is intended in this particular research to identify people and organizations whose artistic work is economically undervalued but culturally important: "Grassroots arts play [...] a key role in sustaining and promoting community identity", concludes McCabe; "they may well have a crucial role in helping communities to celebrate and survive in hard, as well as good, times".[66] But they still don't interrogate that term.

As is clear, the focus of many scholars is to sidestep the (ecological) implications of their so-called ecological approach in order to embrace considerations of community and corresponding dynamics of social capital. They do this using an ecological approach—goes the claim—because of its conceptually networked rather than hierarchical configuration. Ironically, though, the word "grassroots" is used in this context precisely to encode a hierarchical dimension: grassroots activity is implicitly "lower" than other types of activity, and all manner of other metaphors ("on the ground"[67]; "below the radar"[68]; "at the coal face")[69] infer the same positioning.

But if the trend of invoking the term "grassroots" has been to ignore its wider implications, there is one striking exception: *Grassroots Postmodernism*, by Gustavo Esteva and Madhu Suri Prakesh. Here, a playful extension of the ecology metaphor is put to good use, and the authors borrow from the Indian environmentalist Vandana Shiva to summon up their imagery: "Freedom starts with a seed", she says; "Seeds are life itself and the embodiment of culture".[70] The aim of grassroots activities, as Esteva and Prakesh see them, is to use these seeds "to cultivate hope". They see the potential for this hope as deeply embedded not only in communities but also in their "cultural soil". "The people's experiences", they suggest, are "rooted in particular soils", "with roots that tap deep into the genius of their traditions". Yet there is a danger, brought about by the global pressures of the contemporary era: "Without the tending of human hands", they write, "top soil is being blown away, along with the stories, rituals and practices that make cultural soil". This, they claim, results in "two kinds of modern deserts: physical and cultural". In this respect, they evoke a need, calling for communities to "regenerate the soil of their cultures".[71]

This playful rendering of the analogy—invoking a message strikingly similar to that of *The Lorax*—is unusual, and therefore intensely refreshing; and because it is unusual, the potency of the ecological metaphor springs out. Not that theirs is a book that expressly uses an ecological approach: that term is never mentioned, and nor is the grassroots momentum that Esteva and Prakesh explore targeted towards any ecological or environmental agenda. The top soil they speak of (like the cultural desert) is simply a metaphor in itself—even if in this instance its use is well-informed by a knowledge of ecological processes (they're right about the top soil blowing away to leave only desert: "soil degradation is very likely one of the greatest pathways for biological extinction", warn the experts).[72] But we believe that the ecology metaphor can be taken further, bringing the power of metaphorical association to bear on that cultural desert itself in order to "rewild" it and restore the environment of the arts.

From Economies to Ecologies

The general thrust of all of these studies recognizes the dynamic explored in our previous book, *Economies of Collaboration*: the fact that working together (collaboration) seems to generate excess gain (more than the sum of the parts); the collaborative focus is on the work of communities, and as we have seen, those communities are often identified as "grassroots". Morris et al. make the claim that such grassroots initiatives seem to work far better than state- or regionally-directed top-down projects, and on a community scale, they see this invisible gain in terms that are echoed throughout the literature: as "social capital".[73] That term is one that we will find recurring throughout our present study into *Ecologies of Community*, though in this particular book, it is not the concept of social capital that we are addressing per se.

Instead, our focus is in exploring the widespread appropriation of this "ecological approach" or "ecological perspective" in both scholarly and popular discussions of community. Krasny and Tidball define this term succinctly in their accounts of civic ecology, in which they explicitly apply "an ecological perspective to understand how people, practices, communities, and the environment interact".[74]

We might not expect the grassroots organizations themselves to explore the implications of these ecological metaphors, even if their focus is on ecological matters; nevertheless, even in scholarly appraisals of the work of such cultural groups, much of which purports to examine grassroots practices in ecological terms, there is virtually no critical engagement with the terms "grassroots" and "ecology".

Turning to the metaphors of ecology in this book is our way of tracking the connections and interconnections of multiple feeders (and predators) to arts practices, and in this we see a continuation of the economic focus of our previous study. Where we previously explored economics terms such as "invisible cost", "externalities", and "Wikinomics", this study will reflect on ecological terms such as "ecological succession", "disturbance", and "rewilding". We will not be making the argument that the dynamics are somehow *different* in this study, or even that they are no longer examples of economics. Rather, we are applying a different prism to our focus, enabling different perspectives to emerge. By applying an ecological understanding of these concepts to cultural case studies, we will enrich the way in which more generalized applications of vocabulary like "ecosystem" and "grassroots" have been engaged in the context of culture.

Aboveground-Belowground

It's worth demonstrating an example of our methodology by considering the concept of *aboveground-belowground ecology* in relation to Holden's three-sphere model of the ecology of culture ("publicly funded culture, commercial culture and homemade culture").

Aboveground-belowground ecology has emerged since the late 1980s, and as Richard D. Bardgett and David A. Wardle suggest, it has become a focus of study thanks to two previously distinct areas coming together: community ecology and ecosystem ecology.[75] The basic premise is simple: "that understanding plant growth and performance requires integration of both aboveground and belowground biotic interactions".[76] In short, an ecosystem cannot be understood as hermetically sealed, but must be understood in terms of activities within a wider community network which have effects on the system: specifically, those which are underground and therefore ostensibly "hidden" from view. As one study puts it, "It is now well established that linkages between aboveground and belowground communities play a fundamental role in regulating the structure and function of terrestrial ecosystems".[77]

The most obvious way in which these dynamics operate in nature is through the mediation of plants, which act as conduits for nutrients in the soil to be passed on to grazers, who then defecate to enrich the soil and lead to the regrowth of more plants. But it doesn't stop there: that is, in effect, a closed loop ecosystem in itself. In more sophisticated interactions, the plant will make use of the activities of belowground communities to indirectly influence aboveground communities, and vice versa. "Plants mediate multiple interactions between belowground and aboveground heterotrophic communities that have no direct physical contact", write Ohgushi et al.[78] Ciska Veen and her colleagues give some examples: "herbivores and pathogens may prime or reduce parasitoid or pollinator attraction by modifying plant volatile emissions or shoot tissue quality", they write; "In turn, aboveground herbivores may stimulate root exudation, which can increase the activity of the soil community and thereby ecosystem processes such as carbon and nitrogen cycling".[79] This is, in essence, an example of the economics of unintended consequences, something we drew attention to in our previous study in terms of "invisible cost" and "invisible gain". And while we may accept it as a passive feature of systems, understanding its processes might enable us to factor the unintended into our intentions. Put simply: if we can measure the value of unintended consequences, we can harness their mechanics and activate their power to yield more than the sum of the parts. In ecological terms, the cost–benefit analysis can be recognized as more complex, since interactions "can be of antagonistic, synergistic, or neutral nature", and can affect different players to different degrees and extents.[80] Indeed, as

Judith L. Bronstein illustrates, the possibilities with any biotic interaction are many: both species can benefit from an interaction, as is the case with bees pollinating flowers (mutualism); one species can benefit while the other suffers (predation, parasitism); both can suffer, as they would in a fight (competition); one of the species can remain unaffected while the other benefits (a tree offering a nesting site for a bird—commensalism); one can be unaffected while the other suffers (a tree's shadow inhibiting the growth of a smaller plant—amensalism); or, in rare cases, both parties might remain unaffected by their interaction (neutralism).[81] The added complexities with aboveground-belowground interactions are that the species don't necessarily directly interact; that the influences of the different species might be imperceptible to study even if they are there—as is the case with the impact of some micro-organisms; and that current knowledge about belowground life is itself rather undeveloped ("We know more about the movement of celestial bodies than about the soil underfoot", Leonardo Da Vinci famously said).[82] Of course, at the heart of the system is a core player (the plant) that exists in both domains. Roots belowground directly feed the leaves, buds, flowers, and seeds that grow aboveground. This is not really what is of interest to aboveground-belowground ecology, and the same will be true when we apply this analogy to the ecology of culture. Even so, its clear framing and associative inferences resonate with other worldviews such as those of first nations, from which we can learn, and whose cultural engagement with an ecological perspective is profound: "My love of the standing people (trees) is not only in my admiration of their immense beauty that I can see, but in their foundation that I cannot", writes Jennifer Berneda Grenz.[83]

However, there are other ways we might explore the aboveground-belowground analogy. Given our critique of John Holden's model of the ecology of culture, it's appropriate to take his example of "three highly interactive spheres" ("publicly funded culture, commercial culture and homemade culture"),[84] and expand that to conceptualize some of the aboveground-belowground interactions.

There's an inevitable question we face in this modelling, as to where we position the ground level in our analogy: in the natural model, the distinction between above- and belowground is easily understood and materially defined, which is not the case in culture. For our purposes, we are going to assume that the aboveground analogue is the activity that remains manifest or visible: in other words, all three of Holden's spheres of interaction exist in the "aboveground" space, each involving arts activities, an infrastructure that directly supports those activities, and protagonists who participate. We are aware of the producers, practitioners, and audiences; venues, training schools, and media organizations; wider but still obvious influences such as funding bodies and marketing processes; and associated sectors such as hospitality, accommodation, and transport that support and enable cultural interplay. These are all "visible" elements of the cultural scene, and therefore known knowns—hermetically definable expansions of discrete biomes whose dynamics have been explored extensively.

What lies "belowground" is more subtle, and (as in nature) perhaps more difficult to discern—and it's not even as simple as merely detecting an as-it-were subterranean level of activity. Significantly, the study of aboveground-belowground ecology recognizes that its interactions do not take place purely within a spatial dynamic (between above- and belowground, and/or vice versa), but that they also involve multiple other spatial and temporal scales: processes that operate locally, further afield, and at a distance; processes that are influenced by migratory movements, and interspecies interaction; and processes that operate over various spans of time including those driven by circadian rhythms, seasonal fluctuation, biotic life-cycles, or long-term events such as evolution and climate change. The aboveground-belowground dynamic, in short, is a multi-dimensional spatiotemporal ecological system in itself.

One of the most compelling appeals for this ecological feature to be taken seriously is to be found not in a dense scientific study, but in Isabella Tree's popular bestseller, *Wilding*.[85] This is her account of the way she and her husband have converted their farm in the southeast of England into a rewilding project. Her experiences have been transformational, and her learning about ecological processes has been profound. More than anything, her project has revealed to her just how substantially our environment is affected by aboveground-belowground relations, in this case, driven by a deterioration of soil.

"Over the years modern farming has reduced soil to [...] 'dirt'", writes Tree; "a sterile medium in which plants struggle to grow without artificial fertilizers".[86] Dependency on artificial fertilizers and the agricultural use of heavy machinery has not only caused considerable environmental damage, but has also reduced nutrients in the soil to such an extent that the food we eat has been depleted of some nutrients by up to 90%

compared to pre-war levels.[87] In other words, because the (belowground) soil has become degraded, the health of the human population (aboveground) is affected. "It is a self-perpetuating cycle of destruction and chemical dependence", she laments; "Without soil organisms and soil structure to retain them water and nutrients leach away, and the soil compacts and becomes prone to erosion".[88]

One solution to this devastation—aside from reducing the use of chemical fertilizers and intensive farming techniques—would be to re-introduce earthworms to the agricultural land. Earthworms, "aerate the soil", "facilitate the movement and storage of water", and through the growth of bacteria and fungi they create, they "mineralize nutrients in the soil, breaking down soluble and insoluble organic matter into inorganic forms that plants can use". Indeed, she writes, the activity of worms "can change the very nature of the soil"; "earthworm presence increases crop yields by 25 per cent on average, and above-ground biomass by 23 per cent".[89]

Such a belowground remedy would not only transform the fertility of the soil and the nutritional value of the crops produced—with the aboveground effect of transforming the health of the population (whether or not they ever set foot on the soil)—but it would also lead to a systemic impact on the planet. "Improving the structure of our soils and returning unproductive agricultural land to permanent pasture could be a crucial weapon in the battle against rising levels of CO_2", Tree appeals, citing the astounding claims of one environmental organization: "if organic matter in the world's farmed soils was increased by as little as 1.6 per cent, the problem of climate change would be solved". Her conclusion is forceful: "The great concerns of our time—climate change, natural resources, food production, water control and conservation, and human health—all boil down to the conditions of the soil".[90]

Like all profound statements, there is a certain obviousness to this insight, and to our twenty-first-century eco-sensitive mindset, the observation is hardly startling (though the fact that so little is being done about it may be). But it's extraordinary that we often remain rather blinded to "the great concerns of our time". "This is the problem", remarks one of Tree's consultants; "We never think of what's going on below ground".[91] And if this is true of our attitude towards the environment, it is equally the case with our approach to culture, even when it is viewed in terms of ecology. As such, it is instructive to take on board some of the learning from aboveground-belowground ecology in considering the ecology of culture. And it is appropriate for our purposes to interrogate further the AHRC report we've already been critiquing, in order to assess its limitations and to learn how we might reconceptualize the cultural terrain.

Remarkably, considering his championing of an ecological approach, Holden's discussion of the cultural sector avoids the use of the term "grassroots" almost entirely. There is one reference to the term in his 2004 article "Capturing Cultural Value",[92] though it is not used at all in *The Ecology of Culture*. Instead, Holden's analysis refers to "the homemade sector", which as he explains, "extends from traditional 'amateur' and voluntary activities such as am-dram and community choirs through to the uploading of self-produced music and images onto websites".[93] It's seen as a conduit for generating both content and talent for further "up" the tree. "Amateur or youth company settings provide a route into both the funded and commercial spheres", he writes. In fact, here Holden sidesteps into yet another different metaphor: one of a life-cycle. In this analogy, "the 'birth' part of culture lies in the creative process", he reports, noting that "this is still the most fragile part of the ecosystem"; "too little money goes into the early stages of development". Seen at this level, the problem can seem exclusively driven by money and opportunity: "Austerity is hitting the amateur world", he writes; "Some homemade culture only survives because of access to cheap public space or free internet platforms"; and, in a related observation, "'the privileged few are tightening their grip on the arts'".[94] To be sure, these are all pressing concerns, but perhaps from an aboveground-belowground perspective, the problem and the possibilities for addressing it might be revealed as slightly different. In fact, this early phase in providing content for the cultural sector is far from its "birth". Still, this allows us to refer back to the problem noted by Isabella Tree's consultant: "We never think of what's going on below ground".[95] But what exactly is the belowground in this context? What is the economic aboveground-belowground interplay? Or to put it in Holden's terms, how are the skills, interest, and access of the homemade sector "made" in the first place?

In fact, with a little thought and self-reflection we can probably all offer an initial answer: they're made in the day-to-day accumulation of knowledge, interest, and support that comes from growing up; they're made through interactions with "family, friends and fools"[96]; and to a large extent, they're made invisibly—outside the formal structures of education, government policy, or institutional recognition. We see this vividly in bringing up our own kids. One of our daughters, for example, is an aspiring singer. She's still at school, studying arts-based subjects for her post-16 qualifications. There are several options for her once she's completed these exams, including vocational training to nurture her talent. That would be a conventional route into the sort of "commercial sector" Holden identifies. In the meantime, though, Evi has been generating interest from music producers thanks to her "homemade" videos on social media sites, which she has been uploading now for several years. She's got a "home-grown" skill for editing together these short films, allowing her to showcase her singing, and accumulating impressive numbers of followers on sites like Instagram. And it's turning into a small-scale enterprise as various clothing or make-up companies approach her to promote their brands, knowing that they have a captive audience of youngsters following Evi's posts. But Evi's acumen surely hasn't grown out of thin air: no doubt some of her know-how has been picked up through formal channels at school, where one of her subjects is Music Technology; no doubt it's also helped that she's been exposed to arts-making practices in the family home ever since she was born, since her parents work in similar industries, championing a creative mindset in all of their kids. And these sorts of things are the conditions that have prepared the soil, so to speak, well in advance of her work in the home-made sector taking root. From a very early age, we encouraged Evi to take private violin lessons, and though she didn't keep these up, they introduced her to the basics of music, and to the possibilities of learning to play an instrument. Since then, she's taught herself how to play guitar and piano. More substantially, she's attended a weekend performing arts club since she was 10, and before that, a gymnastics club. Week in, week out, she's worked with dozens of other kids on basic song and dance routines, picking up social as well as performance skills; sharing aspirations in a community of peers; and learning the thrill of performance in end of term shows which the club stages at one of the local theatres. All of these activities—the violin lessons, the gymnastics, and the performing arts club—have taken place outside of her formal education. They've been facilitated by professionals, but in the context of a "grassroots" set-up. Just as importantly, her ability to benefit from this formative training has been made possible because other people have invested their energies, their time, and their money: the classes aren't cheap; there are branded hoodies and dance-gear to buy every year; not to mention tickets for the shows themselves; and above all, every Saturday, it's an hour-long round trip to take Evi and her best friend to the performing arts club at the other side of town. It makes sense to car-share, so we do—sometimes it's us, sometimes other parents, sometimes grandparents. It's no big deal, but there's still a cost in time, a cost in fuel, and a cost in emotional and psychological support. These may seem like mundane contributions to Evi's possible future success—but such elements are fundamental to the often-unrecognized infrastructure that supports the arts, not just in our privileged, middle-class, middle-England community, but in countless other communities around the world. "In addition to covering the cost of instruments and lessons, parental involvement/support can take the form of shared musical experiences, home supervision, support for and assistance in children's concerts, provision of materials, and communications with instrumental teachers", reports another author, writing about Hong Kong.[97] It would be easy to extend this list: parents invest time ferrying children to and fro; consoling them when they flunk an audition; balancing their hopes and dreams with the responsibilities of homework. And it's probable that most arts professionals the world over will have been supported by some sort of unrecognized economic contribution of this kind.[98]

In other words, the seeds needed to even "birth" activity in Holden's terms (in the homemade just as much as the publicly-funded or commercial sectors) need to have been planted in fertile soil, given complex nutrients to take root, and then carefully tended, protected and supported. All of these are activities that take place, so to speak, "belowground". Such activity is fundamental to the aboveground presence of the industry, and if we are truly to reflect an "ecology of culture", we need to register such complexities.

Faced with the onslaught of multiple threats to the sustainability of discrete natural communities—global challenges such as climate change, land use intensification, biological invasion, and the loss of biodiversity—"an integrative above- and belowground perspective is critical", note scientists trying to roll back the often man-made damage that has been done to our planet.[99] Their typical argument is that a return to trusting natural systems—a "rewilding" of our resources—is not only crucial for salvaging sustainability, but also beneficial economically and politically if we are to rescue the communities of the world. And the same might be said for the arts.

Holden's acknowledgement of this "belowground" infrastructure is minimal at best. His starting point for "homemade" culture—"traditional 'amateur' and voluntary activities such as am-dram and community choirs through to the uploading of self-produced music and images onto websites"[100]—for instance, assumes a certain degree of aptitude, interest, and access among participants and communities, without accounting anywhere in his "Ecology of Culture" model for the development or provision of these basic needs. Without that extensive landscape of amateur investment, out of which isolated individuals emerge having gained the requisite skillset to subsequently professionalize the transactional economics of their labour, the professional marketplace itself—indeed, the professional market—would be fundamentally depleted to the extent of being non-operational.

Ecologies of Community in Performance

Our aim in this book is to explore the dynamics of community through the prism of an ecological perspective, focusing on the arts.

Following this introduction, we spend three chapters outlining the three principal concepts in our title: ecology (Chapter 2), community (Chapter 3), and performance (Chapter 4). These are all extensive areas in their own right, so our focus sticks to the ways we see them intersecting and interrelating. We'll expand on a basic understanding of ecology to contextualize some of the theories to which we later turn in the book; then we will consider some of the dominant thoughts about community that inform our subsequent case study chapters; finally, we'll explain our perspective on performance, noting a wealth of existing literature associating performance with ecology. As we've already hinted, though, we feel there is much further to go in exploring that association—and it will be in the second section of the book that we take on that challenge.

The second section turns to case studies of artistic practice that enable us to synthesize our theoretical triangulation of ecologies, communities, and performance. Each chapter in this section will focus on a particular term from ecology, honing in on case studies from fine art, poetry, theatre, video making, and music to clarify our thoughts.

Our first case study chapter will focus on the artistic community of St Ives in Cornwall. We'll be considering the concept of *ecological succession*, the phenomenon whereby an ecosystem changes in identity over time. This particular ecosystem changed from a fishing and mining community to a celebrated artist colony and then an international tourist destination. Each of these successional phases has been characterized by a different *population-community* make-up and different *process-functional* dynamics, while the passage of each succession has been marked by particular *mechanisms*, *stages*, and *trajectories*. We'll take a look at the artist as a *keystone species* in the town, picking up on a well-known ecological concept to see how it bears fruit in our critical analogy. We'll then consider the tourist as an *invasive species*, and the town of St Ives itself as an isolated or *island ecosystem*, vulnerable to change or introductions, and therefore prone to develop into an *imbalanced biome*. In particular, we'll note the unsettling impact that Tate St Ives had on the town, in both magnifying its international artistic status, and tipping the balance of its ecosystem. We'll track the fortunes of the town in subsequent years as it began to suffer a decline in its ecosystem and a collapse of its community infrastructure. It's only been through the development of robust *ecosystem stewardship* in the last two decades that the town has seen a change and the complexities of its *ecosystem health* have returned.

Chapter 6 will focus on the ecological concept of the *charismatic* or *flagship species*, noting that our fascination with certain plants or animals can serve as an important way for environmental issues to gain profile. At the same time, we will acknowledge that high-profile environmentalists trade on their celebrity status to become what we'll identify as *charismatic leaders*. We'll focus in particular on Greta Thunberg, the young Swedish campaigner who dominated environmental headlines between 2018 and 2023. We'll look at how she built a significant community of social media followers without appearing to claim privileged status; a performance of leadership that is unassuming, and which stages horizontal relations of environmental efficacy. Yet while her profile brought attention to issues of global warming, and her visibility helped other environmental campaigners to take a stand, these lesser-known figures nevertheless became marginalized, and their issues therefore were often precluded from the debate. Here, we'll consider the work of one such activist, Vanessa Nakate from Uganda, recognizing that her messaging brings slightly different nuances and impulses to the climate change arena. Promoting a message of girls' education and family planning, Nakate shows us that we need to look beyond the most highly-profiled in order to fully grasp the complexities of global environmentalism. In terms of applying this ecological concern to the landscape of the arts, we'll turn to UK arts funding models, and the UK's cultural fascination with Shakespeare, to interrogate the upsides and downsides of charismatic emphasis. We'll consider what we've learned from Vanessa Nakate and those activists kept in Greta's shadow when it comes to taking an *ecosystem* rather than a *single species approach* to the conservation of the arts.

Chapter 7 takes the ecological concept of the *seedbank* and explores its manifestations. We look at two examples of environmentally-aware collections, and we use these to acknowledge the role of archiving as a conservation approach. The first example is the Champion Tree Project run by American environmentalist David Milarch. This is devoted to cloning examples of the most majestic tree species, and preserving them in newly-planted nurseries to ensure their survival. The second example is the artist Katie Paterson's *Future Library*, a longitudinal project inviting a celebrated author each year to donate an unread manuscript to her archive. Gradually the archive expands, and alongside it grows a newly-planted woodland which will provide the paper on which the manuscripts are to be printed after their century of hibernation. Our focus on these archives is by way of recognizing the ecological premise of the *seedbank*, first as a way in which trees and plants ensure genetic continuity; second as a way in which national governments can keep a stock of vital crops for sustainability; and third as a way in which local communities participate in seed-sharing initiatives for food security. Here, the dynamics of community, ecology, and performativity coalesce in our reading of Milarch, Paterson, the seedbanks, and the archives.

Some communities do not have the opportunities to even think about food security, as the topic of Chapter 8 will reveal. We'll be considering migrant and refugee populations housed in detention centres, where any sense of community is fractured and in transition. Dislocated from their homelands and from communities of origin, these individuals face risk, danger, and uncertainty, even if a new community emerges through their shared experience. Our case study will be the Kurdish-Iranian poet Behrouz Boochani, whose book *No Friend But the Mountains* (2018) documents his experiences while detained at Manus Regional Processing Centre. As an expression of incarceration, this award-winning publication—along with his documentary *Chauka, Please Tell us the Time* (2016)—reveals deeply personal insights about the bonding of a community on Manus Island. We discover that his experiences of migration and detention reveal potent dynamics of community, solidarity, and estrangement. One of the things that makes Boochani's book so significant is his writing process. In this chapter, we compare his communicative process with the ecological operations of *mycorrhizae*. We will learn how these fungal growths enable networks of communication to form between trees, and we will use this as a way of understanding how Boochani managed to communicate with the outside world. Prohibited from communicating beyond the fence, he smuggled his writing out in fragments to an accomplice using WhatsApp. We compare his communications using phone technology with the mycorrhizal communications of the ecosystem's trees, and we find in his writing a symbol of hope conveyed to other communities outside the camp.

The experiences on Manus Island and in other global centres of detention demonstrate how horrifically people, government, and organizations can treat those who are unhomed or seeking sanctuary. But in Chapter 9 we encounter a manifestation of community that shows quite the opposite, as residents of the tiny town of Gander in Newfoundland rallied to support refugees stranded during the terror of 9/11. As American airspace was closed, planes from all over the world were grounded for the best part of a week. Some headed for the little-used Gander Airport in Newfoundland, whose local community was faced with accommodating and hosting over 6000 passengers and almost 500 crew. It's a story that has become well-known, thanks to its unlikely adaptation into an award-winning Broadway musical, *Come From Away*. Recognizing this event in ecological terms as the equivalent of a ***disturbance***, we will consider how communities on the ground and among the passengers faced an unprecedented ordeal. Through the lens of disturbance, we will see not only how the communities responded, but also how their sense of togetherness and belonging was emboldened, strengthened, and ultimately, rewarded. Not only does the story celebrate the power of community, but the show exudes the spirit of community in its very presentation. It also bears comparison to the concept of ***ecological memory*** in terms of how an ecosystem finds its resilience through vestiges that it recalls from the past. *Come From Away* is a heartwarming riposte to the devastation of 9/11, which brings our book towards its conclusion with a powerful demonstration of what community and art can achieve.

And if that is the case—if art can speak when words seem insufficient (Chapter 9, *Come From Away*); if it can reach out past impediments that are made to detain (Chapter 8, Boochani's *No Friend But The Mountains*); if it can witness the preservation of our communities into the future (Chapter 7, *Future Library*); if it can galvanize a spirit of community across countries and borders (Chapter 6, Thunberg's *No One is too Small to Make a Difference*); and if it can help to rebuild ecosystems that are in decline (Chapter 5, the artist colony of St Ives)—we are convinced that our capacity to express ourselves through the arts is perhaps the most valuable social capital we have. Yet the arts are themselves imperilled as world governments persist with their neoconservative agendas that leave them devalued, denuded, and overlooked: "a cultural desert", to pick up once more on the phrase of Esteva and Prakesh. Again, we are struck by a parallel with environmentalism, so in our conclusion we turn once more to ecology for our message of hope.

We end by echoing the recent call by many ecologists and commentators to save our ecosystems, eloquently and compellingly stated in Isabella Tree's book, *Wilding* (2019).[101] We'll map some of the projects and initiatives that have been adopted by conservationists in order to rewild the environment, and we'll suggest that a comparable rewilding of the arts is desperately in need. We've seen what the arts can do, just as we've seen what nature is capable of; and yet we've also seen how both the arts and nature are taken for granted, exploited, and abused. And in proposing this solution, our call to action echoes the plaintive appeal of the redemptive Once-ler with whose story we began:

> Plant a new Truffula. Treat it with care. Give it clean water. And feed it fresh air. Grow a forest. Protect it from axes that hack. Then the Lorax and all of his friends may come back.

Notes

1. Dr Seuss (1971), *The Lorax*, New York: Random House.
2. Emma Marris (2011), "In Retrospect: *The Lorax*", *Nature*, 476:7359, pp. 148–149, p. 149.
3. Marianne E. Krasny and Keith Tidball (2015), *Civic Ecology: Adaptation and Transformation from the Ground Up*, Cambridge, MA: The MIT Press, p.xiii.
4. R.K. Bangert et al. (2008), "Genetic structure of a foundation species: scaling community phenotypes from the individual to the region", *Heredity*, 100, pp. 121–131, p. 123.
5. R.J. Putman (1994), *Community Ecology*, London: Kluwer Academic Publishers, p. 2.
6. Peder Anker (2011), "Plant Community, Plantesamfund", in A. Schwarz and K. Jax (eds), *Ecology Revisited*, Dordrecht: Springer, pp. 325–331.

7. See for example Frans de Waal (1982), *Chimpanzee Politics: Power and Sex Among Apes*, London: Jonathan Cape; D.W. Stephens and J.R. Krebs (1986), *Foraging Theory*, Princeton, NJ: Princeton University Press; Edward Osborne Wilson (1971), *The Insect Societies*, Cambridge: Belknap Press.
8. Bryan S. Turner, "Introduction", in Émile Durkheim, *Professional Ethics and Civic Morals*, Abingdon and New York: Routledge Classics, p.x.
9. Nancy L. Rosenblum (1998), *Membership and Morals: The Personal Uses of Pluralism in America*, Princeton: Princeton University Press.
10. R.D. McKenzie (1924), "The Ecological Approach to the Study of the Human Community", American Journal of Sociology, 30:3, pp. 287–301.
11. Robert Ezra Park (1936), "Human Ecology", *American Journal of Sociology*, 42:1, pp. 1–15.
12. Gerald L. Young (1994), "Human Ecology", in Ruth A. Eblen and William R. Eblen (eds) (1994), *The Encyclopedia of the Environment*, Boston: Houghton Miflin, p. 339.
13. Frederick R. Steiner (2002), *Human Ecology*, Washington: Island Press, p. 167.
14. Steiner, *Human Ecology*, pp. 62, 59.
15. Mark E. Warren (2001), *Democracy and Association*, Princeton: Princeton University Press, p. 12.
16. Warren, *Democracy*, pp. 43–45.
17. Warren, *Democracy*, p. 29.
18. Einar Haugen (1972), *The ecology of language*, Stanford, CA: Stanford University Press.
19. Simon Cottle (2000), "Rethinking news access", *Journalism Studies*, 1:3, pp. 427–448.
20. Henry Jenkins (2001), "Convergence? Diverge!", *Technology Review*, June, p. 93.
21. Jude Valdez (1988), "The Entrepreneurial Ecosystem: Toward a Theory of New Business Formation", *Proceedings of the Small Business Institute Director's Association*, San Antonio: University of Texas, http://sbida.org/Resources/Documents/Proceedings/1988%20Proceedings.pdf#page=102, pp. 106–113; James F. Moore (1996), *The Death of Competition: Leadership and*

Strategy in the Age of Business Ecosystems, New York: HarperBusiness; G. Kandiah and S. Gossain (1998), "Reinventing value: The new business ecosystem", *Strategy & Leadership*, 26:5, pp. 28–33.

22. B.A. Huberman (1988), *The Ecology of Computation*, Amsterdam: North-Holland.
23. John T. Pardeck (1988) "An Ecological Approach for Social Work Practice", *The Journal of Sociology & Social Welfare*, 15:2, pp. 133–142.
24. Erich Hörl and James Edward Burton (eds) (2017), *General ecology: The new ecological paradigm*, London: Bloomsbury Publishing, 2017.
25. Arts Council England (2014), *This England: How Arts Council England uses its investment to shape a national cultural ecology*, Manchester: Arts Council England.
26. Tom Fleming and Andrew Erskine (2011), *Supporting Growth in the Arts Economy*, London: Arts Council England.
27. Richard Naylor et al. (2016), *Arts Council England: Analysis of Theatre in England*, London: Arts Council England. See also Anon. (2016), *Analysis of Theatre in England: A Strategic Overview*, London: Arts Council England, https://www.artscouncil.org.uk/sites/default/files/download-file/Analysis%20of%20Theatre%20in%20England%20report%20 %20Arts%20Council%20response.pdf. Accessed 6 November 2020.
28. Ed Vaizey (2011), "Foreword", in Tom Fleming & Andrew Erskine (2011), *Supporting Growth in the Arts Economy*, London: Arts Council England, p. 3.
29. John Holden (2015), *The Ecology of Culture: A Report commissioned by the Arts and Humanities Research Council's Cultural Value Project*, Swindon: Arts and Humanities Research Council, p. 2.
30. Holden, *Ecology*, p. 2.
31. Holden, *Ecology*, p. 3.
32. Holden, *Ecology*, pp. 3–4.
33. Holden, *Ecology*, p. 2.
34. Holden, *Ecology*, p. 33.
35. Michael Carolan (2019), "Capitalizing on financing ecologies: The world making properties of peer-to-peer lending through everyday entrepreneurship", *GeoForum* 102, pp. 17–26.
36. Holden, *Ecology*, p. 5.

37. Holden, *Ecology*, pp. 2–3; Holden borrows the last two of these phrases from Ann Markusen et al. (2011), *California's Arts and Cultural Ecology*, San Francisco: James Irvine Foundation, p. 10.
38. Holden, *Ecology*, p. 24.
39. Holden, *Ecology*, pp. 16, 30, 25, 24.
40. Holden, *Ecology*, p. 15.
41. Holden, *Ecology*, p. 18. Strangely, one of the report's most considered applications of ecological thinking, in which Holden looks more widely at a "cultural environment" that incorporates the heritage sector, the church, and broadcasting, is dismissed as "not strictly part of the ecology of culture"(p. 23); here, we'd suggest he misses a trick.
42. Dominic Symonds (2017), *Broadway Rhythm: Imaging the City in Song*, Ann Arbor: University of Michigan Press.
43. Amos H. Hawley (1986), *Human Ecology: A Theoretical Essay*, Chicago: University of Chicago Press, p. 4.
44. Paul Makeham, Bree Hadley, and Joon-Yee Kwok (2012), "A 'Value Ecology' approach to the performing arts", *M/C Journal* 15:3, pp. 1–12, p. 2.
45. Jen Harvie (2015), "Funding, philanthropy, structural inequality and decline in England's theatre ecology", *Cultural Trends*, 24:1; Christopher Leveille (2019), "'...Aaand Scene!': Moving Towards a Future for the Independent Theatre Landscape in Toronto", Masters thesis, OCAD University, Toronto; Cohen Ambrose (2019), "Ecologies of Experience: John Dewey, Distributed Cognition, and the Cultural-Cognitive Ecosystem of Theatre-Training Settings", *Theatre Symposium*, 27, pp. 80–91; Bree Hadley (2017), "Disability theatre in Australia: a survey and a sector ecology", *Research in Drama Education: The Journal of Applied Theatre and Performance*, 22:3, pp. 305–324.
46. D.M. Post and E.P. Palkovacs (2009), "Eco-evolutionary feedbacks in community and ecosystem ecology: interactions between the ecological theatre and the evolutionary play", *Philosophical transactions of the Royal Society of London. Series B, Biological sciences*, 364: 1523, pp. 1629–1640.
47. Paul W. Speer and Joseph Hughey (1995), "Community organizing: An ecological route to empowerment and power", *American Journal of Community Psychology*, 23: 5, pp. 729–749, p. 746.

48. J.G. Kelly (1986), "An ecological paradigm: Defining mental health consultation as a preventive service", in J.G. Kelly and R.E. Hess (eds.) (1986), *The ecology of prevention: illustrating mental health consultation*, New York: Haworth Press, pp. 1–36.
49. Marcus B. Weaver-Hightower (2008), "An Ecology Metaphor for Educational Policy Analysis: A Call to Complexity", *Educational Researcher*, 37:3, pp. 153–167.
50. Carlos A. Scolari (2012), "Media Ecology Exploring the Metaphor to Expand the Theory", *Communication Theory*, 22:2, pp. 204–225.
51. David Horton Smith (1997), "The International History of Grassroots Associations", *International Journal of Comparative Sociology*, 38:3–4, pp. 189–216, p. 189.
52. Sarah E. Dempsey (2009), "NGOs, Communicative Labor, and the Work of Grassroots Representation", *Communication and Critical/Cultural Studies*, 6:4, pp. 328–345, p. 329.
53. Frances Kunreuther (2011), "Grassroots Associations", in Michael Edwards (ed.) (2011), *The Oxford Handbook of Civil Society*, pp. 55–67, p. 56.
54. Dempsey, "NGOs", p. 331.
55. Susanne Ricee (2020), "How do Grassroots Organizations Work? What and Why is it Important?", *Diversity for Social Impact*, 5 August, https://diversity.social/grassroots/#gsc.tab=0, Accessed 11 December 2020.
56. Marianne E. Krasny (2018), *Grassroots to Global: Broader Impacts of Civic Ecology*, Comstock Publishing Associates.
57. Marianne E. Krasny and Keith Tidball (2014), *Greening in the Red Zone*, Dordrecht, Heidelberg, New York: Springer; Marianne E. Krasny and Keith Tidball (2015), *Civic Ecology: Adaptation and Transformation from the Ground Up*, Cambridge, MA: The MIT Press.
58. Krasny and Tidball, *Greening*, p. 6.
59. Krasny and Tidball, *Civic Ecology*, p. xiii.
60. Krasny and Tidball, *Greening*, p. 4.
61. John C. Morris et al. (2013), *The Case for Grassroots Collaboration: Social Capital and Ecosystem Restoration at the Local Level*, Lanham, Boulder, New York, Toronto & Plymouth: Lexington Books, p.vi.
62. Kunreuther, "Grassroots Associations", p. 56.

63. Angus McCabe and Jenny Philmore (eds) (2017), *Community Groups in Context: Local Activities and Actions*, Bristol: Policy Press, p. 156.
64. Angus McCabe (2012), "The importance of grassroots community arts groups in big society", *The Guardian*, 17 January, https://www.theguardian.com/voluntary-sector-network/community-action-blog/2012/jan/17/grassroots-community-arts-groups. Accessed 26 November 2020.
65. Angus McCabe, Jenny Phillimore and Hilary Ramsden (2012), *The role of grassroots arts activities in communities: A scoping study*, Swindon: AHRC, p. 3.
66. McCabe, "The importance".
67. Anon. (n.d.), "The art of peace: Interrogating community devised arts based peacebuilding", *UKRI* website, https://gtr.ukri.org/projects?ref=AH%2FS001484%2F1. Accessed 13 December 2020.
68. H. Fyfe et al. (2011), "The role of grassroots arts activities in communities: a scoping study", Birmingham: Third Sector Research Centre, p. 48.
69. Emma Harvey (n.d.), "How Does the Music Industry Help the Sector?", *Trinity Community Arts,* https://www.trinitybristol.org.uk/about/news/2020/supporting-the-sector. Accessed 13 December 2020.
70. Gustavo Esteva, and Madhu Suri Prakash (2014), *Grassroots Postmodernism: Remaking the Soil of Cultures* (second edition), London: Zed Books, p.xvi, quoting Sarah Van Gelder (2013), "Freedom Starts With a Seed" (interview with Vandana Shiva), *Yes! Magazine*, 30 November.
71. Esteva and Prakesh, *Grassroots*, pp.xx, 192, 198, 53, 82.
72. Timothy Beach, Sheryl Luzzadder-Beach and Nicholas P. Dunning (2019), "Out of the Soil: Soil (Dark Matter Biodiversity) and Societal 'Collapses' from Mesoamerica to Mesopotamia and Beyond", in Partha Dasgupta, Peter H. Raven and Anna L. McIvor (2019), *Biological Extinction: New Perspectives*, Cambridge: University of Cambridge Press, pp. 138–174, p. 161.
73. Morris et al., *The Case*, p.xxi.
74. Krasny and Tidball, *Civic Ecology*, p.xiii.

75. Richard D. Bardgett and David A. Wardle (2010), *Aboveground-Belowground Linkages: Biotic Interactions, Ecosystem Processes, and Global Change*, Oxford: Oxford University Press, p. 4.
76. G. F. Veen et al. (2019), "Applying the Aboveground-Belowground Interaction Concept in Agriculture: Spatio-Temporal Scales Matter", *Frontiers in Ecological Evolution* 7:300, pp. 1–12, p. 2.
77. Richard D. Bardgett (2018), "Linking Aboveground-belowground Ecology: A Short Historical Perspective", in Takayuki Ohgushi, Susanne Wurst and Scott N. Johnson (eds) (2018), *Aboveground-Belowground Community Ecology*, Cham: Springer, pp. 1–17, p. 1.
78. Takayuki Ohgushi, Susanne Wurst and Scott N. Johnson (2018), "Current Knowledge and Future Challenges of Aboveground and Belowground Community Ecology", in Ohgushi et al. (2018), *Aboveground-Belowground Community Ecology*, pp. 345–361, p. 352.
79. Veen et al., "Applying", pp. 3–4.
80. Ohgushi et al., "Current Knowledge", p. 352.
81. Judith L. Bronstein (1994), "Our Current Understanding of Mutualism", *The Quarterly Review of Biology*, 69, pp. 31–51, p. 32.
82. Bardgett and Wardle (2010), *Aboveground-Belowground*, p. 214.
83. Jennifer Berneda Grenz (2020), "Healing the Land by Reclaiming an Indigenous Ecology: A journey exploring the application of the Indigenous worldview to invasion biology and ecology", Ph.D. thesis, Vancouver: University of British Columbia, p. 2.
84. Holden, *The Ecology*, p. 7. McCarthy et al. construct a similar conceptualization relating to the US: "the three components of local communities' organizational ecology" are made up of "their arts infrastructures; the support systems upon which they depend; and the sociodemographic, economic, and political environments in which they operate"; and Makeham et al. establish another relating to Australia: "a complex range of rhizomatic relationships between production, distribution, and consumption infrastructure and how they influence each other within a sector such as the performing arts". Kevin McCarthy, Elizabeth Heneghan Ondaatje, and Jennifer L. Novak (2007), *Arts*

and Culture in the Metropolis: Strategies for Sustainability, Santa Monica: RAND Corporation, p.xiii; Makeham et al., "A 'Value Ecology' approach".

85. Isabella Tree (2019), *Wilding: The Return of Nature to a British Farm*, London: Picador.
86. Tree, *Wilding*, p. 275.
87. Tree, *Wilding*, p. 281.
88. Tree, *Wilding*, p. 275.
89. Tree, *Wilding*, pp. 277–9.
90. Tree, *Wilding*, pp. 287–9.
91. Tree, *Wilding*, p. 19.
92. John Holden (2004), *Capturing Cultural Value: How Culture has Become a Tool of Government Policy*, London: Demos, p. 10.
93. Holden, *The Ecology*, p. 6.
94. Holden, *The Ecology*, pp. 14–17, quoting respondent Nick Cohen.
95. Tree, *Wilding*, p. 19.
96. See our discussion of crowd-sourced funding from "family, friends and fools" (the "3Fs") in Karen Savage and Dominic Symonds (2018), *Economies of Collaboration in Performance: More Than the Sum of the Parts*, Basingstoke: Palgrave, pp. 239–277.
97. Wai-Chung Ho (2011), "Parental Support and Student Learning of Musical Instruments in Hong Kong", *Visions of Research in Music Education*, 19, pp. 1–53, p. 3.
98. See Stacy Wolf (2020), *Beyond Broadway: The Pleasure and Promise of Musical Theatre Across America*, New York: Oxford University Press.
99. Ohgushi et al., "Current Knowledge", p. 345.
100. Holden, *The Ecology*, p. 7.
101. Tree, *Wilding*.

Bibliography

Ambrose, Cohen (2019), "Ecologies of Experience: John Dewey, Distributed Cognition, and the Cultural-Cognitive Ecosystem of Theatre-Training Settings", *Theatre Symposium*, 27, pp. 80–91.

Anon. (2016), *Analysis of Theatre in England: A Strategic Overview*, London: Arts Council England.

Anon. (n.d.), "The art of peace: Interrogating community devised arts based peacebuilding", *UKRI* website, https://gtr.ukri.org/projects?ref=AH%2FS001484%2F1. Accessed 13 December 2020.

Arts Council England (2014), *This England: How Arts Council England uses its investment to shape a national cultural ecology*. Manchester: Arts Council England.

Bangert, R.K., E.V. Lonsdorf, G.M. Wimp, S.M. Shuster, D. Fischer, J.A. Schweitzer, G.J. Allan, J.K. Bailey, and T.G. Whitham (2008), "Genetic structure of a foundation species: scaling community phenotypes from the individual to the region", *Heredity*, 100, pp. 121–131.

Bardgett, Richard D. and David A. Wardle (2010), *Aboveground-Belowground Linkages: Biotic Interactions, Ecosystem Processes, and Global Change*, Oxford: Oxford University Press.

Beach, Timothy, Sheryl Luzzadder-Beach and Nicholas P. Dunning (2019), "Out of the Soil: Soil (Dark Matter Biodiversity) and Societal 'Collapses' from Mesoamerica to Mesopotamia and Beyond", in Partha Dasgupta, Peter H. Raven and Anna L. McIvor (2019), *Biological Extinction: New Perspectives*, Cambridge: University of Cambridge Press, pp. 138–174.

Bronstein, Judith L. (1994), "Our Current Understanding of Mutualism", *The Quarterly Review of Biology*, 69, pp. 31–51.

Carolan, Michael (2019), "Capitalizing on financing ecologies: The world making properties of peer-to-peer lending through everyday entrepreneurship", *GeoForum*, 102, pp. 17–26.

Cottle, Simon (2000), "Rethinking news access", *Journalism Studies*, 1:3, pp. 427–448.

Dempsey, Sarah E. (2009), "NGOs, Communicative Labor, and the Work of Grassroots Representation", *Communication and Critical/Cultural Studies*, 6:4, pp. 328–345.

Durkheim, Émile (2019), *Professional Ethics and Civic Morals*, Abingdon and New York: Routledge Classics.

Eblen, Ruth A. and William R. Eblen (eds) (1994), *The Encyclopedia of the Environment*, Boston: Houghton Miflin.

Esteva, Gustavo and Madhu Suri Prakash (2014), *Grassroots Postmodernism: Remaking the Soil of Cultures* (second edition), London & New York: Zed Books.

Fyfe, H. A. McCabe, J. Milling, J. Phillimore, H. Ramsden, and R. Simpson (2011), "The role of grassroots arts activities in communities: a scoping study", Birmingham: Third Sector Research Centre.

Grenz, Jennifer Berneda (2020), "Healing the Land by Reclaiming an Indigenous Ecology: A journey exploring the application of the Indigenous worldview to invasion biology and ecology", Ph.D. thesis, Vancouver: University of British Columbia.

Hadley, Bree (2017), "Disability theatre in Australia: a survey and a sector ecology", *Research in Drama Education: The Journal of Applied Theatre and Performance*, 22:3, pp. 305–324.

Harvey, Emma (n.d.), "How Does the Music Industry Help the Sector?", *Trinity Community Arts,* https://www.trinitybristol.org.uk/about/news/2020/supporting-the-sector. Accessed 13 December 2020.

Harvie, Jen (2015), Funding, philanthropy, structural inequality and decline in England's theatre ecology, *Cultural Trends*, 24:1.

Haugen, Einar (1972), *The ecology of language*, Stanford, CA: Stanford University Press.

Hawley, Amos H. (1986), *Human Ecology: A Theoretical Essay*, Chicago: University of Chicago Press.

Ho, Wai-Chung (2011), "Parental Support and Student Learning of Musical Instruments in Hong Kong", *Visions of Research in Music Education*, 19, pp. 1–53.

Holden, John (2004), *Capturing Cultural Value: How Culture has Become a Tool of Government Policy*, London: Demos, p. 10.

Holden, John (2015), *The Ecology of Culture: A Report commissioned by the Arts and Humanities Research Council's Cultural Value Project*, Swindon: Arts and Humanities Research Council.

Huberman, B. A. (1988), *The Ecology of Computation*, Amsterdam: North-Holland.

Jenkins, Henry (2001), "Convergence? Diverge!", *Technology Review*, June, p. 93.

Kandiah, G. and Gossain, S. (1998), "Reinventing value: The new business ecosystem", *Strategy and Leadership*, 26:5, pp. 28–33.

Kelly, J. G. and R. E. Hess (eds.) (1986), *The ecology of prevention: illustrating mental health consultation*, New York: Haworth Press.

Krasny, Marianne E. (2018), *Grassroots to Global: Broader Impacts of Civic Ecology*, Comstock Publishing Associates.

Krasny, Marianne E. and Keith Tidball (2014), *Greening in the Red Zone*, Dordrecht, Heidelberg, New York: Springer.

Krasny, Marianne E. and Keith Tidball (2015), *Civic Ecology: Adaptation and Transformation from the Ground Up*, Cambridge: The MIT Press.

Kunreuther, Frances (2011), "Grassroots Associations", in Michael Edwards (ed.) (2011), *The Oxford Handbook of Civil Society*, pp. 55–67.

Leveille, Christopher (2019), "'...Aaand Scene!': Moving Towards a Future for the Independent Theatre Landscape in Toronto", Masters thesis, OCAD University, Toronto.

Makeham, Paul, Bree Hadley, and Joon-Yee Kwok (2012), "A 'Value Ecology' approach to the performing arts", *M/C Journal*, 15:3, pp. 1–12.

Markusen, Ann, Anne Gadwa, Elisa Barbour and William Beyers (2011), *California's Arts and Cultural Ecology*, San Francisco: James Irvine Foundation.

Marris, Emma (2011), "In Retrospect: *The Lorax*," *Nature*, 476:7359, 11 August, pp. 148–149.

McCabe, Angus (2012), "The importance of grassroots community arts groups in big society", *The Guardian*, 17 January.

McCabe, Angus, Jenny Phillimore and Hilary Ramsden (2012), *The role of grassroots arts activities in communities: A scoping study*, Swindon: Arts and Humanities Research Council.

McCabe, Angus and Jenny Philmore (eds) (2017), *Community Groups in Context: Local Activities and Actions*, Bristol: Policy Press.

McCarthy, Kevin F., Elizabeth Heneghan Ondaatje, and Jennifer L. Novak (2007), *Arts and Culture in the Metropolis: Strategies for Sustainability*, Santa Monica: RAND Corporation.

McKenzie, R. D. (1924), "The Ecological Approach to the Study of the Human Community", American Journal of Sociology, 30:3, pp. 287–301.

Moore, James F. (1996), *The Death of Competition: Leadership and Strategy in the Age of Business Ecosystems*, New York: HarperBusiness.

Morris, John C., William Allen Gibson, William Marshall Leavitt, and Shana Campbell Jones (2013), *The Case for Grassroots Collaboration: Social Capital and Ecosystem Restoration at the Local Level*, Lanham: Lexington Books.

Naylor, Richard, Bethany Lewis, Caterina Branzanti, Graham Devlin and Alan Dix (2016), *Arts Council England: Analysis of Theatre in England*, London: Arts Council England.

Ohgushi, Takayuki, Susanne Wurst and Scott N. Johnson (eds) (2018), *Aboveground-Belowground Community Ecology*, Cham: Springer.

Pardeck, John T. (1988), "An Ecological Approach for Social Work Practice", *The Journal of Sociology and Social Welfare*, 15:2, pp. 133–142.

Park, Robert Ezra (1936), "Human Ecology", *American Journal of Sociology*, 42:1, pp. 1–15.

Post, D. M. and E. P. Palkovacs (2009), "Eco-evolutionary feedbacks in community and ecosystem ecology: interactions between the ecological theatre and the evolutionary play", *Philosophical transactions of the Royal Society of London. Series B, Biological sciences*, 364:1523, pp. 1629–1640.

Putman, R. J. (1994), *Community Ecology*, London: Kluwer Academic Publishers.

Ricee, Susanne (2020), "How do Grassroots Organizations Work? What and Why is it Important?", *Diversity for Social Impact*, 5 August, https://diversity.social/grassroots/#gsc.tab=0, Accessed 11 December 2020.

Rosenblum, Nancy L. (1998), *Membership and Morals: The Personal Uses of Pluralism in America*, Princeton: Princeton University Press.

Savage, Karen and Dominic Symonds (2018), *Economies of Collaboration in Performance: More Than the Sum of the Parts*, Basingstoke: Palgrave.

Schwarz, A. and K. Jax (eds), *Ecology Revisited*, Dordrecht: Springer.

Scolari, Carlos A. (2012), "Media Ecology Exploring the Metaphor to Expand the Theory", *Communication Theory*, 22:2, pp. 204–225.

Seuss, Dr (1971), *The Lorax*, New York: Random House.

Smith, David Horton (1997), "The International History of Grassroots Associations", *International Journal of Comparative Sociology*, 38:3–4, pp. 189–216.

Speer, Paul W. and Joseph Hughey (1995), "Community organizing: An ecological route to empowerment and power", *American Journal of Community Psychology*, 23:5, pp. 729–749.

Steiner, Frederick R. (2002), *Human Ecology*, Washington: Island Press.

Stephens, D. W., and Krebs, J. R. (1986), *Foraging Theory*, Princeton: Princeton University Press.

Symonds, Dominic (2017), *Broadway Rhythm: Imaging the City in Song*, Ann Arbor: University of Michigan Press.

Tree, Isabella (2019), *Wilding: The Return of Nature to a British Farm*, London: Picador.

Valdez, Jude (1988), "The Entrepreneurial Ecosystem: Toward a Theory of New Business Formation", *Proceedings of the Small Business Institute Director's Association*, San Antonio: University of Texas.

Van Gelder, Sarah (2013), "Freedom Starts With a Seed", *Yes! Magazine*, 30 November.

Veen, G. F., E. R. J. Wubs, R. D. Bardgett, E. Barrios, M. A. Bradford, S. Carvalho, G. B. De Deyn, F. T. de Vries, K. E. Giller, D. Kleijn, D. A. Landis, W. A. H. Rossing, M. Schrama, J. Six , P. C. Struik, S. van Gils, J. S. C. Wiskerke, W. H. van der Putten, and L. E. M. Vet (2019), "Applying the Aboveground-Belowground Interaction Concept in Agriculture: Spatio-Temporal Scales Matter", *Frontiers in Ecological Evolution*, 7:300, pp. 1–12.

Waal, Frans de (1982), *Chimpanzee Politics: Power and Sex Among Apes*, London: Jonathan Cape.

Warren, Mark E. (2001), *Democracy and Association*, Princeton: Princeton University Press.

Weaver-Hightower, Marcus B. (2008), "An Ecology Metaphor for Educational Policy Analysis: A Call to Complexity", *Educational Researcher*, 37:3, pp. 153–167.
Wilson, Edward Osborne (1971), *The Insect Societies*, Cambridge: Belknap Press.
Wolf, Stacy (2020), *Beyond Broadway: The Pleasure and Promise of Musical Theatre Across America*, New York: Oxford University Press.

CHAPTER 2

Ecology

"*Imagine a world without trees*", writes Jim Robbins in the preface to his book *The Man Who Plants Trees*.[1] It's a stand-alone sentence, offered in a paragraph all of its own, and italicized to boot. Pretty significant. Even so—and even in his book, where it follows a devastating story about the Once-ler-like razing of an entire forest—it's easy to overlook the significance of those five little words. But he's right. Just imagine. What an alien landscape that would be.

In our previous book, *Economies of Collaboration*, we started each chapter with an anecdote about bees. Our argument focused on the invisible economies that we often aren't aware of, and we recognized that the contribution bees make to pollination would cost billions of dollars if we had to pay for it. The book you are reading now is about *ecologies* rather than *economies*, but there are many synergies between the two. And in this book we start each chapter with an anecdote about the planting of trees. This one's inspired, in part, by Robbins' book. "On balance", he writes, "if we had to pay for the services that trees provide, we couldn't afford them. Because trees offer their services in silence, and for free, we take them for granted".[2] At that point he's literally referring to the monetary value of planting trees ("$10 million in energy costs").[3] But they provide in other ways as well:

> Trees feed oxygen and minerals into the ocean; create rain; render mercury, nitrates, and other toxic wastes in the soil harmless; gather and neutralize

K. Savage and D. Symonds, *Ecologies of Community in Performance*,
https://doi.org/10.1007/978-3-031-94343-0_2

> sulfur dioxide, ozone, carbon dioxide, and other harmful air pollutants in their tissue; create homes and building materials; offer shade; provide medicine; and produce a wide variety of nuts and fruits. They sustain all manner of wildlife, birds, and insects with an array of food and shelter. They are the planet's heat shield, slowing the evaporation of water and cooling the earth. They generate vast clouds of chemicals that are vital to myriad aspects of the earth's ecosystems and likely vital to our health and well-being. [...] Even viewed conservatively, trees are worth far more than they cost to plant and maintain.[4]

As the title of his book implies, planting trees becomes something of a fascination for Robbins—though his eponymous "man who plants trees" is actually a figure called David Milarch, creator of the Champion Tree Project and founder of Archangel Ancient Tree Archive, who we will meet in Chapter 7. Through following Milarch on his exploits, Robbins makes a straightforward ecological point: "Planting trees may be the single most important ecotechnology that we have to put the broken pieces of our planet back together".[5]

Robbins takes his own inspiration from the tale of *The Man Who Planted Trees* by Jean Giono, first serialized in *Vogue* magazine in 1954. Giono's short story is about a man who sets out to hike through a valley. It's a "landscape of unparalleled desolation", he reports; "an expanse of bare and monotonous moorland".[6] The villages are in ruins, the houses are roofless, and the few people who remain are hostile and unhappy. But he comes across a shepherd called Elzéard Bouffier who takes him in, and the next day, they go out planting acorns.

> He'd been planting trees in this wilderness for three years. He'd planted a hundred thousand of them. Out of those, twenty thousand had come up. Of the twenty thousand he expected to lose half, because of rodents or the unpredictable ways of Providence. That still meant ten thousand oaks would grow where before there had been nothing.[7]

Our narrator is intrigued, but he leaves, and is soon drafted to serve in the war. Only after five years does he return, to find the shepherd still planting trees. By now the landscape has transformed: water is flowing in the streams, and seeds have grown into willows and meadows, "and some reason for living".[8] Gradually the valley has come back to life: people have returned to the abandoned villages; they've sown vegetables and flowers; built a fountain; planted a lime tree. "It had become a place where one would wish to live", realizes the narrator; "So hope had returned".[9] Each year he comes back, to see the wilderness of his first encounter changed beyond recognition. He marvels at the transformation the old shepherd has brought, to whom "more than ten thousand people must owe their happiness".[10]

Giono's tale is enchanting, and translated into multiple languages, it has captured imaginations around the world. It's spread a very simple tale about ecology, about conservation, and about the little ways in which we can save our world.

But if we're going to take on an ecological perspective in our discussion, we'll need to understand a little about that terrain. So in this chapter, we will encounter a handful of other environmentalists like Bouffier, getting to grips with the notion of *ecology*, and understanding the term and its principles.

Despite the word's Greek roots, ecology is a relatively recent concept from the late nineteenth century. Etymologically, it's closely related to the word "economy", the terms respectively referring to the study of, and the management of, the household. Ernst Haeckel, who first defined "ecology" as "lessons from nature's household", went on to describe it in more detail as "the collected knowledge of relationships between organisms and their surrounding outside world, from which in a wider sense we can get to know everything about the conditions of existence".[11] Over time, as environmental concerns gained significance, a further meaning emerged: recognizing that the planet's resources are finite, "ecology" has begun to reflect "the notion of the earth as a common home which needs care and maintenance".[12] As our awareness of environmental damage has grown, and as threats such as climate change, pollution, rising sea levels, and species extinction increase, ecology has become a practice of "illuminating and solving environmental problems".[13] Thus, a distinction: "scientific ecology", or how our ecosystems work; and "political ecology", or how ecological insight can be used to protect our planet.

Even so, understanding ecology is more complicated than this simple distinction implies, in part because a culture of metaphor has developed around the term. We've encountered the metaphor *grassroots* in our introduction, but the use of metaphor works both ways, so as much as we commonly use other ecological references like *peas in a pod*, *root and branch*, or *sowing the seeds* in everyday speech, concepts such as *ecosystem health*[14] and *the balance of nature*[15] draw on non-ecological language; and just as the word *ecology* draws on a metaphor in its very meaning (lessons from nature's household), it is also used elsewhere for its own metaphorical power (the ecology of the arts/media etc.).

Later we will consider some of the most recent thinking about ecology, but first we'll set out some of the basic ways in which the term is understood. We'll explore how ecology works, focusing on the *ecosystem*. We'll do this through consulting the work of seminal ecologist Eugene P. Odum. Next, we will consider the common perspective that our ecosystem is at risk and that we are at a point of *ecological crisis*. Our reference point here will be Rachel Carson's *Silent Spring* (1962), though the notion of ecological crisis extends far further back than that. We'll see how ecological approaches might help to stave off crisis as we find ways to protect, preserve, or conserve our ecosystem. Here we will focus on American approaches, tracking the contributions of environmentalists John Muir, Gifford Pinchot, and Aldo Leopold, before considering first nation views. In the concluding parts of the chapter, we will consider how ecology has been reimagined. We'll turn to critiques of standard ecological approaches, introducing Arno Naess's concept of *deep ecology*; we will explore the writing of Gregory Bateson and Félix Guattari, both of whom have reconceptualized ecology as a metaphor; and we will finally turn to Timothy Morton's invocation of *dark ecology*.

Ecology: How It Works

Since several seminal texts suggest that *ecosystems* are "the basic functional unit in ecology",[16] it makes sense for us to start there. It also make sense for us to start with a description by biologist Eugene P. Odum, whose text book *The Fundamentals of Ecology* (1953) has been perennially reprinted. In this opening explanation, Odum gives us not only an insight into *ecosystem*, but also other key terms.

> Ecology is largely, but not entirely, concerned with the system levels beyond that of the organism. In ecology, the term *population*, originally coined to denote a group of people, is broadened to include groups of individuals of any one kind of organism. Likewise, *community*, in the ecological sense (sometimes designated as 'biotic community'), includes all the populations occupying a given area. The community and the nonliving environment function together as an ecological system or *ecosystem*.[17]

As experts tell us, "an ecosystem is not a place, or an assemblage of organisms" per se—ecology offers a different set of conceptual terms to designate specific *places* ("biome"; "region"), and *assemblages* ("population"; "community"). Instead, ecologists who study ecosystems select "convenient portions of the natural world".[18] "As long as the major components are present and operate together to achieve some sort of functional stability, even if for only a short time", suggests Odum, "the entity may be considered an ecosystem".[19] Thus "a pond, a lake, a tract of forest or even a laboratory culture (microecosystem) provide convenient units of study", while at the other end of the scale, we can conceptualize the whole of the planet—"The largest and most nearly self-sufficient biological system we know about"[20]—as an ecosystem. Within these areas the "ecosystem consists of the interactions among its components",[21] which will generally be viewed by ecologists as either "population-community" based or "process-functional" based.[22]

"Population-community ecologists tend to view ecosystems as networks of interacting populations", explains Robert O'Neill.[23] Simplistically, these involve two or more species engaging in interactions such as competition or predation, though in most circumstances a number of different populations will interact with each other and with their environmental surroundings. Odum discusses a pond ecosystem comprised of *producer organisms* such as phytoplankton and floating plants, and *consumer organisms* such as animal plankton, insects and fish: "The primary macroconsumers or herbivores feed directly on living plants or plant parts", he writes; "The secondary consumers or carnivores, such as predacious insects and game fish in the pond [...], feed on the primary consumers or on other secondary consumers".[24] Such a *food chain* is extremely simple. Elsewhere, it has been described using other metaphors such as *food web* or *food pyramid*, terms which suggest something rather more complex. Indeed, Aldo Leopold refers to it as "a tangle of chains so complex as to seem disorderly". Even so, "the stability of the system proves it to be a highly organized structure".[25]

As Odum notes, there are at least nine different types of interactions between species in a typical ecosystem, including those we mentioned in our introduction: mutualism, predation, parasitism, competition, commensalism, amensalism, and neutralism.[26] Together, these form the basis of a food web (Fig. 2.1). The notion of a community is evident from such an image, as all the species in this pond ecosystem interact. It's also worth noting that people are included in this diagram: in other words, we are a part of the system.

In scientific terms, we see in this diagram the primary producers of energy (aquatic plants and phytoplankton) being eaten by primary consumers (tadpoles and snails), who then become the energy source for subsequent consumers (fish and birds) until the apex predators (here, the human and the heron) complete the process. A second flow returns nutrients into the food web as their bodies are consumed by detritivores (like worms) or decomposers (like bacteria). From this perspective, the ecosystem is far more than just a snapshot of community interactions. Instead—and this is the view of "process-functional" ecologists—"the science of ecology is one that considers energy flows within a closed system".[27]

Again, we can turn to Odum for an explanation through considering his "universal" model of energy flow. An energy form—usually sunlight, carbon dioxide, or water—enters the system, and is converted through photosynthesis into assimilated energy—oxygen and sugar. This enables both respiration—the emission of oxygen—and production—a combination of growth, storage of excess matter, and excretion of waste. Some of the energy flow remains unused, while respiration returns oxygen and some carbon dioxide to the atmosphere, for the cycle to begin again. Figure 2.2 shows a version of Odum's schema for this process-functional model (on the left hand side) and how it feeds into the population-community (on the right). Odum predicts 80–90% energy depletion through heat loss, with the remaining 10–20% becoming the energy input at the next stage as the producer becomes food for the consumer.[28]

As we see, even a simple ecosystem such as that of a pond can be extremely complex, and it is by no means either completely stable or completely self-contained. So although many ecologists treat ecosystems as if they are "closed-box"[30] systems,[31] it's clear from the pond example that "all ranks of ecosystems are open systems, not closed ones"; "Energy and matter continually escape from them in the course of the processes of life, and they must be replaced if the system is to continue to function".[32] Equally, the ecosystem will be affected by extrinsic events: weather events and seasonal fluctuation bring changes to the microclimate; long-term build-up of sediment causes changes to the water profile; other species migrate into the ecosystem; incoming microsystems are carried in by streams. In other words, even a pond is an example of a tremendously complex ecosystem. "Its ecology includes all events", Paul Shepard writes:

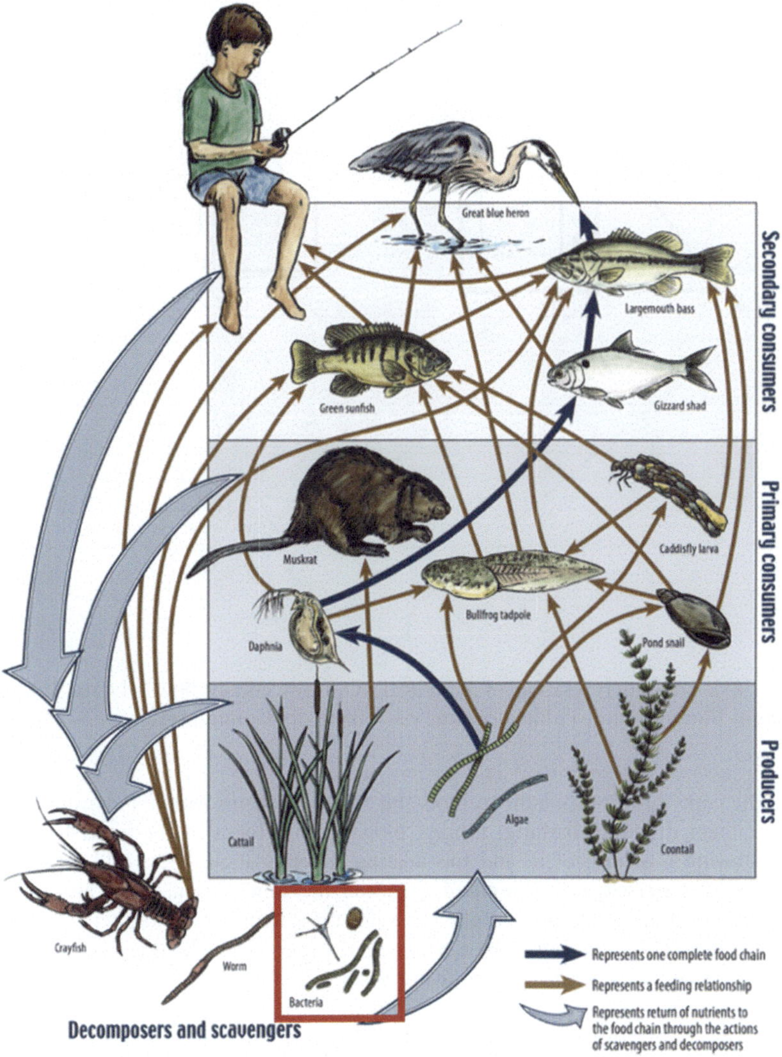

Fig. 2.1 A pond food web showing nutrients and primary producers at the bottom and various levels of consumers. Kestin Schulz, Mariya W. Smit, Lydie Herfort and Holly M. Simon. The image (amended) was provided courtesy of the Missouri Department of Conservation. CC BY-SA 4.0. https://creativecommons.org/licenses/by-sa/4.0, via Wikimedia Commons

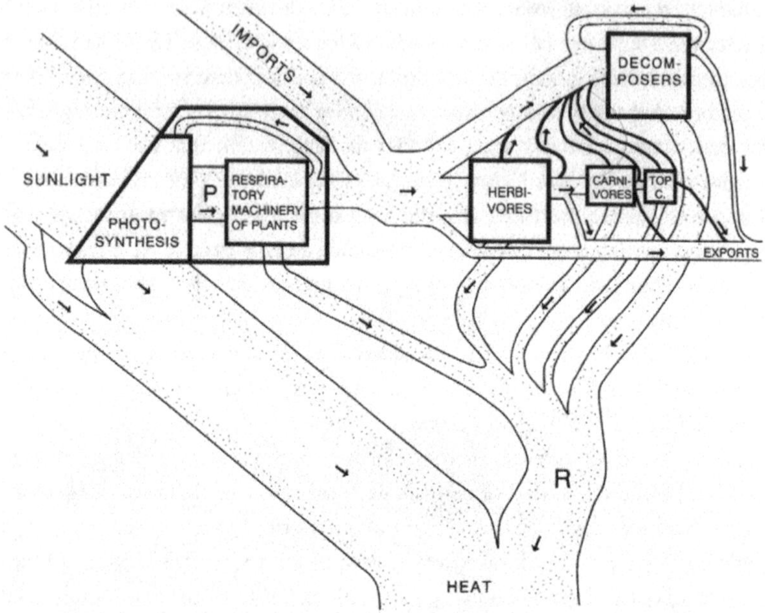

Fig. 2.2 An interpretation of Eugene P. Odum's energy flow schema. Image credit: Sholto Maud, Public domain, via Wikimedia Commons[29]

> the conversion of sunlight to food and the food chains within and around it, man drinking, bathing, fishing, plowing the slopes of the watershed, drawing a picture of it, and formulating theories about the world based on what he sees in the pond. He and all the other organisms at and in the pond act upon one another, engage the earth and atmosphere, and are linked to other ponds by a network of connections like the threads of protoplasm connecting cells in living tissues.[33]

Notice how human activities such as bathing (in the pond), ploughing (beyond the pond) and even drawing a picture (of the pond) become in Shepard's description aspects of the ecosystem. Such an open and permeable system is crucial to ecological thinking, and will become increasingly so as we explore more recent perspectives on ecology.

The tendency towards conceptualizing closed-box ecosystems, therefore, is revealed as a rather artificial way of drawing boundaries around some "thing" that is apparently closed off and self-sufficient. Yet the fact that this has become a tendency demonstrates how we have come to think about ecosystems, and more broadly, nature as a whole. For example, the assumption of a closed system and our attempt to understand its workings as such reveals what Frank B. Golley implies is a post-industrial mindset. The idea of ecosystems, which he notes was introduced to ecology in the 1930s,[34] was "a machine theory applied to nature"[35]—in other words, a human conceit designed to understand, manage, and control the processes of nature. Or, to cite another analogy, Nobel Prize winner Frederick Soddy's "sun-soil-food" summation of energy flow ("the internal energy of life")[36] characterizes an external input to the system from solar energy, and an external output from the system in agricultural provision, an "environmental instrumentalism"[37] that defines our (human) relationship with nature through an *agrilogistical* metaphor. In yet another analogy, which O'Neill and his colleagues point out, "to deal with ecosystems as systems requires an overextension of the organism concept"—in other words, a tendency to characterize or even anthropomorphise the ecosystem, as James Lovelock has arguably done with his concept of the earth-system *Gaia*. "Indeed, in the past, many authors have written as though the community or ecosystem *was* an organism", O'Neill et al. observe[38]; such an approach "implies that there is some scale-independent 'thing' out 'there' that is self-controlled", they warn.[39]

Let's mix some metaphors ourselves: to conceptualize ecology is to imagine countless threads interweaving, merging, blending, and competing in a web of sustenance, vitality, and life. In turn, to evoke "ecology" as a metaphor—whether that is in the ecology of the arts or the media ecology—is to appropriate such conceptual complexity. Yet as we have also seen, other comparisons between ecology (specifically: ecosystems) and machines (systems), resources (food), and humans (organisms) adopt analogues or metaphors, partly to enable clarity in our understanding, but partly to assume some sort of human-centred control or ownership over the way we understand and thereby handle nature.

Crisis, Cause, and Solution

It will come as no surprise to learn that such control—man's influence on our ecosystems—has been far from benign, leading to growing warnings of emergency from organizations such as the United Nations,[40] Greenpeace,[41] and the Red Cross.[42] Experts cite a myriad of concerns, from global warming to plastic pollution, and from deforestation to overfishing, as some of the biggest contributors to a planetary crisis,[43] each of which is human-induced and directly or indirectly damaging the ecosystems of the earth. In order to avoid reaching a catastrophic tipping point, such experts call for us (at the very least) to prevent global temperatures from rising more than 1.5°C from pre-industrial levels, to reduce emissions of methane and carbon dioxide into the atmosphere, and to conserve and restore natural spaces.[44]

Environmentalists point to the publication of Rachel Carson's book *Silent Spring* (1962) as our awakening to this crisis.[45] A compelling account of how we have polluted our planet with agricultural fertilizers and pesticides, Carson's book served as a stark indictment of how industrial progress precipitated disaster. "The most alarming of all man's assaults upon the environment", she wrote, "is the contamination of air, earth, rivers, and sea with dangerous and even lethal materials".[46] By the 1970s, even politicians were raising the alarm: "The threat to the environment is the number one crisis facing this nation and the world", wrote US Senator Gaylord Nelson.[47] In popular culture, writers as diverse as Dr. Seuss (*The Lorax*, 1971), John Wyndham (*The Kraken Wakes*, 1953), and Ernest Callenbach (*Ecotopia Emerging*, 1981) issued fervent warnings. And in academic circles, it was not just the voices of natural scientists that were adding to the fray: "As a direct consequence of our agricultural and industrial technology, we are rapidly causing to degenerate our sole habitation, that narrow strip of soil, air and water—the biosphere—in which we live and move and have our being", cried Australian philosopher John Passmore.[48] He cited "the four major ecological problems" as "pollution, the depletion of natural resources, the destruction of species, [and] over-population".[49] More recently—as we will consider in Chapter 6—the mantle of spokesperson for the environment has been taken up by prominent campaigners such as Al Gore and Greta Thunberg, and more radically by activist groups from Greenpeace to Extinction Rebellion.

Yet these late twentieth-century warnings were by no means the first. As long ago as 1864, George Perkins Marsh had warned that "The destruction of the forests, the drainage of lakes and marshes, and the operations of rural husbandry and industrial art, have unquestionably tended to produce great changes in the hygrometric, thermometric, electric, and chemical condition of the atmosphere".[50] Nor was the challenge confined to America: "If prompt and energetic measures are not taken, it will be almost possible to designate the precise moment when the French Alps will become a desert", wrote Jules Clavé in 1881[51]; "clear-sighted men already foresee a crisis",[52] he warned in 1882, noting how deforestation had denuded the dense woodlands of Russia. In short, "every country, all the world, is met with the threat of an oncoming crisis",[53] wrote Fairfield Osborn in 1948; "blind to the need of co-operating with nature, man is destroying the sources of his life".[54]

In the light of such crises, the global call has been for us to take responsibility: high-profile figures have lobbied governments, and in many countries, political parties, government departments, NGOs, charities, and foundations are now dedicated to environmental concerns. Yet the responsibility to care for our planet does not lie exclusively with formal groups. In the words of British campaigner Guy Shrubsole, whether we see ourselves as eco-warriors, amateur environmentalists, or just responsible citizens, "We, the public, have to step up to become nature's stewards and whistleblowers, its last and best line of defence".[55]

Although there are numerous sceptics, climate deniers and environmental abusers, there is a consensus that the human race should take action. Whether that means *saving from* (preservation: "the saving of species and wildernesses from damage or destruction") or *saving for* (conservation: "the saving of natural resources for later consumption")[56] is a finer point. But a glance at the history of this debate in its American context will be instructive in identifying some of the attitudes and approaches. At the heart of the debate is an interesting set of questions, which scholars, philosophers, artists, and environmentalists have struggled with for the last two hundred years: "why it matters when wildernesses and wild species are destroyed"; "whether, and how far, we should think of ourselves as having a duty to conserve resources for posterity", and whether we should "leave the future to look after itself".[57]

Stewardship: Three US Examples

Histories of American conservation typically focus on the contribution of three larger-than-life figures: the "revered great-grandfather of conservation"[58] John Muir (1838–1914); the "Father of American Forestry" Gifford Pinchot (1865–1946)[59]; and "champion of the wilderness"[60] Aldo Leopold (1887–1948), who is "regarded by many as the founder of the conservation movement in the US".[61] The approaches each had to environmentalism were quite different: Muir aimed "to preserve the American wilderness as a sanctuary for spiritual renewal"[62]; Pinchot pledged "to conserve resources once thought limitless for the continued prosperity of the American nation and the continued growth of American industry"[63]; and Leopold moved towards a theory of "ecological conscience", whereby "conservation is a state of harmony between men and land".[64] These different stances have become guiding principles for different approaches: ecocentrism (Muir), resource conservation (Pinchot), and the land ethic (Leopold).

John Muir (1838–1914)

John Muir was a Scot whose family emigrated to America when he was a boy. He fell in love with the beauty of his natural surroundings, especially once he moved to California, and he was encouraged by friends hearing his evocative descriptions of the land to commit them to writing. His articles gained huge national popularity, and the income supported his otherwise meagre earnings from shepherding and working in sawmills. This gave him leverage with politicians to persuade them to enact legislation to protect against the encroachment of the overgrazing sheep (who he called "hoofed locusts"),[65] the lumber mills that were devastating the sequoia forests, and a burgeoning tourist industry, in part enthused by Muir's own writings. "All sorts of human stuff is being poured into our valley this year", he lamented as early as 1870.[66] In lobbying for the area round Yosemite Valley to be dedicated as a national park, "he hoped both to save humans *for* the wilderness and the wilderness *from* humans".[67]

Muir's approach to the environment was observing the natural beauty of the land, advocating for ordinary people to engage with nature, and lobbying for the government to protect its resources, to preserve what he saw as the transcendence of God's gift to the earth. "The value of these forests in storing and dispensing the bounty of the mountain clouds is infinitely greater than lumber or sheep", he wrote in 1901[68];

> In their natural condition, or under wise management, keeping out destructive sheep, preventing fires, selecting the trees that should be cut for lumber, and preserving the young ones and the shrubs and sod of herbaceous vegetation, these forests would be a never failing fountain of wealth and beauty.[69]

Congress passed the Yosemite National Park Act in 1890, and two years later, Muir founded the Sierra Club, which remains an influential lobbying group. Its mission is still "to explore, enjoy, and protect the wild places of the earth; to practice and promote the responsible use of the earth's ecosystems and resources; to educate and enlist humanity to protect and restore the quality of the natural and human environment; and to use all lawful means to carry out these objectives".[70]

"Any fool can destroy trees", Muir once wrote; "God has cared for these trees, [...] but he cannot save them from fools".[71] In 1908, the area of woodland north of San Francisco was named in his honour as Muir Woods, and subsequently several other parks, associations, and natural features have taken his name, including Mount Muir in the Sierra Nevada and Muir Glacier in Alaska. Following his death in 1914, his collected writings were published to mark the foundation of the US National Park Service in 1916.

Gifford Pinchot (1865–1946)

In many ways Muir's approach to conservation was a vocation, but "in the early years of the new century this informal, amateur approach yielded to the modernist trends of professionalisation and bureaucratization".[72] If there was one man who embodied the spirit of professionalizing conservation, it was Gifford Pinchot.

Following a formative education in Europe, Pinchot returned to the US with the express desire to establish American forestry, and to transform the use of woodlands from an ad hoc free-for-all to a regulated, well-managed system. His commitment to careful research into trees, forests, management practices, and policy, and his practical experimentation in the field, enabled him to turn forestry into a science. Meanwhile, his political aptitude and ability to connect with people in power managed to effect real change in federal policy. He considered his early work at Biltmore—George Washington Vanderbilt's private estate in North Carolina—to be "the beginning of practical Forestry in America",[73] and his publication of *The White Pine* (1896) as initiating "the systematic study of American trees" by foresters.[74] Thanks to his work with the Forest Commission in the late 1890s, he was able to push for a slew of new forest reserves in 1897, and then became Chief of the Forestry Division in 1898. He formed the Society of American Foresters in 1900, and gained the support of Theodore Roosevelt from the beginning of his presidency. The transfer of the nation's public woodland to a newly instituted Forest Service in 1905, together with the publication of formal guidelines for forestry in *The Use of the National Forest Reserves* (1905) was a seminal moment. This "marked the beginning of sound, clean, and effective Government forest administration in America",[75] and ended years of poor woodland use, commercial abuse, and general ignorance of forestry needs. Finally, in 1908, the Roosevelt administration established the Commission on the Conservation of Natural Resources. Pinchot was appointed chairman.

Pinchot's lifelong belief was one of "utilitarian conservationism".[76] He did not agree with what he called the "denudatics" (or with Muir, or for that matter the Lorax) that forests should be completely left alone. That, he felt, would be "marching up a blind alley"; instead, "the job was not to stop the ax, but to regulate its use".[77] He believed "that natural resources such as lumber, coal, and water should be sustainably used".[78] He recognized that "trees could not only be protected but *managed* for sustained yields", explains Fox; "thus in his own person Pinchot embodied the transition from amateur protection to scientific management".[79]

In his biography, *Breaking New Ground* (1947), written a year before his death, Pinchot set out his views, comparing forestry to farming:

> To grow trees as a crop is Forestry. Trees may be grown as a crop just as corn may be grown as a crop. [...] A well-handled farm gets more and more productive as the years pass. So does a well-handled forest.[80]

Meanwhile, his views on conservation, which had been fundamental in building national knowledge and policy, especially between 1896 and 1910, were also made clear. Conservation was articulated as "the use of the natural resources for the greatest good of the greatest number for the longest time".[81] Pinchot was unequivocal:

> The Conservation of natural resources is the key to the future. It is the key to the safety and prosperity of the American people, and all the people of the world, for all time to come. The very existence of our Nation, and of all the rest, depends on conserving the resources which are the foundations of its life. That is why Conservation is the greatest material question of all.[82]

Pinchot's words were influential, and in just a short amount of time, America's attitudes towards its land were transformed. He relied on the backing of powerful leaders to effect such gains, and acknowledged

Roosevelt's support in *Breaking New Ground*.[83] Roosevelt reciprocated, writing in his own autobiography, "Gifford Pinchot is the man to whom the nation owes most for what has been accomplished as regards to the preservation of the natural resources of our country".[84]

Aldo Leopold (1887–1948)

Leopold, inspired by Pinchot, enrolled in the graduate programme at Yale School of Forestry before joining the US Forest Service in 1909. There, he developed a lasting legacy—his concept of *wilderness*. "I mean a continuous stretch of country preserved in its natural state", he wrote, "open to lawful hunting and fishing, big enough to absorb a two weeks' pack trip, and kept devoid of roads, artificial trails, cottages, or other works of man".[85] Where some (like Pinchot) feared that preserving wildernesses would be equivalent to "bottling up" their resources so that nobody could benefit,[86] and where others[87] assumed that the very idea of wilderness referred to wasteful land that had no purpose, Leopold's view was that such at-risk landscapes—prime areas for hunting, fishing, canoeing, and hiking—could help restore America's sense of freedom.[88]

Leopold's revisioning of land-use from exploitation to recreation was in part a way of smuggling land protection into the national conversation through more palatable human interests. On the other hand, when viewed as part of his holistic vision of the land, it reads human recreation as a part of the ecosystem: "Plants, animals, men, and soil are a community of interdependent parts, an organism", he stated; and as such "the ecological paradigm thus began to change from nature-as-organism to nature-as-society".[89]

More than this, Leopold remonstrated against the use of the land for economic gain, as had been Pinchot's vision: "a system of conservation based solely on economic self-interest is hopelessly lopsided", he chided; "It tends to ignore, and thus eventually to eliminate, many elements in the land community that lack commercial value, but that are [...] essential to its healthy functioning"[90]—elements that we now know bring benefits to mental wellbeing and to society. In a first step towards preserving such areas as sanctuaries, he successfully lobbied for Gila Wilderness, a 500,000-acre tract of New Mexico, to be designated the world's first wilderness area in 1924.

In 1933 he was appointed Professor of Game Management at the University of Wisconsin. Through observation, and influenced by European approaches, he refined America's understanding of how to handle big game. His insights were driven by a critique of existing practice: "wolves and lions are cleaned out of a wilderness area in the interest of big-game management", he observed, but "the big-game herds (usually deer or elk) then increase to the point of overbrowsing the range".[91] To remedy this, he developed ideas that would influence today's rewilding initiatives, "restoring big wilderness based on the regulatory roles of large predators",[92] and "creating wildlife corridors between existing reserves, linking them up to create wildernesses large enough to support native carnivores such as wolves and bears".[93]

Beyond advocacy for wilderness, Leopold would also lay the foundations of contemporary attitudes: he deplored what he called "land sickness": "the disappearance of plants and animal species without visible cause [...], and the irruption of others as pests",[94] but objected even more to the solutions of the day: "The practices we now call conservation", he wrote—referring to localized or short-term treatments using fertilizers or pest control solutions—"are, to a large extent, local alleviations of biotic pain. They are necessary, but they must not be confused with cures".[95] Rather than firefighting problems, he sought preventative measures by maximizing the underlying vitality of the land. "Health is the capacity of the land for self-renewal", he believed; "Conservation is our effort to understand and preserve this capacity".[96] In the end, he would critique both government approaches to conservation (which didn't understand local needs), and the short-sightedness of landowners (who couldn't see beyond their personal gain).[97] Instead, he saw solid land management or husbandry as vital for preserving America's land, cultural identity, and even its social make-up.

In Leopold's view, conservation should aspire to "a state of harmony between men and land",[98] and he promoted a "focus on human/nonhuman *relations* and the central role of *citizenship*".[99] This was the basis of his "ecological conscience" or "land ethic". "All ethics so far evolved rest upon a single premise: that the individual is a member of a community of interdependent parts", he argued; "The land ethic simply enlarges the boundaries of the community to include soils, waters, plants and animals, or collectively: the land".[100] By encouraging ordinary citizens to complement the conservation professionals at the heart of reform, he precipitated many of the environmental initiatives we now take for granted. Towards the end of his life, in 1947, Leopold became president of the Ecological Society of America, by which point he had distilled his views: "In short, a land ethic changes the role of *Homo sapiens* from conqueror of the land-community to plain member and citizen of it".[101]

To some extent the stances of these three individuals map onto the thematic development of American sensibilities to the land over the past 150 years: Muir reflected America's pioneer spirit, a lone figure benignly surveying the vast wilderness of the continent, yet embodying the conquering doctrine of manifest destiny; Pinchot represented the utilitarian and anthropocentric values of the "Progressive Age", in which the resources of the land were leveraged for the benefit of the people; and Leopold, especially in his later writings, became increasingly socially-conscious, championing the values of ecosystem health and citizen environmentalism.

Yet they also map onto another clear American sensibility that in lots of ways is profoundly more problematic. We noted earlier that histories of American conservation "typically focus" on these three figures. It won't have escaped notice that they are all white men, and in the US context, these are inheritors of the European settler legacy, the "dominant paradigm" in US and Western thought,[102] and a worldview that has been promulgated by "entities that operate through a lens of extractive personal gain, the control and maintenance of such gain, and the physical, spiritual, and epistemic superiority of the settler".[103] Scholars and researchers are nowadays acutely aware that other voices have historically been marginalized from the history books; what is equally problematic is the fact that this erasure has meant that "other" perspectives and worldviews have been overlooked.

When it comes to narratives about ecology and histories of the land, this is a concern since many cultural positions beyond those of the global North acknowledge a deep spiritual and existential affinity with the more-than-human world. In terms of ecological awareness in the North American context, looking beyond the dominant paradigm means engaging with the "lifeway" of first nation indigenous communities,[104] which Jennifer Grenz describes as "knowledge acquired from our deep relationship with the places we are from. Intergenerational knowledge, passed through our lineages about plants, animals, places, and things we did. Ancestral knowledge that is also simply 'in' us". Yet she also acknowledges that "While it has become trendy to incorporate our traditional knowledge into ecology, it is simply not enough".[105] Other scholars note similar concerns: "Science and scientists have in the past and continue at present", notes one report, to "impose harm on indigenous peoples"; "discount indigenous knowledge"; and "inappropriately reproduce, apply, or other wise use information derived from indigenous knowledge".[106] To Grenz, it is important for us to turn to an indigenous worldview "not to result in alternative perspectives on current scientific understanding", nor as "a simple exercise of integration of traditional knowledge into western scientific methodologies". Instead, "The application of the indigenous worldview pushes us into an entirely different world of research".[107]

Of course, there is not a homogenous indigenous worldview; "there are over five hundred distinct cultural traditions which still maintain sacred relations with the diverse terrains and life forms of the North American continent", John A. Grim concedes—and that's just one region.[108] Nevertheless, he presents as a representative example the worldview of the Apsaaloke people which "orients a person toward an affective, symbolic, mythopoeic, private, and ethical relationship between the human and the sacred realms of the earth and the sky". Crucially, it is not "a scientific-conservationist attitude which seeks to manage nature so as to preserve it for future use"[109]; in this respect, it departs from the dominant narratives of North American ecological awareness, especially those following the *preservation* or *conservation* assumptions of Muir and Pinchot which have established the principles of forestry and land management. First nation communities call on us "to decolonize ecological restoration so that we can heal the land together",[110] and refer variously to doctrines of respect, relationality, and reciprocity,[111] or respect, reciprocity, and confidentiality[112]; to "deep listening and engagement"[113]; and even to "a 'two-eyed seeing' approach", which moves beyond mere assimilation to adopt "an ethic and framework of knowledge coexistence and complementarity".[114]

But even if we amalgamate indigenous worldviews to a general understanding, what might that be?

Commonly, the perspective of indigenous peoples towards the land is taken to be one of synergy and holism: "Indigenous peoples often view nature as a densely tangled web of interrelated elements bound together in 'relationship' as opposed to a collection of discrete unrelated objects", write Robinson, Gellie, and Redvers.[115] It is "often guided by a land-use ethic held by indigenous people that supports interactions with nature in a respectful manner and in a way that allows for all creatures to coexist".[116] Although there are studies that show "many ancient societies did not live in harmony with their environment",[117] indigenous communities typically view themselves as possessing "higher levels of ecological consciousness" than settler communities.[118] And "ways of knowing" are also markedly different from Western epistemologies: indigenous knowledge is "abstract, qualitative, inclusive, intuitive, diachronic, and is formed from communal knowledge gained over time through practice and application", writes Hoagland.[119] In relation to ecology, Gazing Wolf et al. write, it includes "keen observations of Earth's physical processes and cycles of change recorded over millennia"; its own "theory and application of ontology, cosmology, eschatology, axiology, praxeology, and epistemology"; its own "standards of hypothesis testing, replication, data management, research ethics, and peer review"; and its own "mechanisms of intergenerational knowledge transmission guided by elders through the modalities of experiential learning, storytelling, art, music, and ceremony".[120] Even articulating indigenous knowledge can challenge settler epistemologies: Peter Cole, for example, in advancing his "academic take on 'Indigenous Traditions and Ecology'", presents his thoughts in a style that is "closer to the oral style of my Nation than standard academic format, more in the storytelling mode". His work is published as how we might perceive a journal article, though it reads more like how we might perceive a poem: "For me language is topology topography geo/graphy", he writes/says.[121] And why not? This is a way of "querying the scholarly norms, simplistic divides and cosmological assumptions that have provided impetus for the establishment, and continuance, of westernization over several thousand years"[122]; in short, it's a way to resist the dominant paradigm and what Jennifer Grenz calls out as its "ideology".[123]

Nowadays, judges Hoagland, "we have surpassed the traditional ecological knowledge dark ages; a time when individuals like Aldo Leopold [...] created poor, stereotypical images of Native Americans and traditional ecological knowledge".[124] Though their narratives might form the foundations of our historiographies about the land and our principles of protecting it, the work of Muir, Pinchot, and Leopold must be seen within a context—an *ecological* context, we might say.

Man and Nature

As we can see, the relationship between humans and nature begins with our instincts, commitments, and approaches to the land. Yet one of the challenges to taking on a more ecological relationship with the planet is that we tend to conceptualize the human species as distinct from the world around us: we think *anthropocentrically*. It was "the growth of the natural sciences in the seventeenth and eighteenth centuries", writes Anna Bramwell, that "produced a separation between ourselves and the world".[125] Others broadly concur, ascribing the separation to the development of modern scientific thinking, underpinned by the anthropocentric philosophies of major world religions,[126] and a procession of political decisions that divorced us increasingly from the land. According to ecologists like George Sessions, this has led to the belief "that modern civilized humans are *literally* separated from the rest of Nature in an artificial technological 'second nature' of their own creation".[127] "Civilization has 'transcended' or 'evolved out of' Nature", they believe; "and is thus not subject to evolutionary/ecological laws".[128] Either way, the distinction between "human" on the one side and "nature" on the other—the idea that humans, somehow, are not a part of nature—has become entrenched, leading to the perception of a nature/culture divide: Nathaniel Rivers refers to "a troubling environmental rhetoric that places humans outside of nature"[129]; "'nature' only has meaning when set in opposition to human works", writes anthropologist Philippe Descola[130]; while philosopher Bruno Latour writes of "the *background* of the majestic cycles of nature, against which human history had always stood out".[131] Elsewhere, performance scholar Baz Kershaw—whose work we will consider in Chapter 4—writes of "the separation of humans from their environment via 'nature'",[132] suggesting that "nature" itself is the artificial construct, distanced conceptually from human beings, yet constructed by human concerns. That he and other writers use scare quotes or capitals to parenthesize the word is a familiar ploy, and elsewhere, scholars even anthropomorphize or feminize the term into Mother Nature,[133] Dame Nature,[134] or Great Mother,[135] as if to further distance it. Some even seem unable to use the word, preferring phrases like the "Great Beyond"[136]; the "Great Outdoors"[137]; "Gaia"[138]; or "the Great Big Green Thing Out There".[139]

By way of interrogating "Nature", Elinor Fuchs and Una Chaudhuri consider the term "landscape" as it has emerged from the field of fine art, recognizing that not only does it place nature in the service of people, but it also privileges the sense of sight.[140] "Nature" is something that we gaze upon, that we passively survey, and over which we hold dominion, a perspective set up by hundreds of landscape works of art, either implicitly in the way they frame the viewer, or explicitly in the way they image human figures into those landscapes (see Fig. 2.3). Chaudhuri tracks the development of this privileged positioning from the inception of perspective in the fifteenth century, and goes on to consider how landscape's role in art moved progressively to become a discussion of social relations.[141]

Such views are also traced by Phil Macnaghten and John Urry in their book *Contested Natures*, where they argue that a certain sense of "environmental realism" in the imaging of "nature" led to an "idealism" projecting key values and relationships, and a subsequent "instrumentalism" of our behaviours vis à vis our natural surroundings.[142] They suggest that "the classical or Enlightenment mentality" depicted a "'savage' wilderness" that "had to be tamed, subdued and cultivated" (consider the storm scene from Shakespeare's *King Lear*); in contrast, "the more Romantic sense" reflected "reverence towards untouched and unmanaged wilderness" (think Walt Whitman); while as the twentieth century broke, images turned towards "'nature' as the 'unspoilt' other, as embodied in the relics, customs and mystery of the English countryside" (T.S. Eliot's "The Waste Land"). This was a nature that was "vulnerable, threatened by urban growth and industrial expansion, and in need of the state for protection in the collective interest".[143] In other words, "there is no singular nature as such, only natures. And such natures are historically, geographically, and socially constituted".[144] Through such machinations, writes John Adams, "the landscape gaze becomes a symbolic act of conquest and colonization",[145] and nature something to be "subdued and controlled".[146]

One of the most prolific scholars of ecology in the last few years has been the philosopher Timothy Morton, who has explored this topic in depth. "'Nature' is an arbitrary rhetorical construct", he writes, "empty of independent, genuine existence behind or beyond the texts we create about it".[147] And beyond that conceptual phenomenon, the actual *thing* we refer to as "nature" is "always 'over yonder', alien and alienated"[148]; "a reified thing in the distance, under the sidewalk, on the other side where the grass is always greener, preferably in the mountains, in the wild".[149] Morton turns to examples from Romantic literature to evidence

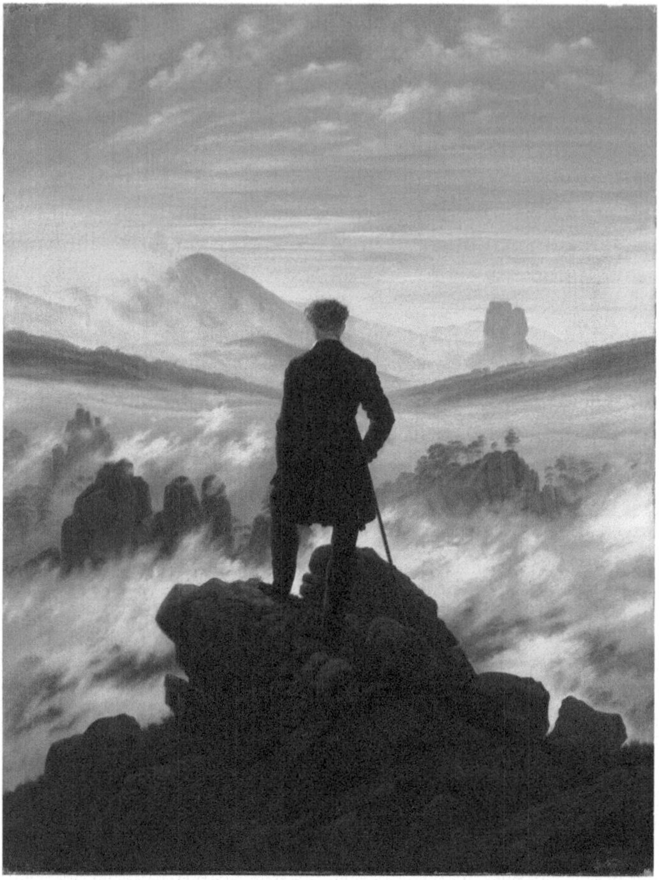

Fig. 2.3 Caspar David Friedrich (c.1818), "The hiker above the Sea of Fog" ("Der Wanderer über dem Nebelmeer"). Image credit: Caspar David Friedrich, Public domain, via Wikimedia Commons

what he calls "ambient poetics", which have played a part in the construction of this *thing*, "putting something called Nature on a pedestal and admiring it from afar".[150] He refers to a trope of representation that he calls "ecomimesis",[151] found extensively throughout the arts, especially in the Romantic period, and derived from the anthropocentric fascination of our human perspective: "The more you have of landscapes that convey the subjectivity of the implied viewer", Morton argues, "the more you have of hills, trees and water".[152] It's a trope found in painting, but also in music, poetry, and literature: the "often first-person rendering of situatedness 'in'", as he explains it[153]; the "as I write" type of expression.[154] Think of Wordsworth's "I wandered lonely as a cloud" (1804); Constable's "The Haywain" (1821); Grieg's "Morning" from *Peer Gynt* (1875), or virtually any imagery that evokes sentiment, beauty, or sublimation in an ostensibly "natural" setting. Through what Morton calls its "poetics of ambience", which he dismisses as "the materially embodied, meaningless, and inconsistent kernel of 'idiotic enjoyment' that sustains an otherwise discursive ideological field",[155] ecomimesis and its rhetorical accomplices, ecorhapsody,[156] and ecodidacticism,[157] "generate [...] a fantasy of nature as a surrounding atmosphere, palpable but shapeless".[158] This, he suggests, is an ideology of transcendence: "Nature becomes supernatural", he writes[159]; "a transcendental term in a material mask".[160]

At times Morton weaves his extensive discussion about *dark ecology* into a Marxist context, for two reasons. First, because for him such anthropocentric subjectivity is problematically ideological, imprinting an uncontested power dynamic onto the situation. "Ideology is not just in your head", he writes; "It's in the way some things appear 'natural'—rolling hills and greenery—as if the Industrial Revolution had never occurred".[161] In some ways, this is standard Marxist positioning. But the second reason Morton invokes Marx is rather more critical, because he sees Marx's relationship with the nonhuman, non-social world as typifying our general attitude towards ecological matters—bending to anthropocentric mystification just as much as the ideologically naive. "What such a Marxism calls *nature*", he observes, "is not actual trees and Arctic foxes but trees and foxes as they are metabolized by human economic relations". Put more simply: "Use value isn't 'what things really are for', but 'what things are for humans'".[162] And as an example? The observation (ironic on many levels) that "objections to wind farms and solar arrays are often based on arguments that they 'spoil the view'".[163]

And this is causing problems: "The smoothly functioning human world is now malfunctioning precisely because of this gap between our world, with its anthropocentric access modes, and actual reality".[164] In other words the positioning of the natural world as *something separate from us* ("Nature"; "over yonder") has caused a disturbance in the force. He is dismissive—or at least ambivalent—about typical gestures towards environmentalism, which, he acknowledges, are "broad and inconsistent"[165]; after all, many of those environmental gestures themselves position activism outside of or separate from the environment (such as going online to offset your carbon footprint by paying for a tree to be planted).[166] He is particularly critical of urgent calls to action: "All too often the siren song of the beautiful soul these days comes in the form of a call to act, *now!*", he laments, "which induces the guilt that cripples genuine action".[167] And as for sustainability: "what exactly is being sustained?", he asks; "'Sustainable capitalism' might be one of those contradictions in terms along the lines of 'military intelligence'".[168] Such a separation of nature is also "getting in the way of properly ecological forms of culture, philosophy, politics, and art", he maintains[169]—in other words, art that could be making a difference to our treatment of the natural world.

"The upshot of the radical Western split between people and nature", writes Edward Grumbine, is that as long as we consider nature "as a collection of resources for humans, [we] inhabit a world that categorically denies the full range of symbiotic relationships that may exist between people and wilderness".[170] In the words of Thomas Berry, "To be viable, the human community must move from its present anthropocentric norm to a geocentric norm of reality and value".[171]

> We must begin by accepting the fact that the life community, the community of all living species, is the greater reality and the greater value, and that the primary concern of the human must be the preservation and enhancement of this larger community.[172]

In the end, "The aesthetics of Nature truly impedes ecology", Morton writes[173]; "Ecology, if it means anything at all, means being *without* nature".[174]

Deep Ecology

This brings us to a concept introduced in the 1970s by the Norwegian Arne Naess (1912–2009): *deep ecology* is a form of ecological philosophy, or "ecosophy",[175] that responds to what Naess viewed as the inadequacy of *shallow ecology*, which he considered "anthropocentric" ("It views humans as above or outside of nature, as the source of all value, and ascribes only instrumental, or use value to nature"). By contrast,

> Deep ecology does not separate humans from the natural environment, nor does it separate anything else from it. It does not see the world as a collection of isolated objects but rather as a network of phenomena that are fundamentally interconnected and interdependent.[176]

The distinction between the shallow and the deep is crucial, as Naess clarifies: where shallow ecology represents a limited ambition for environmentalism—merely a "fight against pollution and resource depletion", with a central objective of sustaining "the health and affluence of people in the developed countries"[177]—deep ecology goes far further, "problematizing"[178] existing approaches to ecology like preservation and conservation, "cultivating ecological consciousness",[179] and even engaging in direct action.[180] "The essence of deep ecology—as compared with the science of ecology, and with what I call the shallow ecological movement", Naess explains, "is to ask deeper questions. The adjective 'deep' stresses that we ask why and how, where others do not".[181]

In taking up his ideas, Bill Devall and George Sessions summarized Naess's ecosophy in the 1980s by invoking two principal concepts or "ultimate norms": "Self-realization",[182] which Naess explains as moving beyond the ego, the social self, and even the metaphysical self to reach "the ecological self",[183] and "biocentric equality"[184] which reflects "the equal right [of all living beings] to live and blossom".[185]

The first norm, "Self-realization", situates a notion of "our selves", as Naess put it, "in, and of, Nature".[186] "The study of our place in the Earth household includes the study of ourselves as part of the organic whole", explain Devall and Sessions, distinguishing between "self" (lowercase) and "Self" (capitalized): "A nurturing nondominating society can help in the 'real work' of becoming a whole person", they write; "The 'real work' can be summarized symbolically as the realization of 'self-in-Self' where 'Self' stands for organic wholeness".[187] This understanding clearly confronts the "shallow" views that we are distinct from nature, and that we have the right to control or exploit it. "Ultimately, the recognition of value inherent in all living nature stems from the deep ecological awareness that nature and the self are one", another adherent of Naess opined.[188]

The second norm, "biocentric equality", according to Devall and Sessions, "is intimately related to Self-realization in the sense that if we harm the rest of Nature then we are harming ourselves". They explain further: "This process of the full unfolding of the self can also be summarized by the phrase, 'No one is saved until we are all saved', where the phrase 'one' includes not only me, an individual human, but all humans, whales, grizzly bears, whole rain forest ecosystems, mountains and rivers, the tiniest microbes in the soil, and so on". Thus any sense of human superiority—or worse, dominion—over nature was dismissed:

> Ecological consciousness and deep ecology are in sharp contrast with the dominant worldview of technocratic-industrial societies which regards humans as isolated and fundamentally separate from the rest of Nature, as superior to, and in charge of, the rest of creation.[189]

Naess's view of ecology exploded the boundaries of any ecosystem: "Through the extension of our understanding of the ecological context, it will ultimately be possible to develop a sense of belonging with a more expansive perspective: *ecospheric belonging*", he wrote; "We are a part of the ecosphere just as intimately as we are a part of our own society".[190] In many respects, deep ecology follows the tenets of spiritual philosophies such as Taoism, Buddhism, and first nation cultures, whose evocation of the Circle of Life is familiar (if often simplistically presented): "In the Circle of Life", according to indigenous American lore, "every being is no more, or less, than any other. We are all Sisters and Brothers. Life is shared with the bird, bear, insects, plants, mountains, clouds, stars, sun".[191] But that sense of belonging is key: just as "Family belonging, the tie of kinship, has a material basis in perceived togetherness and cooperation", so too does our belonging in the world: "an interdependence for the benefit of all"; "*a cognitive basis for a sense of belonging*".[192] And if this is the case—that we are all intimately interdependent with all living things—then by extension, any harm that befalls any other thing is a harm also done to ourselves.

If this transforms our understanding, it also behooves us to transform the way we think, calling for "*the necessity of a substantial change in economic, social, and ideological structures*",[193] as Naess put it. In place of the shallow environmental responses of conservation and preservation, an alternative approach—or rather, an alternative mindset—emerged: he called this "moral extensionism", though, more substantially, he invoked the idea of "ecological sensibility": "noninterference with natural processes", "limited intervention to repair environmental damage", and "a style of coinhabitation that involves the knowledgeable, respectful, and restrained use of nature".[194]

For those such as Paul Shepard who have taken up the call, such a sensibility is fundamental to our (mature) understanding of the world, our behaviour within it, and our very existence. "Man is in the world and his ecology is the nature of that in-ness", he wrote; and "ecological thinking [...] requires a kind of vision across boundaries" (one that encapsulates both Self-realization and biocentric equality). As such, the self becomes "ennobled and extended rather than threatened as part of the landscape and the ecosystem, because the beauty and complexity of nature are continuous with ourselves".[195] Such a conceptually disorienting view may challenge the quotidian image of ourselves, though we can see in numerous spiritual insights a version of this holism ("We are from the soil, we live of the soil, we shall return to the soil"; or even "ashes to ashes, dust to dust"). It's in the spirit of such holism that terms such as *ecology* and *ecosystem* most dynamically come into play as metaphors.

Ecology of Mind

We've seen how understandings of ecology have incorporated metaphor before, not only in our core definition of the word itself (lessons from nature's household), but also in phrases like *food web* and *ecosystem health*, and in comparisons of ecosystems to machines or organisms. We've also seen how the concept of *ecosystem* has been transferred to other domains such as the arts or media, to evoke the dynamic interrelations of those fields. But there is one further ecological metaphor we need to introduce. It comes from the evolutionary psychologist Gregory Bateson, and has since been taken up by other writers, not least the philosophical psychoanalyst Félix Guattari.

Bateson's *ecology of mind* suggests that we can imagine the ways in which ecology works as similar to the ways in which the mind works: "it is the notion that ideas are interdependent, interacting, that ideas live and die", he suggests; "You've got the sort of complicated, living, struggling, cooperating tangle like what you'll find on any mountainside with the trees, various plants and animals that live there".[196] So far, so good; this offers a neat analogy between one thing (the mind) and another (the mountainside). Bateson goes further with other aspects of the metaphor: now that we have accepted his basic premise, he introduces other terms from ecology, teasing us with more remote connections: "How do ideas interact? Is there some sort of natural selection which determines the survival of some ideas and the extinction or death of others? […] What are the necessary conditions for stability (or survival) of such a system or subsystem?"[197] He's stretching the metaphor, but beneath the playful suggestions their substance is rooted in ecological knowledge. For example, he draws attention to the difference between an idea being situated in a context (i.e., against a background or landscape, so to speak), and an idea being a part of a whole system (i.e., ecological).[198] We can see a direct correlation between this and the ways in which we frame ourselves with/in nature: is my thought (the idea), and by extension my human sensibility captured against the background of the landscape (as in that image by Caspar David Friedrich), or is it a part of the messy tangle of the mountainside? Bateson's answer is simple: "We are not outside the ecology for which we plan", he suggests; "we are always and inevitably a part of it".[199]

Like other thinkers, Bateson starts from the recognition of "ecological crisis", whose cause lies "in the combined action of (a) technological advance; (b) population increase; and (c) conventional (but wrong) ideas—about the nature of man and his relation to the environment".[200] Those wrong ideas are key: providing the preface to Bateson's book, Mark Engel states that "The essence of all our problems is bad thinking".[201]

The bad thinking, Bateson suggests, convinces us that "a) it's us *against* the environment; b) it's us *against* other men; c) it's the individual (or the individual company, or the individual nation) that matters; d) we *can* have unilateral control over the environment and must strive for that control; e) we live within an infinitely expanding 'frontier'; f) economic determinism is common sense; g) technology will do it for us".[202] In other words, like other critiques, Bateson dismisses the assumption that we can control nature, the problematic nature/culture divide such an assumption causes, and the misguided pursuit of economic growth through industrial progress. All of this "is not the only possible human way", he urges; *"It is conceivably changeable"*[203]—but "the only medicine", warns Engel, "is better thinking".[204]

Our route to better thinking is through changing the way we live to become more ecological, adopting a sensibility that is in better synergy with our environment, and allowing that mindset to form a conceptual map of how we understand knowledge. Bateson's recommendations are straight from the 1970s environmentalist's playbook: "consume unreplaceable natural resources only as a means to facilitate necessary change", he urges; "depend upon the energy income which Spaceship Earth derives from the sun"; and "us[e] as energy sources only photosynthesis, wind, tide, and water power".[205] In short, as Peter Harries-Jones summarizes, "the solution to ecological issues must be ecological".[206]

In his 1989 book *The Three Ecologies*, the philosopher Félix Guattari made a very similar point when he claimed that ("as in painting or literature") "by their very essence analytic cartographies extend beyond the existential Territories to which they are assigned".[207] Guattari opens with an epigraph from Gregory Bateson, and his theorizing of three ecologies—mental, social, and environmental—leans heavily on Bateson's thinking. He starts with what might be called a metaphorical "sidestep" (conceptualizing the functions of the mind, society, and the environment as ecologies), but the real substance of his idea comes when he loops what has been revealed through the metaphor back into its context.

Guattari also starts from a point of crisis, though the "ecological disequilibrium" he evokes applies more broadly ("human modes of life [...] progressively deteriorating"; "kinship networks" and "neighbourhood relations [...] reduced"; "domestic life [...] poisoned by the gangrene of mass-media consumption"; "family and married life [...] 'ossified'").[208] "Now more than ever", he writes, "nature cannot be separated from culture". So he issues a call to arms:

> in order to comprehend the interactions between ecosystems, the mechanosphere and the social and individual Universes of reference, we must learn to think 'transversally'.[209]

As we have seen, the recognition that the way we currently think is flawed, and therefore the invitation to change the way we think, has been central to many of the concepts we have so far explored; Guattari's call to think 'transversally' is one more example. He clarifies that our current epistemology derives from Enlightenment thinkers, whose assumption of rationality is a simplification, and he reiterates Bateson's conceptual understanding of the "ecology of ideas". Ecology is "an ongoing complex system, open-ended for slow change",[210] and the ecology of ideas "cannot be contained within the psychology of the individual"; instead, it "organizes itself into systems or 'minds', the boundaries of which no longer coincide with the participant individuals".[211]

Where Guattari extends Bateson's thinking is in articulating the contributions of not only a mental ecology but also a social ecology and an environmental ecology to an overall ecosophy. We need, he argues, "a nascent subjectivity; a constantly mutating socius; [and] an environment in the process of being reinvented".[212] Fundamentally, the way we imagine (and operate with/in) these spheres needs to be reconceptualized; and more specifically, the way that we lock ourselves into the "rhythms and refrains" of their constructs. We need to resist "the mythic and phantasmatic lure of psychoanalysis", for example, but also similarly repetitive discourses in "education, health, culture, sport, the arts, the media, and fashion".[213] Reorienting the way we think is also about reunderstanding the way in which those processes work—not according to the signifying assumptions about which we have been taught, but according to a different logic: "it is a logic of intensities, of auto-referential existential assemblages engaging in irreversible durations"—in short, an "eco-logic".[214]

Again, Guattari advocates the facility of art to contribute to such an evolution: "Are not the best cartographies of the psyche, or, if you like, the best psychoanalyses, those of Goethe, Proust, Joyce, Artaud and Beckett, rather than Freud, Jung and Lacan?", he asks[215]; at the very least, the "ways of operating [of the mental ecosophy] will be more like those of an artist" than a psychiatrist.[216] And it is the operating that is key: the "workings of art" rather than "works of art", as Mark Nash puts it; "the slide from product to process".[217]

Dark Ecology

Our final stage of the journey through the philosophical enquiries of *ecology* returns to Timothy Morton and his concept of "dark ecology". Here, the idea that ecology can be associated with the mind moves beyond the realms of metaphor.

Morton begins with a familiar refrain: "The current global crisis demands that we wake up and smell the total coffee", he writes characteristically.[218] But he too goes beyond the environment: "One of the things that modern society has damaged, along with ecosystems and species and the global climate", he writes, "is thinking".[219] We think only from a human perspective, as if we are the most important things in the world. And while we are caught in that anthropocentrism, we will continue to molest the world, continue to destroy it. Thus to "wake up and smell the total coffee", we need to find a new way of thinking, something to jolt us out of our habitual mindsets and recalibrate our ways. He labels this new thinking "the ecological thought", describing it as "a virus that infects all other areas of thinking",[220] and "the thinking of interconnectedness".[221] "The ecological thought is a thought about ecology, but it's also a thinking that is ecological".[222] And he means this in multiple ways: it's ecological in the sense that it thinks (cares for) our environs, and it's also ecological in the sense that it works like an ecosystem.

Morton's stance is that humans have not only contributed to damaging our environment, but that we have also habituated a way of thinking which prevents us from seeing our culpability. We have severed *the real* from *reality*, he suggests,[223] and now we arrogantly assume that "only humans [...] bestow meaning" on and in the world.[224] This hasn't been a recent phenomenon. "The Severing", as he calls it, occurred as we developed *agrilogistics*, "a specific logistics of agriculture that arose in the Fertile Crescent and that is still plowing ahead".

> Humans in Mesopotamia established villages with granaries. The storage and selection of grain pushed the harvested plants to evolve. Humans selected grain for its tastiness, ease of harvesting and other criteria favoured by the agrilogistical program.

This was "an agricultural programme so successful that it now dominates agricultural techniques planet-wide".[225]

We have already seen that Morton is not alone in making such a claim, which extends from previous commentaries about the nature/culture divide, and the foundational precepts of deep ecology. "Scaled up the evolutionary pressure was substantial", he writes; "Agrilogistics led rapidly to patriarchy, the impoverishment of all but a very few, a massive and rigid social hierarchy, and feedback loops of human–nonhuman interaction such as epidemics".[226] In the end, we papered over this "original trauma"[227] with what scientists more typically refer to as *shifting baseline syndrome*. "The 'landscape' look of agriculture", he writes, "is the original 'greenwashing'"[228]:

> It's in the way gigantic fields of rapeseed extend everywhere. It's in the way huge lonely front lawns extend a meaningless one-size-fits-all statement about individuality. It's in the way malls have gigantic parking lots, and housing lots have giant McMansions without so much as a garden.[229]

Yet contemporary manifestations of agrilogistics are only the most recent indications of our destruction. "Agrilogistics has been effective since the beginning", Morton writes,[230] and may at first have seemed relatively innocuous. We began to adapt the earth's resources to our own ends with analogue technologies like clothes and shoes to protect us against the elements (destroying animals and plants in the process); soon we fashioned walking sticks and spectacles to extend our bodily functions (carved from trees and mined from the earth's minerals); before long we were digging canals and laying roads. These may all have indicated "progress", but they came at the expense of the planet we live with/in. By the time we started fencing off territory, claiming ownership, and erecting borders, the damage was existential. "Humans did it, not jellyfish and not computers", he concedes; "But humans did it with the aid of beings that they treated as prostheses: nonhumans such as engines, factories, cows, and computers". And that is not all, because although "no one decided in 1790 to wreck the planet by emitting carbon dioxide and related gases", as he puts it, the damage has become anthropocentrically inevitable and relentless thanks to "viral ideas about agricultural logistics living rent-free in minds".[231] In other words, anthropocentrism has taken root in the very way we think. Nowadays, smart technologies derived from industrial-scale quarrying and global oil extraction fit on our wrists and into our pockets, turning individuals into superconnected components of a massive machine that has cybernetically repurposed the world. Once we have irreparably desolated our immediate surroundings, it's the anthropocentric habit to simply move on. "You turn your farmland into a desert, and you move west",[232] Morton objects.

But there is a way out, he eventually reasons—even if the solution requires us to strip back layers of well-meaning but ultimately doomed false starts. It is difficult to condense Morton's prolific and complex writings into a meaningful precis as we have tried to do here, but one of the reasons his work resonates with the impetus for our book is because he frequently turns to art (both "high" and "low") to witness glimpses of our escape. Ecological thinking "isn't just to do with the sciences of ecology", he insists; "[it] is to do with art, philosophy, literature, music, and culture".[233] "Art forms have something to tell us about the environment because they can make us question reality".[234]

> Art's ambiguous, vague qualities will help us think things that remain difficult to put into words. Reading poetry won't save the planet. Sound science and progressive social policies will do that. But art can allow us to glimpse beings that exist beyond or between our normal categories.[235]

To begin with, as we have seen, Enlightenment poets sought to address our vandalism of the environment by reclaiming its virtues, by placing it as central subject in their work. Nature was held up as either a "thing" of sublime aesthetic transcendence, or a background against which we were positioned. But this was not enough: "The 'thing' we call Nature becomes, in the Romantic period and afterward, a way of healing what modern society has damaged", Morton writes; yet "'Nature' is an arbitrary rhetorical construct, empty of independent, genuine existence behind or beyond the texts we create about it".[236] We may have idolized our natural surroundings, but we still did this with anthropocentric hubris.

Rather than succumbing to the ambient "blur"[237] of this type of representation—"ecomimesis"—Morton encourages a spirit of "ecocritique"[238] or "ecognosis"[239]: an ecological awareness that he likens to feminist theory[240] or queer theory[241] in their attempts to see beyond the shibboleths of patriarchy or heteronormativity. Instead, "ecological awareness is a detailed and increasing sense, in science and outside of it, of the innumerable interrelationships among lifeforms and between life and non-life", Morton writes[242]; it is a way of us beginning to recognize our interconnectedness with the planet; "it is something", he claims, "like co-existing"[243]—and according to him, such recognition helps us to abolish the idea "that we are living in an environment".[244]

In the most succinct account of his thesis, *Dark Ecology*,[245] Morton lays out his thoughts in three "threads", as if they are part of a web, or in his terms, a "mesh".[246] In the First Thread, he introduces dark ecology and indicates that ecognosis works on a far bigger scale, spatially and temporally—the scale of what he later calls "hyperobjects".[247] In the second thread he queries the conceptual boundaries that we construct around "things" like Nature, ecosystems, humans, and even our thoughts: "Ecosystems are made not only of trees, rock formations, and pigs (seemingly 'external' to the human)", he writes, "but also of thoughts, wishes, fantasies (seemingly 'inside' our heads)".[248] In order to extract ourselves from what he calls the "strange loop", we need to recalibrate this interconnectedness, seeing the earth's resources not as a means of survival or progress, but as a symbiosis of living (and non-living) things. So in the third thread he shows us a route map of sorts, revealing how we might attain a state of ecognosis: "we can't get to ecognostic society through further agrilogistic manipulation", he writes[249]; instead, we must descend through environmental guilt, shame, melancholy, and horror, and into the realm of toys ("meaningful collaborations between the arts, the humanities, and engineering, rather than the mutual suspicion that reigns today")[250]; at last, we must swim down through the hollow[251] of the "uncanny valley" to reach sadness, longing, and finally joy, "deep in the interior of life".[252] "The whole point", he proposes, "is to fight one's way back from the brink (species-cidal and suicidal) toward the comedy".[253] For Morton, this is a journey not towards *transcendence*, but instead to *subscendence*: "Play is subscendence", he clarifies, "connecting me with the Lego brick, the lichen, the activist network, the microbiome, the melting glacier".[254]

Throughout his writing, Morton returns to examples of art that in his judgement have gone some way towards doing this. For example, he suggests that Mary Shelley's *Frankenstein* is ecological "*not* because it compels us to care for a preexisting notion of nature, but because it questions the very idea of nature"[255]—contributing to the second thread of his thesis. He notes that artists such as Raphael depict an "'impossible' viewpoint" such as is "a cornerstone of the ecological thought"; and more than this, that a writer like Milton "achieves the ecological thought in form as well as in content",[256] paving the way for further re-representations of the ecosphere. In case we should think that it is only the culturally validated canon that affords the ecological thought, Morton also turns to popular culture, contemporary art, and even kitsch. He frequently refers to "Once in a Lifetime" by *Talking Heads*,[257] he has collaborated with Laurie Anderson, Yoko Ono, Olafur Eliasson, Jennifer Walshe and Björk, and he cites experiences at Glastonbury as formative.[258] Of course those artists may just reveal Morton's taste, though he also says "all art is ecological".[259] As he writes, "art sometimes gives voice to what is unspeakable elsewhere, either temporarily—one day we will find the words—or intrinsically—words are impossible".[260] In one of his shorter essays, Morton suggests that we have entered a new post-Romantic artistic era (the "Age of Asymmetry") in which our artistic expression is guided not by our human-anthropocentric imperative, but by a symbiotic "collaboration between humans and non-humans". He notes this, rather ironically, by calling upon our artistic use of prostheses: "when you write a poem you are making a deal with some paper, some ink, wordprocessing software, trees, editors and air (and more)".[261] Does this mean that for Morton there are some uses of planetary resources that are compatible with the ecological thought? As if by response, he tells us that "Ecology isn't about 'resources' [...]. 'Resources' is one of those ideas of something 'over yonder' that the ecological thought deletes".[262] Rather, ecological art is "art that includes its environment(s) in its very form".[263]

"Art in the Asymmetric Phase seems to have three properties", Morton explains, listing these as 1) Demonic force; 2) Hypocrisy; and 3) Collaboration between humans and nonhumans.[264]

First, when he refers to art as "an attunement to a demonic force coming from the nonhuman and permeating it", he is borrowing the "demonic" idea from Plato, but referring to electromagnetic, gravitational, and quantum waves that are all around us. Ecological art, he believes, works best when the artist tunes into these waves, something that Felix Hess's *Air Pressure Fluctuations* (2001) does quite literally, but something that, more obliquely, Morton understands us all to be doing all the time, filtered through objects in the world. "I never hear the wind in itself, only the wind in the chimney, the wind in the trees", he explains, picking up on Heidegger in recognizing that sound and therefore music is an interaction with sound waves manipulated through space, and visual art is caused by distortions of light bouncing on to our retinas, and so on. "We tune to the environment, then we tune to our tuning", he writes; "Tuning in this respect is attuning the art object (voice, breath, instrument) to these physical waves, quite literally".[265] Perhaps we can recognize in this a corollary to the common citing by artists of concepts like inspiration; the sense that (in Elgar's words) "music is in the air all around us", or (according to Bob Dylan), "The song was there before me, before I came along".[266]

Morton's second discussion of "hypocrisy" is split into two sections. Here he suggests that the hypocrisy of ecological art is defined by a) weakness and b) irony. He notes that art (i.e. representation), one stage removed from reality, can only be a pale imitation of that reality, and therefore weak (hence "asymmetrical"). Meanwhile, irony is something that we cannot avoid but must embrace. As an example, he cites the idea of Jonah realizing that he is inside the digestive system of the whale, alive yet at the same time being eaten. Of course, Jonah's is a singular experience, but one to which we can relate by analogy: we are always a part of some bigger being—a "hyperobject" such as the universe, global warming, or capitalism—that we can only experience through a "sensual impression".[267] For us, the impression resonates: we are the centre of our own attention and feel subjectively to be supremely significant; though we are in fact an insignificant moment of that hyperobjective existence, as ecological art causes us to realize. Thus "irony is coexistence without center or edge", as he puts it.[268]

In the context of these explanations, understanding Morton's third point, about ecological art as a collaboration between humans and nonhumans, becomes more straightforward. If we are constantly in the wash, so to speak, of the electromagnetic waves of the universe, and if artistry means us synchronizing our tuning to them as closely as we can, we are then in a type of collaboration with the world around us. "Beauty is what happens when an object and its tuning fit so snugly that they fuse together, in a kind of loving extinction", Morton explains,[269] and these objects can be (and are) anything and everything. "More and more, art in the Asymmetric Phase tries to come as close to that magic bullet as possible".[270]

Morton's work can be philosophically rewarding, though it's perhaps rather less useful in terms of practical solutions to environmental crisis. Still, for us to understand the myriad complexities of ecological thought, we need a whole community of thinkers, and for the purposes of this study, we need to explore the intellectual terrain of other disciplines and areas. In the next chapter, we will turn to the second of our theoretical priorities, the field of *community*, and we will start with a field trip—join us as we hook up with a small community to plant a tree.

Notes

1. Jim Robbins (2012), *The Man Who Plants Trees*, London: Profile Books, p. xvi.
2. Robbins, *The Man*, p. xix.
3. Later: "It's estimated that over a fifty-year lifetime, a tree provides $162,000 in ecosystem services, including $62,000 in air pollution control and $31, 250 in soil erosion, and that is a very conservative estimate because things like atmospheric cooling and protection from ultraviolet radiation cannot be valued". Robbins, *The Man*, p. 200.
4. Robbins, *The Man*, pp. xviii–xix.
5. Robbins, *The Man*, p. 5.
6. Jean Giono (1996), *The Man Who Planted Trees* (trans. Barbara Bray), London: The Harvill Press, p. 8.
7. Giono, *The Man*, pp. 16–17.
8. Giono, *The Man*, p. 24.

9. Giono, *The Man*, p. 32.
10. Giono, *The Man*, p. 34.
11. Ernst Haeckel (1866), *Allgemeine Entwickelungsgeschichte der Organismen*, Berlin: Verlag von Georg Reimer, pp. 235, 286.
12. Phil Macnaghten and John Urry (1998), *Contested Natures*, London: SAGE Publications, p. 147.
13. Oswald J. Schmitz (2007), *Ecology and Ecosystem Conservation*, Washington, DC: Island Press, p. 1.
14. Steward T.A. Pickett, Jurek Kolasa, and Clive G. Jones, (2007), *Ecological Understanding: The Nature of Theory and the Theory of Nature* (second edition), Amsterdam: Elsevier, p. 142.
15. Pickett et al., *Ecological Understanding*, p. 196.
16. Eugene P. Odum and Gary W. Barrett (2004), *Fundamentals of Ecology* (fifth edition), USA: Cengage Learning, p. 8. See also Francis C. Evans (1956), "Ecosystem as the Basic Unit in Ecology", *Science*, 123:3208, pp. 1127–1128; Thomas P. Burns (1992), "Ecosystem: a powerful concept and paradigm for ecology", *Bulletin of the Ecological Society of America*, 73:1, pp. 39–43; Frank B. Golley (1993), *A history of the ecosystem concept in ecology: more than the sum of the parts*, New Haven: Yale University Press.
17. Odum and Barrett, *Fundamentals*, p. 5.
18. US Department of the Interior Bureau of Land Management (n.d.), "Chapter Introduction: Ecology of the Table Rocks", *Table Rocks Curriculum*, https://www.blm.gov/sites/blm.gov/files/tablerocks_ecology-intro.pdf. Accessed 16 July 2024.
19. Odum and Barrett, *Fundamentals*, p. 9.
20. Odum and Barrett, *Fundamentals*, p. 5.
21. US Department, "Chapter Introduction".
22. Robert V. O'Neill et al (1986), *A hierarchical concept of ecosystems*, Princeton: Princeton University Press, p. 1.
23. O'Neill et al., *A Hierarchical*, pp. 8–9.
24. Odum and Barrett, *Fundamentals*, p. 28.
25. Aldo Leopold (1970), *A Sand County Almanac: with essays on conservation from Round River*, New York: Oxford University Press, p. 252.

26. Neutralism, competition (direct interference), competition (resource use), amensalism, commensalism, parasitism, predation, protocooperation, mutualism. Odum and Barrett, *Fundamentals*, p. 283.
27. Anna Bramwell (1989), *Ecology in the 20th Century: A History*, New Haven: Yale University Press, p. 4.
28. Eugene P. Odum (1968), "Energy flow in ecosystems: a historical review", *American* Zoologist, 8:1, pp. 11–18, pp. 11–12; see also Odum, *Fundamentals*, pp. 63–77.
29. Original source: Howard T. Odum (1971), *Environment, Power, and Society*, New York: Wiley-Interscience.
30. Pickett et al., *Ecological Understanding*, p. 11.
31. "They try to study ecosystems that are as far as possible self-contained, and they divide each one into discrete compartments ('strata'), such as the tree canopy, dead wood and soil". Philip J. Stewart (2000), "Eddies in the flow: towards an inclusive ecology". In Duncan Poore (ed.) (2000), *Where Next? Reflections on the Human Future*, Kew: Royal Botanic Gardens, pp. 271–288, p. 274.
32. Evans, "Ecosystem", p. 1127.
33. Paul Shepard (1995), "Ecology and Man—a Viewpoint", in George Sessions (ed.) (1995), *Deep Ecology for the Twenty-First Century*, Boston: Shambhala, pp. 131–140, p. 134.
34. Arthur George Tansley (1935), "The Use and Abuse of Vegetational Concepts and Terms", *Ecology*, 16:3, pp. 284–307.
35. Golley, *A History*, p. 2.
36. Bramwell, *Ecology*, p. 84.
37. Macnaghten and Urry, *Contested Natures*, p. 2.
38. O'Neill et al., *A Hierarchical*, pp. 13–14.
39. O'Neill et al., *A Hierarchical*, p. 54.
40. United Nations (2022), "What Is The Triple Planetary Crisis?", *Climate Change*, 13 April, https://unfccc.int/news/what-is-the-triple-planetary-crisis. Accessed 18 August 2024.
41. Bea Amador and Sudhanshu Malhotra (2023), "The climate emergencies of 2023", *Greenpeace*, 15 December, https://www.greenpeace.org/international/story/64537/the-climate-emergencies-of-2023/. Accessed 18 August 2024.
42. Red Cross (2024), "Everyone has a role to play in tackling the climate crisis. Words alone are not enough", *Red Cross Climate*

Centre, 15 August, https://www.climatecentre.org/14389/everyone-has-a-role-to-play-in-tackling-the-climate-crisis-words-alone-are-not-enough/. Accessed 18 August 2024.

43. Earth.Org (2024), "15 Biggest Environmental Problems of 2024", *Earth.Org*, https://earth.org/the-biggest-environmental-problems-of-our-lifetime/. Accessed 18 August 2024.
44. United Nations (2024), "Facts about the climate emergency", *United Nations Environment Programme*, https://www.unep.org/facts-about-climate-emergency. Accessed 18 August 2024.
45. Phoebe Weston (2024), "No birdsong, no water in the creek, no beating wings: how a haven for nature fell silent", *The Guardian*, 16 April, https://www.theguardian.com/environment/2024/apr/16/nature-silent-bernie-krause-recording-sound-californian-state-park-aoe. Accessed 18 August 2024.
46. Rachel Carson (2000), *Silent Spring*, Dublin: Penguin Random House, p. 5.
47. Gaylord Nelson (1970), "Foreword: What Kind of Legacy?", in Walt Anderson (ed). (1970), *Politics and environment: a reader in ecological crisis*, Pacific Palisades: Goodyear, pp. xi–xvi, p. xiv.
48. John Passmore (1974), *Man's Responsibility for Nature; ecological problems and Western traditions*, New York: Scribner, p. 3.
49. Passmore, *Man's Responsibility*, p. ix.
50. George Perkins Marsh (1864), *Man and Nature: Physical Geography as Modified by Human Action*, New York: Scribner, p. 13.
51. Clavé was citing a report to the French government from 1853. Jules Clavé (M. J. Cleve) (1881), "Forest-Culture in Alpine Ravines", *Popular Science Monthly*, 19. https://en.wikisource.org/wiki/Popular_Science_Monthly/Volume_19/October_1881/Forest-Culture_in_Alpine_Ravines#cite_note-1. Accessed 17 August 2024.
52. "Les esprits clairvoyants en sont déjà à prévoir une crise qui pourrait bien être terrible". Jules Clavé (1862), *Études sur l'économie forestière*, Paris: Guillaumin, p. 261.
53. Fairfield Osborn (1948), *Our Plundered Planet*, Boston: Little, Brown and Company, p. 201.
54. Osborn, *Our Plundered Planet*, p. 37.
55. Guy Shrubsole (2024), "Stewardship", in Nick Hayes and Jon Moses (eds) (2024), *Wild Service: Why Nature Needs You*, London: Bloomsbury, pp. 47–57, p. 57.

56. Passmore, *Man's Responsibility*, p. 73.
57. Passmore, *Man's Responsibility*, p. x, p. 79.
58. Harry Jenkinson (2024), "Kinship", in Hayes and Moses, *Wild Service*, pp. 81–94, p. 81.
59. Thomas M. Bonnicksen and Diana M. Burton (2003), "Forest Policy Development in the United States", in Raymond A. Young and Ronald L. Giese (eds) (2003), *Introduction to Forest Ecosystem Science and Management* (third edition), Hoboken: John Wiley & Sons, pp. 5–37, pp. 17–18.
60. Harold C. Bradley (1951), "Aldo Leopold: Champion of the Wilderness", *Sierra Club Bulletin*, 36:5, pp. 14–18.
61. Katherine L. Gross (2001), "Address of the Past President. It's a Good Time to Get Involved: Perspectives on the ESA, from Aldo Leopold to Today", *Bulletin of the Ecological Society of America*, 82:2, pp. 112–18, p. 113.
62. Michael B. Smith (1998), "The Value of a Tree: Public Debates of John Muir and Gifford Pinchot", *The Historian*, 60:4, pp. 757–778, p. 759.
63. Smith, "The Value", p. 759.
64. Leopold, *A Sand*, p. 243.
65. John Muir (1916), *My First Summer in the Sierra*, Boston: Houghton Mifflin, p. 56.
66. William Frederic Bade (1924), *The Life and Letters of John Muir*, Boston: Houghton Mifflin, p. 221.
67. Stephen Fox (1985), *The American Conservation Movement: John Muir and his Legacy*, Madison: University of Wisconsin Press, p. 58.
68. John Muir (1901), *Our National Parks*, Boston: Houghton Mifflin, p. 329.
69. Muir, *Our National*, p. 360.
70. The Sierra Club (2024), "Policies: Mission Statement", *Sierra Club website*, https://www.sierraclub.org/policy. Accessed 13 August 2024.
71. Muir, *Our National*, pp. 364–365.
72. Fox, *The American*, p. 108.
73. Gifford Pinchot (1947), *Breaking New Ground*, New York: Harcourt, Brace and Company, p. 50.
74. Pinchot, *Breaking*, p. 306.
75. Pinchot, *Breaking*, p. 265.

76. Char Miller (2001), *Gifford Pinchot and the Making of Modern Environmentalism*, Washington: Island Press, p. 121.
77. Pinchot, *Breaking*, pp. 27–29.
78. Miller, *Gifford Pinchot*, p. 4.
79. Fox, *The American*, p. 111.
80. Pinchot, *Breaking*, p. 31.
81. Pinchot, *Breaking*, p. 326.
82. Pinchot, *Breaking*, p. 355.
83. Pinchot, *Breaking*, p. 383.
84. Theodore Roosevelt (1914), *Theodore Roosevelt: an Autobiography*, New York: The Macmillan Company, p. 409.
85. Aldo Leopold (1921), "The Wilderness and its Place in Forest Recreational Policy", *Journal of Forestry*, 19:7, pp. 718–721, p. 719.
86. Leopold, "The Wilderness", p. 718.
87. Passmore, *Man's Responsibility*, pp. 107–110.
88. Curt Meine (2010), *Aldo Leopold: His Life and Work*, Madison: University of Wisconsin Press, p. 197.
89. Susan L. Flader and J. Baird Callicott (1991), "Introduction", in Susan L. Flader and J. Baird Callicott (eds) (1991), *The River of the Mother God: and Other Essays by Aldo Leopold*, Madison: University of Wisconsin Press, pp. 3–31, p. 6. The quote is from the text of Leopold's lecture "The Arboretum and the University" (1934).
90. Leopold, *A Sand*, p. 251.
91. Leopold, *A Sand*, p. 268.
92. Michael Soulé and Reed Noss (1998), "Rewilding and biodiversity: complementary goals for continental conservation", *Wild Earth*, 8, pp. 19–28, p. 22.
93. Graham Lawton (2018), "The Call of Rewilding", *New Scientist*, 240:3199, pp. 34–38, p. 34.
94. Leopold, *A Sand*, p. 272–273.
95. Leopold, *A Sand*, p. 274.
96. Leopold, *A Sand*, p. 258.
97. Passmore, *Man's Responsibility*, p. 187.
98. Leopold, *A Sand*, p. 243.
99. Teena Gabrielson and R. McGreggor Cawley (2010), "Plain member and citizen: Aldo Leopold and environmental citizenship", *Citizenship Studies*, 14:5, pp. 605–615, p. 612.

100. Leopold, *A Sand*, p. 239.
101. Leopold, *A Sand*, p. 240.
102. Jennifer Berneda Grenz (2020), "Healing the Land by Reclaiming an Indigenous Ecology: A journey exploring the application of the Indigenous worldview to invasion biology and ecology", Ph.D. thesis, Vancouver: University of British Columbia, p. 10.
103. Joseph Gazing Wolf et al (2024), "Centering Indigenous Knowledges in ecology and beyond", Frontiers in Ecology and Environment, 22:7, e2776, p. 2.
104. John A. Grim (1993), "Native North American Worldviews and Ecology", in Mary Evelyn Tucker and John A. Grim (eds), *Worldviews and Ecology*, Lewisburg: Bucknell University Press, pp. 41–54, pp. 42–43.
105. Grenz, "Healing", p. 1.
106. Tyler D. Jessen et al (2022), "Contributions of Indigenous Knowledge to ecological and evolutionary understanding", *Frontiers in Ecology and the Environment*, 20:2, pp. 93–101, p. 99.
107. Grenz, "Healing", p. 15.
108. Grim, "Native North American", p. 41.
109. Grim, "Native North American", p. 45.
110. Grenz, "Healing", p. 14.
111. Grenz, "Healing", p. 3.
112. Jessen et al., "Contributions", p. 100.
113. Jake M. Robinson, Nick Gellie, and Nicole Redvers (2021), "Traditional ecological knowledge in restoration ecology: a call to listen deeply, to engage with, and respect Indigenous voices", *Restoration Ecology*, 29:4, e13381, p. 2.
114. Jessen et al., "Contributions", p. 99.
115. Robinson, Gellie, and Redvers, "Traditional ecological knowledge", p. 2.
116. Serra Jeanette Hoagland (2017), "Integrating Traditional Ecological Knowledge with Western Science for Optimal Natural Resource Management", *IK: Other Ways of Knowing*, 3:1, p. 2.
117. Fikret Berkes and Mina Kislalioglu Berkes (2009), "Ecological complexity, fuzzy logic, and holism in indigenous knowledge", *Futures*, 41:1, pp. 6–12, p. 6.
118. Gazing Wolf et al., "Centering", p. 1.
119. Hoagland, "Integrating", p. 2.

120. Gazing Wolf et al., "Centering", p. 1.
121. Peter Cole (1998), "An Academic Take on 'Indigenous Traditions and Ecology'", *Canadian Journal of Environmental Education*, pp. 100–115, p. 100.
122. Amba J. Sepie (2017), "More than Stories, More than Myths: Animal/Human/Nature(s) in Traditional Ecological Worldviews", *Humanities*, 6:78, p. 2.
123. Grenz, "Healing", p. 16.
124. Hoagland, "Integrating", p. 9.
125. Bramwell, *Ecology*, p. 30.
126. George Sessions (1995), "Ecosophism and the Anthropocentric Detour", in Sessions, *Deep Ecology*, pp. 156–183.
127. George Sessions (1995), "Introduction (Part Two)", in Sessions, *Deep Ecology*, pp. 97–103, p. 99.
128. Sessions, "Introduction (Part Two)", p. 97.
129. Nathaniel A. Rivers (2018), "Deep ambivalence and wild objects: Toward a strange environmental rhetoric", in Joshua Gunn and Diane Davis (eds) (2018), *Fifty Years of Rhetoric Society Quarterly: Selected Readings 1968–2018*, New York: Routledge, pp. 322–342.
130. Philippe Descola (2013), *Beyond Nature and Culture* (trans. Janet Lloyd), Chicago: University of Chicago Press, p. 8.
131. Bruno Latour (2017), *Facing Gaia: Eight Lectures on the New Climatic Regime*, London: Polity Press, pp. 92–93.
132. Baz Kershaw (2007), *Theatre Ecology: Environments and Performance Events*, Cambridge: Cambridge University Press, p. 316.
133. Dennis E. Jelinski (2010), "On the notions of mother nature and the balance of nature and their implications for conservation", in Daniel G. Bates and Judith Tucker (eds) (2010), *Human ecology: Contemporary research and practice*, New York: Springer, pp. 37–50.
134. Simon Schaffer (2013), "Dame Nature Cares Nothing for Us", in Pasquale Gagliardi, Anne Marie Reijnen, and Philipp Valentini (2013), *Protecting Nature, Saving Creation: Ecological Conflicts, Religious Passions, and Political Quandaries*, New York: Palgrave Macmillan, pp. 43–61.
135. Thomas Berry (1994), "The Viable Human", in Sessions, *Deep Ecology*, pp. 8–18, p. 16.

136. Marc Boglioli (2015), "Into the Great Beyond: Ontology and the Non-Human", *Reviews in Anthropology*, 44:4, pp. 225–236.
137. Ridvan Askin (2021), "Narrating the 'Great Outdoors'", in Marco Caracciolo, Marlene Karlsson Marcussen and David Rodriguez (2021), *Narrating Nonhuman Spaces: Form, Story, and Experience beyond Anthropocentrism*, New York: Routledge, pp. 201–220.
138. James Lovelock (2016), *Gaia*, Oxford: Oxford University Press.
139. Stewart, "Eddies", p. 274.
140. Elinor Fuchs and Una Chaudhuri (eds) (2002), *Land/Scape/Theater*, Ann Arbor: University of Michigan Press.
141. Una Chaudhuri, "Land/Scape/Theory", in Fuchs and Chaudhuri, *Land/Scape/Theater*, pp. 11–29, p. 11.
142. Macnaghten and Urry, *Contested Natures*, pp. 1–2.
143. Macnaghten and Urry, *Contested Natures*, pp. 34–38.
144. Macnaghten and Urry, *Contested Natures*, p. 15.
145. John Adams (2005), "Performative Locations: Wilderness Space and Place in Early Film", in Gabriella Giannachi and Nigel Stewart (eds) (2005), *Performing Nature: Explorations in Ecology and the Arts*, Oxford: Peter Lang, pp. 85–101, p. 88.
146. Macnaghten and Urry, *Contested Natures*, p. 7.
147. Timothy Morton (2007), *Ecology Without Nature: Rethinking Environmental Aesthetics*, Cambridge: Harvard University Press, pp. 21–22.
148. Timothy Morton (2010), *The Ecological Thought*, Cambridge: Harvard University Press, p. 5.
149. Morton, *Ecological Thought*, p. 3.
150. Morton, *Ecology Without*, p. 5.
151. Morton, *Ecology Without*, p. 77.
152. Timothy Morton (2012), "Art in the Age of Asymmetry: Hegel, Objects, Aesthetics", *Evental Aesthetics*, 1, pp. 121–142, p. 130.
153. Timothy Morton (2013), *Hyperobjects: Philosophy and Ecology After the End of the World*, Minneapolis, University of Minnesota Press, p. 5.
154. Morton, *Hyperobjects*, p. 194.
155. Morton, *Ecology Without*, p. 67.
156. Morton, *Ecology Without*, p. 56.
157. Morton, *Ecology Without*, p. 61.
158. Morton, *Ecology Without*, p. 77.

159. Morton, *Ecology Without*, pp. 18–19.
160. Morton, *Ecology Without*, p. 14.
161. Morton, *Hyperobjects*, p. 106.
162. Timothy Morton (2016), *Dark Ecology: For a Logic of Coexistence*, New York: Columbia University Press, p. 26.
163. Morton, *Hyperobjects*, p. 105.
164. Timothy Morton (2017), *Humankind; Solidarity with Nonhuman People*, London: Verso, p. 74.
165. Morton, *Ecology Without*, p. 9.
166. Though Morton is equally dismissive of the more embedded approaches of "ecophenomenology". Morton, *Hyperobjects*, p. 18.
167. Morton, *Hyperobjects*, p. 92.
168. Morton, *Hyperobjects*, p. 111.
169. Morton, *Ecology Without*, p. 1.
170. Edward Grumbine (1995), "Wildness, Wise Use, and Sustainable Development", in Sessions, *Deep Ecology*, pp. 376–396, p. 377.
171. Thomas Berry (1995), "The Viable Human", in Sessions, *Deep Ecology*, pp. 8–18, p. 8.
172. Berry, "The Viable Human", p. 10.
173. Morton, *Hyperobjects*, p. 105.
174. Morton, *Ecology Without*, p. 204.
175. See Arne Naess (1973), "The Shallow and the Deep, Long-Range Ecological Movement", *Inquiry*, 16, pp. 95–100; Stephan Bodian (1982), "Simple in Means, Rich in Ends: An Interview with Arne Naess", in Sessions, *Deep Ecology*, pp. 26–36, p. 27; Arne Naess (2001), *Ecology, Community and Lifestyle: Outline of an Ecosophy* (trans. David Rothenberg), Cambridge: Cambridge University Press, p. 32.
176. Fritjof Capra (1995), "Deep Ecology: A New Paradigm", in Sessions, *Deep Ecology*, pp. 19–25, p. 20.
177. Naess, "The Shallow", p. 95.
178. Arne Naess (1995), "Deepness of Questions and the Deep Ecology Movement", in Sessions, *Deep Ecology*, pp. 204–212, p. 205.
179. Bill Devall and George Sessions (1985), *Deep Ecology*, Layton: Gibbs Smith, p. 8.
180. Devall and Sessions, *Deep Ecology*, p. x.
181. Bodian, "Simple in Means", p. 27.

182. Devall and Sessions, *Deep Ecology*, p. 66.
183. Arne Naess (1995), "Self-Realization: An Ecological Approach to Being in the World", in Sessions, *Deep Ecology*, pp. 225–239, p. 226.
184. Devall and Sessions, *Deep Ecology*, p. 67. Naess uses the rather more lofty phrase, "biospherical egalitarianism": Naess, "The Shallow", p. 95.
185. Naess, "The Shallow", p. 96.
186. Naess, "Self-Realization", p. 226.
187. Devall and Sessions, *Deep Ecology*, pp. 66–67.
188. Fritjof Capra (1995), "Deep Ecology: A New Paradigm", in Sessions, *Deep Ecology*, pp. 19–25, p. 20.
189. Devall and Sessions, *Deep Ecology*, pp. 65–68.
190. Naess, *Ecology, Community*, pp. 168, 165.
191. George Sessions (1995), "Ecosophism and the Anthropocentric Detour", in Sessions, *Deep Ecology*, pp. 156–183, p. 158, citing American Indian philosophy.
192. Naess, *Ecology, Community*, p. 168.
193. Naess, "Deepness of Questions", p. 211.
194. John Rodman (1995), "Four Forms of Ecological Consciousness Reconsidered", in Sessions, *Deep Ecology*, pp. 121–130, pp. 125–126.
195. Paul Shepard (1995), "Ecology and Man—a Viewpoint", in Sessions, *Deep Ecology*, pp. 131–140, pp. 131–132.
196. Gregory Bateson (1991), *A Sacred Unity: Further Steps to an Ecology of Mind*, New York: Cornelia and Michael Bessie Books, p. 265.
197. Gregory Bateson (1976), *Steps to an Ecology of Mind*, New York: Ballantine Books, pp. xv–xvi.
198. Bateson, *Steps*, pp. 338–339.
199. Bateson, *Steps*, p. 504.
200. Bateson, *Steps*, p.488.
201. Mark Engel (1971), "Preface", in Bateson, *Steps*, pp. vii–viii, p. viii.
202. Bateson, *Steps*, p. 492.
203. Bateson, *Steps*, p. 493.
204. Engel, "Preface", p. viii.
205. Bateson, *Steps*, pp. 495–496.

206. Peter Harries-Jones (2002), *A Recursive Vision: Ecological Understanding and Gregory Bateson*, Toronto: University of Toronto Press, p. 211.
207. Félix Guattari (2018), *The Three Ecologies*, London: Bloomsbury, p. 40.
208. Guattari, *Three Ecologies*, p. 27.
209. Guattari, *Three Ecologies*, p. 43.
210. Bateson, *Steps*, p. 494.
211. Guattari, *Three Ecologies*, p. 54.
212. Guattari, *Three Ecologies*, p. 68.
213. Guattari, *Three Ecologies*, pp. 38–39.
214. Guattari, *Three Ecologies*, p. 44.
215. Guattari, *Three Ecologies*, p. 37.
216. Guattari, *Three Ecologies*, p. 35.
217. Nash is picking up on the ideas of Iain Chambers. Mark Nash (2023), "Three Ecologies: Conversations with Newton Harrison", in *Field: A Journal of Socially Engaged Art Criticism*, 23, https://field-journal.com/issue-23/three-ecologies-conversations-with-newton-harrison/#:~:text=Of%20course%20the%20challenge%20is,very%20different%20notion%20of%20subjectification. Accessed 22 August 2024.
218. Morton, *Ecological Thought*, p. 40.
219. Morton, *Ecological Thought*, p. 3.
220. Morton, *Ecological Thought*, p. 2.
221. Morton, *Ecological Thought*, p. 7.
222. Morton, *Ecological Thought*, p. 7.
223. Morton's actual words: "The Severing is a foundational, traumatic fissure between, to put it in stark Lacanian terms, *reality* (the human-correlated world) and *the real* (ecological symbiosis of human and nonhuman parts of the biosphere)". Morton, *Humankind*, p. 13.
224. Timothy Morton (2012), "The Oedipal Logic of Ecological Awareness", *Environmental Humanities*, 1, pp. 7–21, p. 9.
225. Timothy Morton (2015), "What is Dark Ecology?", *Changing Weathers*, https://www.changingweathers.net/en/episodes/48/what-is-dark-ecology/#:~:text=In%20this%20essay,%20which%20draws%20on%20his%20book. Accessed 30 September 2024, n.p.
226. Morton, "What is Dark Ecology?", n.p.

227. Chellis Glendinning (1995), "Recovery from Western Civilization", in Sessions, *Deep Ecology*, pp. 37–40, p. 37.
228. Morton, *Hyperobjects*, p. 106.
229. Morton, *Dark Ecology*, p. 154.
230. Morton, *Humankind*, p. 45.
231. Morton, *Dark Ecology*, p. 21.
232. Morton, *Humankind*, p. 45.
233. Morton, *Ecological Thought*, p. 4.
234. Morton, *Ecological Thought*, p. 8.
235. Morton, *Ecological Thought*, p. 60.
236. Morton, *Ecology Without*, pp. 21–2.
237. Morton, *Ecology Without*, p. 15.
238. Morton, *Ecology Without*, p. 13.
239. Morton, *Dark Ecology*, p. 5.
240. Morton, *Dark Ecology*, p. 65.
241. Morton, *Ecology Without*, p. 13.
242. Morton, *Hyperobjects*, p. 128.
243. Morton, *Dark Ecology*, p. 5.
244. Morton, *Hyperobjects*, pp. 128–129.
245. A truncated account of *Dark Ecology* can be found in Morton, "What is Dark Ecology?".
246. Morton, *Ecological Thought*, pp. 28–33.
247. Morton, *Hyperobjects*.
248. Morton, *Dark Ecology*, p. 67.
249. Morton, *Dark Ecology*, p. 120.
250. Morton, *Dark Ecology*, p. 142.
251. Morton, *Dark Ecology*, p. 147.
252. Morton, *Dark Ecology*, p. 157.
253. Morton, *Dark Ecology*, p. 154.
254. Morton, *Dark Ecology*, p. 116.
255. Morton, *Ecology Without*, p. 194.
256. Morton, *Ecological Thought*, p. 23.
257. See Morton, *Being Ecological*, p. 12; Aengus Anderson, Micah Saul, and Timothy Morton (2012), "The Conversation #10—Timothy Morton", in *The Conversation*, 29 May, http://opentranscripts.org/transcript/the-conversation-timothy-morton/. Accessed 2 August 2024.
258. Leslie Carol Roberts (2019), "Eco-Thoughts: An Interview with Timothy Morton (Vol. 1)", *The Believer*, 16 October,

https://www.thebeliever.net/logger/an-interview-with-timothy-morton/. Accessed 22 August 2024.

259. Morton, *Being Ecological*, p. 12.
260. Morton, *Ecological Thought*, p. 12.
261. Morton, "Art in the Age", pp. 137–138.
262. Morton, *Ecological Thought*, p. 122.
263. Morton, *Being Ecological*, p. 52.
264. Morton, "Art in the Age", p. 134.
265. Morton, "Art in the Age", pp. 135–136. A whole chapter of *Being Ecological* is dedicated to the idea of *tuning*.
266. Charles Ford cites both of these examples in Charles Ford (2010), "Musical Presence: Towards a New Philosophy of Music", *Contemporary Aesthetics*, 8:1, n.p. Available at https://digitalcommons.risd.edu/cgi/viewcontent.cgi?article=1235&context=liberalarts_contempaesthetics. Accessed 3 August 2024.
267. Morton, "Art in the Age", p. 134.
268. Morton, "Art in the Age", p. 137.
269. Morton, "Art in the Age", p. 138.
270. Timothy Morton (2012), "From modernity to the Anthropocene: ecology and art in the age of asymmetry", *International Social Science Journal*, 63:207–8, pp. 39–51, p. 49. (This is a slightly revised version of Morton, "Art in the Age").

Bibliography

Amador, Bea, and Sudhanshu Malhotra (2023), "The climate emergencies of 2023", *Greenpeace*, 15 December, https://www.greenpeace.org/international/story/64537/the-climate-emergencies-of-2023/. Accessed 18 August 2024.

Anderson, Aengus, Micah Saul, Timothy Morton (2012), "The Conversation #10—Timothy Morton", in *The Conversation*, 29 May, http://opentranscripts.org/transcript/the-conversation-timothy-morton/. Accessed 2 August 2024.

Anderson, Walt (ed). (1970), *Politics and environment: a reader in ecological crisis*, Pacific Palisades: Goodyear.

Anon. (n.d.), "Act Establishing Yellowstone National Park", https://www.loc.gov/rr/program/bib//ourdocs/yellowstone.html#:~:text=An%20act%20establishing%20Yellowstone%20National,created%20the%20first%20national%20park. Accessed 16 August 2023.

Anon. (n.d.), "Brief History of the National Parks", *Library of Congress*, https://www.loc.gov/collections/national-parks-maps/articles-and-essays/brief-history-of-the-national-parks/#:~:text=Yellowstone%20became%20the%20first%20national,%2C%20Agriculture%2C%20and%20the%20Interior. Accessed 16 August 2023.

Anon. (n.d.), "The Great Outdoors Initiative", https://obamawhitehouse.archives.gov/administration/eop/ceq/initiatives/ago. Accessed 16 August 2023.

Anon. (n.d.), "Yosemite Act, 1864", *National Park Service*, https://www.nps.gov/parkhistory/online_books/anps/anps_1a.htm. Accessed 16 August 2023.

Bade, William Frederic (1924), *The Life and Letters of John Muir*, Boston: Houghton Mifflin.

Bates, Daniel G., and Judith Tucker (eds) (2010), *Human ecology: Contemporary research and practice*, New York: Springer.

Bateson, Gregory (1976), *Steps to an Ecology of Mind*, New York: Ballantine Books.

Bateson, Gregory (1991), *A Sacred Unity: Further Steps to an Ecology of Mind*, New York: Cornelia and Michael Bessie Books.

Berkes, Fikret, and Mina Kislalioglu Berkes (2009), "Ecological complexity, fuzzy logic, and holism in indigenous knowledge", *Futures*, 41:1, pp. 6–12.

Boglioli, Marc (2015), "Into the Great Beyond: Ontology and the Non-Human", *Reviews in Anthropology*, 44:4, pp. 225–236.

Bradley, Harold C. (1951), "Aldo Leopold: Champion of the Wilderness", *Sierra Club Bulletin*, 36:5, pp. 14–18.

Bramwell, Anna (1989), *Ecology in the 20th Century: A History*, New Haven & London: Yale University Press.

Burns, Thomas P. (1992), "Ecosystem: a powerful concept and paradigm for ecology", *Bulletin of the Ecological Society of America*, 73:1, pp. 39–43.

Caracciolo, Marco, Marlene Karlsson Marcussen, and David Rodriguez (2021), *Narrating Nonhuman Spaces: Form, Story, and Experience beyond Anthropocentrism*, New York: Routledge.

Carson, Rachel (2000), *Silent Spring*, Dublin: Penguin Random House.

Char Miller (2001), *Gifford Pinchot and the Making of Modern Environmentalism*, Washington: Island Press.

Clavé, Jules (1862), *Études sur l'économie forestière*, Paris: Guillaumin, p. 261.

Clavé, Jules (M.J. Cleve) (1881), "Forest-Culture in Alpine Ravines", *Popular Science Monthly*, 19. https://en.wikisource.org/wiki/Popular_Science_Monthly/Volume_19/October_1881/Forest-Culture_in_Alpine_Ravines#cite_note-1. Accessed 17 August 2024.

Cole, Peter (1998), "An Academic Take on 'Indigenous Traditions and Ecology'", *Canadian Journal of Environmental Education*, pp. 100–115.

Descola, Philippe (2013), *Beyond Nature and Culture* (trans. Janet Lloyd), Chicago: University of Chicago Press.

Devall, Bill, and George Sessions (1985), *Deep Ecology*, Layton: Gibbs Smith.

Earth.Org (2024), "15 Biggest Environmental Problems of 2024", *Earth.Org*, https://earth.org/the-biggest-environmental-problems-of-our-lifetime/. Accessed 18 August 2024.

Elinor Fuchs and Una Chaudhuri (eds) (2002), *Land/Scape/Theater*, Ann Arbor: University of Michigan Press.

Evans, Francis C. (1956), "Ecosystem as the Basic Unit in Ecology", *Science*, 123:3208, pp. 1127–1128.

Flader, Susan L., and J. Baird Callicott (eds) (1991), *The River of the Mother God: and Other Essays by Aldo Leopold*, Madison: University of Wisconsin Press.

Ford, Charles (2010), "Musical Presence: Towards a New Philosophy of Music", *Contemporary Aesthetics*, 8:1. https://digitalcommons.risd.edu/cgi/viewcontent.cgi?article=1235&context=liberalarts_contempaesthetics. Accessed 3 August 2024.

Fox, Stephen (1985), *The American Conservation Movement: John Muir and his Legacy*, Madison: University of Wisconsin Press.

Gabrielson, Teena, and R. McGreggor Cawley (2010), "Plain member and citizen: Aldo Leopold and environmental citizenship", *Citizenship Studies*, 14:5, pp. 605–615.

Gagliardi, Pasquale, Anne Marie Reijnen, and Philipp Valentini (2013), *Protecting Nature, Saving Creation: Ecological Conflicts, Religious Passions, and Political Quandaries*, New York: Palgrave Macmillan.

Gazing Wolf, Joseph, Danielle D. Ignace, Dominique M. David-Chavez, Lydia L. Jennings, Deondre Smiles, Paulette Blanchard, Ellen Simmons, Diana Doan-Crider, Ruth Plenty Sweetgrass-She Kills, Michelle Montgomery, Melissa K. Nelson, Linda Black Elk, Luke Black Elk, Gwen Bridge, Ann Marie Chischilly, Kevin Deer, Kathy DeerinWater, Trudy Ecoffey, Judith Vergun, Daniel Wildcat, and James Rattling Leaf (2024), "Centering Indigenous Knowledges in ecology and beyond", *Frontiers in Ecology and Environment*, 22:7, e2776.

Giannachi, Gabriella, and Nigel Stewart (eds) (2005), *Performing Nature: Explorations in Ecology and the Arts*, Oxford: Peter Lang.

Golley, Frank B. (1993), *A history of the ecosystem concept in ecology: more than the sum of the parts*, New Haven: Yale University Press.

Grenz, Jennifer Berneda (2020), "Healing the Land by Reclaiming an Indigenous Ecology: A journey exploring the application of the Indigenous worldview to invasion biology and ecology", Ph.D. thesis, Vancouver: University of British Columbia.

Gross, Katherine L. (2001), "Address of the Past President. It's a Good Time to Get Involved: Perspectives on the ESA, from Aldo Leopold to Today", *Bulletin of the Ecological Society of America*, 82:2, pp. 112–118.

Guattari, Félix (2018), *The Three Ecologies*, London: Bloomsbury.

Gunn, Joshua, and Diane Davis (eds) (2018), *Fifty Years of Rhetoric Society Quarterly: Selected Readings 1968–2018*, New York: Routledge.

Haeckel, Ernst (1866), *Allgemeine Entwickelungsgeschichte der Organismen*, Berlin: Verlag von Georg Reimer.

Halderman, Logan, Mitchell Zischke, Fred Whitford, Jarred Brooke, David Osborne, Amy Thompson, Bill Horan, and Jonathan Ferris (2021), "More Than Just A Puddle: Understanding The Pond Ecosystem For Better Management", *Pond Management*, National Oceanic and Atmospheric Administration, https://repository.library.noaa.gov/view/noaa/49649/noaa_49649_DS1.pdf. Accessed 30 September 2024.

Harries-Jones, Peter (2002), *A Recursive Vision: Ecological Understanding and Gregory Bateson*, Toronto: University of Toronto Press.

Hayes, Nick, and Jon Moses (eds) (2024), *Wild Service: Why Nature Needs You*, London: Bloomsbury.

Hirsch, Tim (2005), "Living Beyond Our Means: Natural Assets and Human Well-being: Statement from the Board", *Millennium Ecosystem Assessment*, https://www.millenniumassessment.org/en/index.html. Accessed 7 April 2024.

Hoagland, Serra Jeanette (2017), "Integrating Traditional Ecological Knowledge with Western Science for Optimal Natural Resource Management", *IK: Other Ways of Knowing*, 3:1.

Jessen, Tyler D., Natalie C. Ban, Nicholas XEMŦOLTW Claxton, and Chris T. Darimont (2022), "Contributions of Indigenous Knowledge to ecological and evolutionary understanding", *Frontiers in Ecology and the Environment*, 20:2, pp. 93–101.

Kershaw, Baz (2007), *Theatre Ecology: Environments and Performance Events*, Cambridge: Cambridge University Press.

Latour, Bruno (2017), *Facing Gaia: Eight Lectures on the New Climatic Regime*, London: Polity Press.

Lawton, Graham (2018), "The Call of Rewilding", *New Scientist*, 240:3199, pp. 34–38.

Leopold, Aldo (1921), "The Wilderness and its Place in Forest Recreational Policy", *Journal of Forestry*, 19:7, pp. 718–721.

Leopold, Aldo (1970), *A Sand County Almanac: with essays on conservation from Round River*, New York: Oxford University Press.
Lovelock, James (2016), *Gaia*, Oxford: Oxford University Press.
Macnaghten, Phil and John Urry (1998), *Contested Natures*, London: SAGE Publications.
Marsh, George Perkins (1864), *Man and Nature: Physical Geography as Modified by Human Action*, New York: Scribner.
Meine, Curt (2010), *Aldo Leopold: His Life and Work*, Madison: University of Wisconsin Press.
Morton, Timothy (2007), *Ecology Without Nature: Rethinking Environmental Aesthetics*, Cambridge: Harvard University Press.
Morton, Timothy (2010), *The Ecological Thought*, Cambridge: Harvard University Press.
Morton, Timothy (2012), "Art in the Age of Asymmetry: Hegel, Objects, Aesthetics", *Evental Aesthetics*, 1, pp. 121–142.
Morton, Timothy (2012), "From modernity to the Anthropocene: ecology and art in the age of asymmetry", *International Social Science Journal*, 63:207–208, pp. 39–51.
Morton, Timothy (2012), "The Oedipal Logic of Ecological Awareness", *Environmental Humanities*, 1, pp. 7–21.
Morton, Timothy (2013), *Hyperobjects: Philosophy and Ecology After the End of the World*, Minneapolis, University of Minnesota Press.
Morton, Timothy (2015), "What is Dark Ecology?", *Changing Weathers*, https://www.changingweathers.net/en/episodes/48/what-is-dark-ecology/#:~:text=In%20this%20essay,%20which%20draws%20on%20his%20book. Accessed 30 September 2024.
Morton, Timothy (2016), *Dark Ecology: For a Logic of Coexistence*, New York: Columbia University Press.
Morton, Timothy (2017), *Humankind; Solidarity with Nonhuman People*, London: Verso.
Muir, John (1916), *My First Summer in the Sierra*, Boston: Houghton Mifflin.
Naess, Arne (1973), "The Shallow and the Deep, Long-Range Ecological Movement", *Inquiry*, 16, pp. 95–100.
Naess, Arne (1984), "What is Basic in Deep Ecology?", *Ecophilosophy*, VI, p. 5. http://environmentalethics.info/ecophilosophy.html. Accessed 22 July 2024
Naess, Arne (1986), "The Basics of Deep Ecology", *The Trumpeter*, 21:1, pp. 61–71.
Naess, Arne (2001), *Ecology, Community and Lifestyle: Outline of an Ecosophy* (trans. David Rothenberg), Cambridge: Cambridge University Press.

Nash, Mark (2023), “Three Ecologies: Conversations with Newton Harrison”, in *Field: A Journal of Socially Engaged Art Criticism*, 23, https://field-journal.com/issue-23/three-ecologies-conversations-with-newton-harrison/#:~:text=Of%20course%20the%20challenge%20is,very%20different%20notion%20of%20subjectification. Accessed 22 August 2024.

O’Neill, Robert V., Donald Lee DeAngelis, J. B. Waide, and Timothy F. H. Allen (1986), *A hierarchical concept of ecosystems*, Princeton: Princeton University Press.

Odum, Eugene P. (1968), “Energy flow in ecosystems: a historical review”, *American* Zoologist, 8:1, pp. 11–18

Odum, Eugene P. and Gary W. Barrett (2004), *Fundamentals of Ecology* (fifth edition), USA: Cengage Learning.

Odum, Howard T. (1971), *Environment, Power, and Society*, New York: Wiley-Interscience.

Olson, Mark, Alfonso Arroyo-Santos and Francisco Vergara-Silva (2019), “A User’s Guide to Metaphors In Ecology and Evolution”, *Trends in Ecology and Evolution*, 34:7, pp. 605–615.

Osborn, Fairfield (1948), *Our Plundered Planet*, Boston: Little, Brown and Company.

Passmore, John (1974), *Man’s responsibility for nature; ecological problems and Western traditions*, New York: Scribner.

Pickett, Steward T. A., Jurek Kolasa, and Clive G. Jones, (2007), *Ecological Understanding: The Nature of Theory and the Theory of Nature* (second edition), Amsterdam: Elsevier.

Pinchot, Gifford (1947), *Breaking New Ground*, New York: Harcourt, Brace and Company.

Poore, Duncan (ed.) (2000), *Where Next? Reflections on the Human Future*, Kew: Royal Botanic Gardens.

Red Cross (2024), “Everyone has a role to play in tackling the climate crisis. Words alone are not enough”, *Red Cross Climate Centre*, 15 August, https://www.climatecentre.org/14389/everyone-has-a-role-to-play-in-tackling-the-climate-crisis-words-alone-are-not-enough/. Accessed 18 August 2024.

Roberts, Leslie Carol (2019), “Eco-Thoughts: An Interview with Timothy Morton (Vol. 1)”, *The Believer*, 16 October,

Robinson, Jake M., Nick Gellie, and Nicole Redvers (2021), “Traditional ecological knowledge in restoration ecology: a call to listen deeply, to engage with, and respect Indigenous voices”, *Restoration Ecology*, 29:4, e13381.

Roosevelt, Theodore (1914), *Theodore Roosevelt: an Autobiography*, New York: The Macmillan Company.

Schmitz, Oswald J. (2007), *Ecology and Ecosystem Conservation*, Washington, DC: Island Press.

Sepie, Amba J. (2017), "More than Stories, More than Myths: Animal/Human/Nature(s) in Traditional Ecological Worldviews", *Humanities*, 6:78.

Sessions, George (ed.) (1995), *Deep Ecology for the Twenty-First Century*, Boston: Shambhala, pp. 131–140.

Smith, Michael B. (1998), "The Value of a Tree: Public Debates of John Muir and Gifford Pinchot", *The Historian*, 60:4, pp. 757–778.

Soga, Masashi, and Kevin J. Gaston (2018), "Shifting baseline syndrome: causes, consequences, and implications", *Frontiers in Ecology and the Environment*, 16:4, pp. 222–230.

Soulé, Michael, and Reed Noss (1998), "Rewilding and biodiversity: complementary goals for continental conservation", *Wild Earth*, 8, pp. 19–28.

Suzanne Simard et al (1997), "Net transfer of carbon between tree species with shared ectomycorrhizal fungi", *Nature*, 388, pp. 579–582.

Tansley, Arthur George (1935), "The Use and Abuse of Vegetational Concepts and Terms", *Ecology*, 16:3, pp. 284–307.

The Sierra Club (2024), "Policies: Mission Statement", *Sierra Club website*, https://www.sierraclub.org/policy. Accessed 13 August 2024.

Tucker, Mary Evelyn, and John A. Grim (eds), *Worldviews and Ecology*, Lewisburg: Bucknell University Press.

United Nations (2022), "What Is The Triple Planetary Crisis?", *Climate Change*, 13 April, https://unfccc.int/news/what-is-the-triple-planetary-crisis. Accessed 18 August 2024.

United Nations (2024), "Facts about the climate emergency", *United Nations Environment Programme*, https://www.unep.org/facts-about-climate-emergency. Accessed 18 August 2024.

US Department of the Interior Bureau of Land Management (n.d.), "Chapter Introduction: Ecology of the Table Rocks", *Table Rocks Curriculum*, https://www.blm.gov/sites/blm.gov/files/tablerocks_ecology-intro.pdf. Accessed 16 July 2024.

Weston, Phoebe (2024), "No birdsong, no water in the creek, no beating wings: how a haven for nature fell silent", *The Guardian*, 16 April, https://www.theguardian.com/environment/2024/apr/16/nature-silent-bernie-krause-recording-sound-californian-state-park-aoe. Accessed 18 August 2024.

Young, Raymond A., and Ronald L. Giese (eds) (2003), *Introduction to Forest Ecosystem Science and Management* (third edition), Hoboken: John Wiley & Sons.

CHAPTER 3

Community

So how do you plant a tree? Let's ask an expert.

Japanese botanist Akira Miyawaki has inspired thousands of people with the popular "Miyawaki Method" of tree planting. Concerned about pollution and environmental destruction, he proposed that "environmental protection forests" should be planted in urban areas.[1] Since the 1970s, he has planted "over 40 million trees in 15 countries",[2] establishing a practical, affordable, and efficient approach to improving the environment. The method encourages the development "of a mature natural forest in a comparatively small amount of time",[3] writes Hannah Lewis; instead of the 150–200 years that it would take a conventional mature woodland to develop, a Miyawaki forest can reach maturity in as little as 15–20 years.[4]

One of the most appealing innovations of the Method is that it champions "public involvement in eco-restoration", with grassroots communities participating in a "mini-forest revolution".[5] Like Miyawaki's other initiatives, such as his South-East Asian tree-planting festivals,[6] or his "Great Forest Wall of Tōhoku",[7] this approach emphasizes "crowdforesting"[8]—working with community organizations on mass planting events "to help solve a shared problem that calls for all hands on deck".[9] As Monika Egerer and Michael Suda explain, "an important component of Tiny Forests is the involvement of local residents in the planning, implementation, and maintenance of the forests to ensure that their

K. Savage and D. Symonds, *Ecologies of Community in Performance*, https://doi.org/10.1007/978-3-031-94343-0_3

perspectives, wishes and needs for the design and function of nearby urban greenspaces are considered".[10] When Miyawaki received a Blue Planet Award for his work in 2006, he acknowledged that "the main participants of forest creation are citizens. It is the local citizens of all ages who plant trees, with sweat on their brow and hands in the earth".[11] In particular, notes Kamala Thiagarajan, "the projects aim to connect city dwellers, especially children, with nature".[12] As such, the Miyawaki Method is a process that brings communities together in pursuit of "transnational ecological consciousness, regeneration and community empowerment",[13] and in so doing, builds social capital among participants. "The communal activity of forest planting is a way forward", writes Lewis; "It is a way to revitalise our common lands, communicate hope, and allow each of us to take responsibility".[14]

The first step is to conduct "field surveys of the local vegetation"[15] to identify appropriate species to plant in the area: "The tree species must be chosen from the forest communities of the region", Miyawaki writes, explaining that native trees will respond best to the nutrients in the soil.[16] The topsoil is prepared to a depth of around 20–30 centimetres, the soil is analysed, and where necessary the site is primed with straw or organic fertilizer if required. Meanwhile, seedlings are propagated in pots for up to two years to facilitate their growth. "The major objective is to grow seedlings with strong root systems because the root system is the key to successful survival of the plantings and the stabilization of the soil",[17] explains Miyawaki. Tens of thousands of seedlings can be planted in order to have sufficient coverage for the method. On that sort of scale, the planting is a major operation: "Planting of seedlings occurs during a planting festival", writes Miyawaki, "including local authorities, families, school children and guests".[18] The area is portioned into one metre squares, and participants "are invited to select seedlings from tanks of water and plant the seedlings in the prepared holes in the squares on the site".[19] Egerer and Suda explain: "One main tree species, one secondary tree species and one shrub are randomly planted per square", they note.[20] When transferred to the ground, the seedlings are planted more densely than is usual, in order to cause them to compete with one another, a trick that enhances their vertical growth. For the first two to three years, the soil will need tending, with mulch material added and weeds removed. Once the crowns of the trees spread though, very little sunlight will reach the forest floor, meaning that weed growth reduces. "Three years after planting", writes Miyawaki, "the site basically becomes maintenance free".[21]

The attractions of the Miyawaki Method are both practical and scientific. Practically, it can be carried out in the smallest of spaces ("sometimes the size of a tennis court"),[22] meaning that projects can be organized in available plots of land with relatively few logistical obstacles. This also means that the method appeals to community groups and local neighbourhoods, who often have appropriate parcels of land on their doorstep, and who by banding together can plant the trees in a single day. Scientifically, it is a process that achieves very speedy results, since it bypasses the usual successional stages of ecosystem growth by populating the area with carefully selected climax vegetation from the start. "Miyawaki forests are easy to establish, self-sustaining and require very little maintenance", writes Mike Charkow, while their "high biodiversity planting with species suited to a site means that the trees will be more resilient".[23]

It's true that there are some drawbacks: "labor need is high", acknowledges one report, "and planting costs are quite expensive because of the high plant density required"[24]; another report cites both cost concerns and ecological reservations in calling the Miyawaki Method "wildly unreasonable, illogical and inappropriate"[25]; meanwhile, the Miyawaki Method has also received criticism for its trade in "symbolic capital"[26]—the attempt by companies, organizations, or administrations to raise their green credentials through public shows of environmentalism which may have little substance.

Nevertheless, according to researchers, Miyawaki Forests bring a wealth of ecological and societal benefits. Studies report an exponential increase in general biodiversity,[27] while in urban areas the mini-forests offset greenhouse gas emissions,[28] provide natural flood management,[29] mitigate the adverse effects of urban over-heating,[30] and bring numerous health and wellbeing effects.[31] More than this, the social gains reported from communal tree-planting activities further commend Miyawaki's approach not only as beneficial for individuals, but also for communities—researchers cite "improved public-health outcomes"[32] and in some cases, greater social cohesion.[33] Participants "engaged in future civic activities because of tree planting",[34] claims one account, noting that "neighbors also took better care of their yards, spent more time being physically active outdoors, and spent more time outside in general because of the planting"[35]; another account even reported "increased voter turnout in primary and general elections".[36] Meanwhile, comparable urban tree-planting initiatives list benefits including "protecting human and ecosystem health, improving quality of life in cities, and educating broader publics about the benefits of urban environments".[37] "When people come together for the good of humanity and the earth", states the environmentalist David Suzuki, "we can accomplish great things".[38] And this is indeed an example of coming together: something that's done *by* communities *for* communities and *in* communities.[39] This makes the Miyawaki Method an interesting filter for our own study, which puts community at its heart.

In this chapter, we'll attempt a definition of "community". We'll start in the UK with a state-led push to promote a "big society" and "connected communities", considering how this establishes dynamics of *connection*, *difference*, *boundaries*, and *development* at the heart of the community concept. But we'll recognize that much of the classical theorizing about community has come from elsewhere, so we'll introduce three European thinkers to understand their concepts: Ferdinand Tönnies' ideas of *Gemeinschaft* and *Gesellschaft*; Émile Durkheim's thoughts on *mechanical* and *organic solidarity*, and *collective representation*; and Alexis de Tocqueville's views on the value of *associations* in America. Then we'll move to twentieth-century theories from John Dewey, Robert Nisbet, Timothy Carney, Robert Putnam, and others. We'll consider the lament for a loss of community and recognize contemporary populism in part to be driven by feelings of alienation. In the end, we will turn to the thoughts of "Social Network Theory" to consider *community lost*, *community saved*, and *community liberated*. We'll see in this way of thinking a sort of synergy with our ecological perspective, and we'll recognize the community as an ecosystem of its own.

Defining "Community"

One of the principal slogans branded by the UK's Conservative-Liberal Democrat coalition following the 2010 election invoked "The Big Society", a drive to "encourag[e] greater volunteering and philanthropy", and "to enable 'some of the most dynamic' charities, social enterprises and voluntary groups to take over the running of public services".[40] It was a restatement of conservative commitments to the principles of civil society, a legacy of seventeenth- and eighteenth-century thinkers like John Locke and Edmund Burke, and a decidedly paternalistic initiative. Opposition voices saw the gesture as a money-saving scheme pushing the costs of welfare back onto communities following the financial collapse of 2008.[41] But at the heart of the initiative was an appealing pledge of "community empowerment",[42] and the government gushed breathlessly about its virtues:

> There is a very exciting new approach here. For too long now, government has been something done to people. There is the chance of re-thinking politics to get people in the heart of government, to help them to do things in their own community.[43]

By some, the idea of empowering communities was interpreted as progressive, though complications with the programme were numerous, and others viewed it as just plain cynical.[44] Nor was the idea seen as particularly novel: "Minus the snappy name", noted *The Guardian*, Tony Blair's New Labour had proposed something similar when it offered its "Third Way" policy at the end of the 1990s: "the same mix of localist empowerment and civil society regeneration couched in the rhetoric of community".[45] Like that policy, the "Big Society" initiative started from the recognition of something that was missing ("few people feel empowered"; "people feel they have no influence"; there's a "lack of trust"),[46] and offered reasonable solutions: "close bonds in deprived areas matter", acknowledged the government; "a lack of shared values can make things worse", it acquiesced; "community hubs, including pubs and sports clubs, are important for social contacts", came the rallying cry; "joining a group helps", it concluded—though by now, it seemed something of a plea.[47] The project was destined to languish, and though the Conservatives remained in power for the next 14 years, the term "Big Society" disappeared, even if various ideas were rolled out which might be seen as following its legacy, cynical, or not: the devolution of power to regional mayors; the agenda to "level up" the country; the encouragement of co-production or co-creation initiatives between authorities and communities; and a frequent appeal to mobilize grassroots organizations.

Strangely enough, the Conservative government was rather coy about defining either "society" or "community" during its time in office, though some of the associated Big Society projects attempted to glossarise the terms. One such was *Connected Communities*, a programme led by the UK's Arts and Humanities Research Council from 2010–2019 which aimed "to mobilise the potential" of communities "to enhance" their quality of life through "prosperity, sustainability, health and well-being".[48] The official report on the *Connected Communities* programme, by Alice Mah and Graham Crow (2012), identified four "interrelated and overlapping themes": *connection*, *difference*, *boundaries*, and *development*.[49] These are useful touchstones in conceptualizing "community", or at least in helping us to understand what that word might mean.

The first term, *connection*, is implicit in the idea of community, the report suggests, and is "captured in notions such as 'common bonds', 'common ties', or the idea that 'we belong to one another'".[50] Many accounts of community pick up on similar themes: popular authors write of "shared circumstances",[51] "common attitudes",[52] or "a feeling of fellowship with others"[53]; scholarly writers use terms like "belonging",[54] "interdependence",[55] and "social cohesion".[56] Indeed, there is evidence that increasing connectivity within communities brings positive benefits: "informal and formal networks strengthen communities and improve partnership working", suggests the report.[57] Underlining that significance was the title of the programme itself: "Connected Communities".

Next, *difference* might be seen as the photo-negative of connection, identifying a distinction between those who belong *within* a community and those who remain *without* ("outsiders and insiders"; "them and us").[58] The report acknowledges difference both as positive and negative: on the one hand "a celebration of social and cultural diversity", but on the other "related to social exclusion, conflict and inequality".[59] It also draws attention to difference as a driver of transformation, and recent #MeToo, Trans Rights, and Black Lives Matter campaigns have demonstrated that sort of power, often with explicit emphasis on community in phrases such as "You are not alone", or "Different together".[60] But there is another transformational take on the politics of difference: "the collective capacity of citizens to make a difference", as Peter Block puts it.[61] This is evidenced, for example, in the "37 Ways to Make a Difference in the World" initiative from *Goodgoodgood*, which promotes not only altruistic behaviour, but healthy living, positive thinking, lifelong learning, and environmental awareness. "Show up for your neighbors, the planet, your mental health, and more…", runs its website; "Create positive change in your community and beyond".[62]

The third term highlighted in the report is *boundaries*. "Ideas of connection and difference, inclusion and exclusion, and tolerance and acceptance, both between and within communities", it reflects, "suggest that communities have limits, edges, or boundaries".[63] Suzuki emphasizes a similar point: "Communities are bound together by shared beliefs, values, history and rituals", he writes; but this means that "they have always walked a fine line between inclusiveness and exclusivity".[64] Where some boundaries are protective and safeguarding (such as the confines of a family home), others are exclusive and prohibitive, which is vividly enacted in policies relating to immigration (Trump's wall). In extreme cases, exclusionary practices keeping people out also form boundaries of incarceration keeping people in: inmates thrown together find themselves restricted by the bonds of an uncomfortable forced community. This sense of bonding (as an impediment) is not precisely what Robert Putnam writes about in *Bowling Alone*, where he references bonding as "a kind of sociological superglue",[65] though he too concedes that boundaries circumscribe communities and what matters for them. To Putnam, communities tend to be either internally focused (centripetal) or externally driven (centrifugal),[66] and he refers to these as "bonding" communities that focus on their own interests (for example, fraternity houses), or "bridging" communities that work together to reach out (such as the Rotarians). Thus the conceptual boundary line that demarcates a community can influence profound dynamics of cohesion or division, solidarity or altruism, togetherness or distinction, with all of the mixed connotations of these terms.

The report's final category, *development*, turns its attention to the stated aims of the *Connected Communities* programme: "the more practical issue of how to address issues of sustainability, public policy, community engagement, health, sustainability, welfare, socioeconomic deprivation, and related issues".[67] These aims are clearly well-intentioned attempts to assist communities, though once again, such initiatives are not straightforwardly resolutions. While many community-centred endeavours seek positive ends in empowering communities through participatory practices or co-production, working with communities can be seen as condescending[68]; community practitioners can be seen as virtue-signalling[69]; top-down initiatives can impose unwanted change[70]; neighbourhood improvement can drive out communities[71]; giving communities a voice can lead to backlash[72]; and charitable activities can lead to corruption.[73] Development is by no means a given outcome of any community work, however well-intentioned the initial idea.

With its four touchstones of *connection*, *difference*, *boundaries*, and *development*, the *Connected Communities* report offers a useful overview of some of the conceptual dynamics of "community". Even so, this is still a snapshot rather than a comprehensive analysis—after all, it explicitly refers to community as a "contested concept", reminding us that a big challenge to understanding the term is that it has been used in a variety of competing ways. As long ago as the 1950s, scholars had "identified some 94 definitions of community", reveals Ted Bradshaw,[74] while others admit that "the lack of an accepted definition" can result in "contradictory or incompatible assumptions about community".[75] "'Community' means different things to different people", Putnam notes,[76] and those different people can perceive their own affiliations with community in different ways.

After all, as Robert A. Nisbet observes, the concept of *community* "goes far beyond mere local community".[77] Think of communities of identity such as the gay community, the Christian community, or the Asian community; communities borne from experience, such as the residents of Grenfell Tower in West London, who were brought together psychologically and emotionally following a tragic fire in their block of flats in 2017, or the international community of pilgrims who have walked the Camino de Santiago; consider administrative organizations like the European Community or the Community of Nations; or projects that conceptualize community as a synonym for vulnerability, such as Care in the Community, community outreach, and some forms of community arts. Each of these broad community affiliations enables us to categorize different applications of the term.

Given such scope, perhaps any quest for a definition of *community* is thwarted from the start. This might be especially the case in the contexts of top-down initiatives for change and development: "Simple definitions by their very nature are too static to encompass multiple connotations and fluidity, qualities that abound in communities", writes Vera Zolberg.[78] Indeed, many authors acknowledge the changing dynamics that have caused understandings of community to fluctuate and evolve: the swift recent transition to technology that has created so-called virtual communities[79]; "a weakening of traditional ties among community members due to urbanization and industrialization"[80]; a "loss of the sense of community based on 'place'" which underpins issues of migration and diaspora[81]; even the impossibility of community, which Esposito projects in his writing.[82] There is also a "'darker' side": as much as the idea of community is taken to "represent social belonging, collective well-being, solidarity and support", write Crow and Mah, some community-based dynamics can reflect "social problems", "problem populations", or issues of "exclusion, inequality, oppression and social divisions".[83] For all the well-meaning plans, positive spin, and energetic initiatives of committees or ambassadors, what defines a community can be unique, resistant, and carefully guarded. The concept of community, acknowledges one community reader, "is so ambiguous, so difficult to define".[84]

The Origins of Community Theory

Tracking back to the origins of sociological thinking about community helps us to understand the landscape, and introduces us to three European thinkers: the German scholar Ferdinand Tönnies (1855–1936), the French writer Émile Durkheim (1858–1917), and his compatriot Alexis de Tocqueville (1805–1859).

Ferdinand Tönnies (1855–1936)

Tönnies' classic text on community from 1887 was titled *Gemeinschaft und Gesellschaft*. This picked up on two familiar German terms, broadly translatable as "community" and "society". For him, the distinction between them was clear: "Community means genuine, enduring life together, whereas Society is a transient and superficial thing".[85]

At its core, Tönnies suggests, the bonds of community (*Gemeinschaft*) are a fundamental feature of humanity, and are most evident in the ties between mother and child, the union of married couples, and the connection between siblings.[86] He also notes beyond the "primal unity" of these blood-ties the significance of *place*, which "holds life together on a physical level", and *spirit*, "the binding link on the level of conscious thought",[87] both of which can enhance connection. Indeed, he acknowledges that it is the community of spirit which "makes a truly human community in its highest form",[88] though he also conceptualizes such a "comradeship of minds" (the community of spirit) as "a kind of invisible location, a mystical city and meeting-place".[89]

By contrast, a society (*Gesellschaft*) reflects a group of people who "live peacefully alongside one another, but in this case without being essentially united".[90] These are "separate individuals", he writes[91]; "their wills and spheres of interest interact with each other in manifold different ways, yet they remain independent of one another and lacking in deep intimacy".[92] What brings a society together, Tönnies asserts, is "some common, conscious purpose".[93] Thus there is a clear difference between attachments of the heart and mind: the emotional bond of *Gemeinschaft*, and the intellectual connection of *Gesellschaft*. "*Gemeinschaft* is made up of the union of natural wills", he wrote; "while *Gesellschaft* is made up of the union of rational, arbitrary wills".[94]

Émile Durkheim (1858–1917)

Only a few years later, Tönnies' ideas were developed by another scholar, Émile Durkheim. Writing in his doctoral thesis in 1893, Durkheim conceptualized two types of "social solidarity": that of traditional or "lower societies"[95] which he called "mechanical" (*la solidarité mécanique*); and that of "more advanced peoples",[96] whose relationships he saw as "organic" (*la solidarité organique*).

"Mechanical solidarity" was defined as "a more or less organized society composed of beliefs and sentiments common to all the members of the group".[97] This society was typically made up of families or ancestral clans, or bound together in "a very special kind of intensity"[98] through "collective or common consciousness".[99] In many ways, this resembles Tönnies' *Gemeinschaft*: "Men who feel themselves united, partially by bonds of blood, but still more by a community of interest and tradition, assemble and become conscious of their moral unity", wrote Durkheim.[100] He noted that such assembly might stem from the shared belief system of a religion: villagers or clans would all worship at the same time, eat together, and have their routines determined communally in a shared but straightforward system. "The foundation of society is thus found in a *cogitamus*, in a collective consciousness that expresses itself in the plural, saying 'we'".[101]

"Organic solidarity", on the other hand, is not determined by the similarity of people but by their differences, and by the way in which their different skills can be brought together to contribute to a whole. Individuals are distinguished by their "social activity" or "profession",[102] and society is derived from "the result of their combining together" in a "division of labour".[103] Here, Durkheim uses Marxist language, noting that the set-up "makes each individual an agent of exchange"[104] which thereby "multiplies intra-social relationships".[105] The carpenter, for example, needs a woodcutter to provide wood and a blacksmith to make nails before they can build a table for the village inn; already, there are four people in *organic solidarity* (carpenter, woodcutter, blacksmith, innkeeper). As such, Durkheim's thinking develops Tönnies' *Gesellschaft* to explain how a society forms from *collaboration* (literally: co-labour).

Even so, Durkheim's conceptualization of a collective consciousness would continue to form the basis for his understanding of how a society bonds, though he increasingly turned to an alternative phrase, "collective representation", to explain the psychology of our connection. "A collective representation is a concept or a category of thought held in sufficiently similar form by many persons to allow effective communication", explains Paul Bohannon.[106] It's only through such shared representations, Durkheim argues, that humans are able to make sense of the world. For example, in order for us to discuss "the world" with you, we need a shared understanding of that concept; the same can be said of "time", "society", or an example he uses, "whiteness"[107] (not to mention "ecology", "community", and "performance"). It is our shared recognition and use of such "collective representations" that makes us social beings.

More than this, the ways in which our understanding of society are formed in respect of these representations is in itself reminiscent of the dynamics of ecology:

> Social life is a network of criss-crossing references to the roles and statuses that make it up. Shared representations, such as money, leadership, rights, are themselves the currency of interaction. They are not only created and sustained by these interactions they are, in a large measure, constitutive of them.[108]

As such, we can see a line of thinking between the concept of community ecology, the sociological concept of community, and the ecological way of thinking.

Each of these theorists conceptualizes the distinctions between community and society not only in terms of size, scale, and quality, but also in terms of progress or perceived advancement: we've seen that Durkheim juxtaposes "lower societies" and "advanced peoples", and a similar divide pervades Tönnies' work, according to Steven Brint: "*gemeinschaft* represented the childhood of humanity and *gesellschaft* its maturity".[109] No doubt there is an element of condescension here, and in their nostalgic evocation of simpler times, Brint suggests, the terms *Gemeinschaft* and *la solidarité mécanique* may have encouraged "the tendency of subsequent writers either to romanticize or debunk community".[110] Even so, for all the old-fashioned and rose-tinted romanticism that the term *community* might evoke, and by contrast, for all the dynamics of small-world hierarchy, power, and division that might lurk beneath its manifestations, contemporary sociologists, anthropologists, and even economists have noted generally progressive benefits rather than mere primitivism in the coming together of communities. "As a symbol and aspiration, the idea of community continues to resonate in public discourse", notes Brint, and he links this to "the virtues of civil society": "a place of democratic and communal social relations, set apart from both state and market".[111]

Alexis de Tocqueville (1805–1859)

The third scholar we are introducing as one of the "classical" thinkers in this area is the French aristocrat Alexis de Tocqueville, who recognized such "democratic and communal social relations" at the very heart of American society. Following an extended trip to the US in the early 1830s, De Tocqueville returned to France where he wrote *Democracy in America* (1835–1840). In this, he counselled his compatriots on the virtues of America's democracy, considering its political, social, and civil set-up as the last word in how a nation ought to operate. Notwithstanding its still-active trade in slavery and its abuse of the indigenous population, he saw America as a land steeped in equality and democratic rights; notwithstanding the civil war that broke out only a generation later, he considered America to be so politically stable that it was inoculated against revolution or civil unease. De Tocqueville was undoubtedly somewhat optimistic in his analysis; even so, his thoughts about community and society are worth considering.

For De Tocqueville, community and society offered ways of working together in order to inhabit and enact the virtues of democracy. "The most natural privilege of man", he wrote, "next to the right of acting for himself, is that of combining his exertions with those of his fellow-creatures, and of acting in common with them".[112] He continued: "the most democratic country on the face of the earth is that in which men have in our time carried to the highest perfection the art of pursuing in common the object of their common desires, and have applied this new science to the greatest number of purposes".[113] The "science" that he spoke of was in fact more of a custom—the practice of joining together in community groups or associations: "As soon as several inhabitants of the United States have taken up an opinion or a feeling which they wish to promote in the word", he explained, "they look out for mutual assistance; and as soon as they have found each other out, they combine". He lists a selection of activities driven by such associations, such as building churches, founding hospitals, distributing books, or putting on entertainment. The benefits of such communal collaboration were profound, judged De Tocqueville, since such associations offered not only mutual support and common interest, but also, potentially, the political and cultural influence of a lobbying group: "From that moment they are no longer isolated men, but a power seen from afar", he wrote, "whose actions serve for an example, and whose language is listened to". More than this, De Tocqueville writes effusively about how in America the expansion of the newspaper industry alongside these community associations enabled the dissemination and advocacy of ideas. Bringing such dynamism to the idea of community avoids both the potential insularity of Tönnies' *Gemeinschaft* and the assumed primitivism of Durkheim's *solidarité mécanique*. "The science of association is the mother of science", De Tocqueville opined; "if men are to remain civilized, or to become so, the art of associating together must grow and improve".[114]

Indeed, so fundamental was the effective triangulation of equality, community, and advocacy in De Tocqueville's thinking that he implicitly saw these virtues as the abiding pillars of democratic society: "The more equal the conditions of men become", he explained, "the less strong men individually are, the more easily do they give way to the current of the multitude, and the more difficult is it for them to adhere by themselves to an opinion which the multitude discard". And he went on: "Thus they learn to surrender their own will to that of all the rest, and to make their own exertions subordinate to the common impulse".[115] That the dynamics of community inherently supported democracy (or at least, resisted the danger of the individual) thereby played into De Tocqueville's understanding of society as a whole.

Of course, the three examples we have considered here are all born out of the European tradition, and this in itself is in some ways a rather lopsided way of considering community, since for many other cultures, the concept is both so innate and organic that it hardly needs to be defined, and at the same time, it is perceived as "a notion that has been imposed"[116]; "an externalist word—the creation of an intellectual class alienated from its land and kin".[117] As such, the term "community" itself is not always useful to an indigenous worldview, which may encompass metaphysical and spiritual modes of attachment as well as those linked to place, family, or other forms of identity. Nevertheless, the dynamics that bond groups of people together in such societies may rely on "a similar way of knowing and a similar culture"[118]; a shared cultural negotiation involving context, meaning, and process[119]; "physical groupings and psychological belonging to a particular group(s)"[120]—in other words, many of the dynamics we have encountered in Westernized notions of "community".

Communities in the Twentieth and Twenty-First Centuries

Underpinned by the thinking of these historical philosophies about communities and the way they work, the academic discipline of sociology has flourished throughout the twentieth century. Barry Wellman suggests that "the community question has set the agenda for much of sociology",[121] while Nisbet writes that community is "the most fundamental and far-reaching of sociology's unit-ideas".[122] Sociological theorists such as these have staked their claim to the term "community", but so too have educational theorists such as John Dewey who in the 1920s and 30s imagined community as "an integrative matrix wherein symbiotic relations could exist".[123]

Like De Tocqueville, Dewey asserted that "American democratic polity was developed out of [...] association in local and small centers".[124] However, by the time he was writing, around a hundred years after the Frenchman, America had undergone substantial change. "The machine age in developing the Great Society", he observed, "has invaded and partially disintegrated the small communities of former times".[125] To Dewey it was particularly egregious that this had happened "without generating a Great Community",[126] and it was here that he identified the notion of the communal at the very heart of democratic American life: "democracy is not an alternative to other principles of associated life", he wrote; "It is the idea of community life itself".[127] As the power of earlier local communities waned, leaving society in a "state of blind and aggregate interconnection",[128] Dewey recognized that the moral, emotional, intellectual, and conscious[129] virtues of communal life would need to be restored. This could only happen, he believed, through coming to a shared set of views (literally: a "common sense").[130]

The way of harnessing such common sense would be through learning (education) and dialogue (communication). "There is more than a verbal tie between the words common, community, and communication", he observed; "Men live in a community in virtue of the things which they have in common; and communication is the way they come to possess things in common".[131] But his conclusion was sombre: "Till the Great Society can be converted into a Great Community, the public will remain in eclipse".[132]

Dewey's concern about a loss of community is reiterated throughout twentieth- and twenty-first-century thought. Nisbet's *The Quest for Community* (1953), for example, opens with a chapter titled "The Loss of Community". Here, he establishes the pervasive mood of the twentieth century as one of "*disorganization, disintegration, decline, insecurity, breakdown, instability*",[133] and later, "moral estrangement and spiritual isolation".[134] By 2020, Timothy Carney was recognizing this as *alienation*, "the disease of working class America". "Civil society has dissolved in huge swaths of America", Carney lamented; "Millions of Americans are alienated"; "this is a plague of alienation".[135] Carney's book *Alienated America* focuses on the societal dynamics that led Donald Trump to the White House, thereby recognizing disaffection as a specific driver of populist politics. A hundred years may span the work of these scholars (Dewey to Carney), but along with others, they cite the same crisis (alienation) and the same cause (loss of community).

Dewey attributes the loss of community to a gradual erosion of communal groupings such as the family, the neighbourhood, or the church, "traditional primary relationships" which "have become functionally irrelevant to our State and economy and meaningless to the moral aspirations of individuals".[136] Into that vacuum has stepped the "Political Community"—the organizational machine of a top-down centralized state, which brings authority and guidance, but which in the absence of real community risks becoming a totalitarian entity. For Carney too, "the erosion of society [...] is the disappearance of the middle institutions",[137] which he suggests has been caused by the dual influences of state centralization and individualism prising apart societies and destroying the middle ground. Yuval Levin says something similar: "While individualism pulls at them from below, centralized government pulls at them from above", he writes, "hollowing out the middle spaces where a free society forms its citizens".[138] For it is in this "middle ground" that much of the stability of communities has historically been found, from the associations and societies witnessed by De Tocqueville, and the qualities they seed—"common attitudes",[139] "belonging",[140] "support",[141] and above all, "trust: the glue that holds communities together".[142] Thus "at the root of the most significant problems America faces at home", Levin laments, "is the weakening of our core institutions—family and community, church and school, business and labor associations, civic and fraternal organizations".[143] Community organizations like these, suggests Sandra Navidi, are networking "superhubs" combatting the "rootlessness" around us.[144]

It was just such associations and their decline that Robert D. Putnam investigated in a series of related publications including *Bowling Alone* (2000) and *Better Together* (2003).[145] He observed civic engagement declining across America in all manner of pursuits: "voter turnout, newspaper readership, membership in choral societies and football clubs",[146] and bowling leagues (as the title infers). He noted that over the course of a generation, there had been a significant drop in the number of people engaging with local politics, religious groups, civic organizations, volunteering, and even scouting. "When people don't see the immediate purpose of belonging to institutions, they turn away", suggests Carney; "then they lose the less obvious, less immediate benefits of community, such as the sense of purpose, the modeling, the camaraderie, the hidden safety net, the prods to take on tasks one might never think of on one's own".[147]

According to Putnam, what's most at stake when benefits such as these are lost is a community's *social capital*. "For a variety of reasons", he writes, "life is easier in a community blessed with a substantial stock of social capital". He goes on to explain that such a quality fosters "social trust", "collective benefits", "coordination and communication", "collaboration" and "reciprocity"; it can also "amplify reputations" and "allow dilemmas of collective action to be resolved".[148] The decline he perceives in social engagement comes with a stark consequence: "As the social fabric of a community becomes more threadbare", he warns, "its effectiveness in transmitting and sustaining reputations declines, and its power to undergird norms of honesty, generalized reciprocity, and thin trust is enfeebled".[149] This chimes with Carney: "When you strengthen the vertical bonds between the state and the individual", he notes, "you tend to weaken the horizontal bonds between individuals. What's left is a whole that by some measures is more cohesive, but individuals who are individually all less connected to one another".[150]

Putnam attributes the decline in social engagement to a number of factors causing us to be less sociable: he sees us being hampered by pressures of time and money, which simultaneously cause us to work more and spend less[151]; he notes that the increasing sprawl of our suburbs has led to more and more social segregation[152]; and he observes that we spend more of our time engaging with individualistic technologies such as television[153]; finally, he acknowledges that this has been a generational shift,[154] and that there has been over the same period an associated loosening of the bonds in American families.[155] Although he doesn't claim to have a full understanding of the cause of the decline,[156] he's resolute that "more social capital and civic engagement would improve things".[157] Towards the end of *Bowling Alone*, he issues a call: "Let us find ways", he begs, to raise "the level of civic engagement"; to make the workplace "more family-friendly and community-congenial"; to "spend less time travelling and more time connecting"; to design "more integrated and pedestrian-friendly areas"; to become "more tolerant of the faiths and practices of other Americans"; to spend more time in active engagement rather than "sitting passively alone in front of glowing screens"; to participate more in public life ("running for office, attending public meetings, serving on committees, campaigning in elections, and even voting"); and—significantly for us—"to use the arts as a vehicle for convening diverse groups of fellow citizens".[158] Putnam's call, in essence, maps on to the same sort of thinking we have seen in the Big Society policy and its *Connected Communities* programme.

Putnam proposed 2010 as the date by which his objectives should be achieved, and the election in 2008 of Barack Obama as America's first black President might have indicated a progressive step towards a more communitarian ideal. To be sure, Obama's election campaign focused on "the notion of a common good",[159] and a "common kinship"[160]; he acknowledged that "despite great differences in wealth we rise and fall together",[161] and he recognized that an intrinsic part of the American constitution was the belief that "a common culture, a common faith, and a well-developed set of civic virtues [...] limited contention and strife".[162] In his book *The Audacity of Hope* (2007) Obama reflects on his days in Chicago as a community organizer, writing of "communal values" (family, community, nation), "the glue upon which every healthy society depends". In particular, he notes that "we value the constellation of behaviours that express our mutual regard for one another: honesty, fairness, humility, kindness, courtesy, and compassion".[163] Obama is not unaware of the sort of challenges (alienation) that have been highlighted by scholars and politicians on both sides of the Atlantic, recognizing communities whose "immune system has broken down almost entirely"[164]; "soaring high school dropout rates" and "inadequate health insurance"[165]; and "the nuclear family [...] on the verge of collapse".[166] Yet he invoked "the audacity to believe despite all the evidence to the contrary that we could restore a sense of community to a nation torn by conflict".[167]

Far from restoring communal values, common kinship, and healthy communities, however, the politicians of Western democracies, as history attests, have further alienated vast swathes of their electorate, driving populism. In the UK this led to the divisive wrench of the Brexit vote in 2016, the political instability of Boris Johnson's period as Prime Minister from 2019, and the even more turbulent legacy of his successor Liz Truss in 2022; in America, it led to what Presidential candidate Donald Trump observed as the death of the American Dream in 2016,[168] to a storming of the Capitol Building by Trumpist insurrectionists in 2020, and to a flurry of subsequent legal cases brought against "the Donald" in the courts. Similar chaotic situations have erupted in other Western democracies from France to Brazil. As policy pushers, then, the politicians have been decidedly unsuccessful in combatting societal disintegration and alienation.

In fact, neither political soundbites nor academic theories seem to have worked; neither religious faith nor audacious hope seem particularly practical; and neither learning nor dialogue—despite appearing to be the most strategic and balanced solutions—seem fully effective. As Timothy Carney asserts, you can't just "tell the people to 'choose' not to be left behind, to get better educated, to adapt, and to stop being so backward". By the same token, you can't just put together a wish-list, as Putnam did, and expect the situation to resolve. Carney makes the case for crisis clear: "something deeper is broken".[169]

Contemporary community theorists insist that restoring community is about "a reinvigoration of the middle layers of society and a resuscitation of our mediating institutions".[170] Deneen explains: "Such efforts should focus on building practices that sustain culture within communities, the fostering of household economics, and 'polis life', or forms of self-governance that arise from shared civic participation".[171] Generally, these writers discourage quick-fix cultural cures like "easy religion", "the psychiatrist's office", "the cult", "the movie", "the radio serial", and "popular fiction"; they deride the public's "pathetic dependence on [these] superficial symbols of friendship and association",[172] which are no more than "fantasy avenues to social belonging".[173] Instead, as mostly conservative traditionalists, such writers urge the restoration of family and community values, pointing towards church communities and their practices. "Family, work, community, and religion are what fill the middle layers of our social life", writes Levin[174]; "Institutions like these build living communities around them from the bottom up, and bring order to the lives of those communities and their members".[175]

Once more, though, any of these potential solutions need to start by establishing what "community" means. By now, perhaps, our understanding of community is becoming more nuanced: something "that gives you a role, a place, and a set of relationships and responsibilities to other particular human beings", as Levin puts it; "a mix of dependence on others and obligations to them, and so a connection with specific people with whom you share some meaningful portion of the actual experience of life in common".[176] Or, as Wendell Berry has it, "a rich and varied set of personal relationships, a complex of practices and traditions drawn from a store of common memory and tradition, and a set of bonds forged between a people and a place that—because of its situatedness—is not portable, mobile, fungible, or transferable".[177] Or, according to the sociologist Steven Brint, "dense and demanding social ties", "social attachments to and involvements in institutions", "ritual occasions", "small group size", "perceptions of similarity", and "common beliefs".[178]

Social Network Theory

Brint's understanding of community is a *social network* point of view, something which stems from Durkheim's interest in (labour) relations. Social network theory sees communities formed through "structures of relationships and flows of activities"[179] as opposed to blood ties, kinship, or place. According to Barry Wellman, one of the foremost scholars of social network theory in the 1970s and 80s, such networks have established increasingly through industrialization with its improved transportation and communication, and its general tendency towards urbanization.

Wellman acknowledges that there is a general perception of traditional community values having waned (*community lost*), and he recognizes that this has in part come about because of a general flow of populations towards the cities, where signifiers of dysfunction such as higher crime and greater poverty give the impression that there is no sense of community. However, he notes, some sociologists offer a different perspective—*community saved*, which argues that there are in fact strong pockets of community in inner-city urban environments, especially among populations that are marginalized, ethnic, or poor. In fact, for him the reality involves a third interpretation, *community liberated*, which sees the strength of a contemporary community as being pluralistic, porous, diverse, fluid, and complex: "Instead of being fully incorporated into a single solidary community", he writes, "urbanites are seen as being limited members of multiple social networks, sparsely knit and loosely bounded"[180]; "While such ties may have fewer strands in the relationship than those in which kinship, residence, and work are combined, they are prevalent and important sources of sociability and support".[181]

Perhaps not surprisingly, we see in Wellman's theorizing some of the terms that have become ubiquitous slogans in our more decentralized world—terms like "network" and "web" signify connectivity,[182] and remind us of the various communities we engage with through social media. At the same time, as we have seen in Chapter 2, these are terms that have been widely used in ecology, looping us back to the metaphorical landscape of our study. We can also see in social network theory's "structures of relationships and flows of activities"[183] a link to the sort of process-functional dynamics that we've explored in relation to ecosystems.

Ecosystems of Community

In the context of our discussion, it's easy to see some obvious links between our perspectives on ecosystems and our perspectives on community. Though slightly different language may be used, we can recognize both constructs as (sometimes) conceptually bounded systems, with parallel tensions between closed and open or inside and out. We can also compare many of the relationships and dynamics within the systems, whether we view these as mutualistic, competitive, or even predatory. After all, ecological theory borrows metaphors like *population*, *guild*, and *selection* from other domains, while wider conceptions of ecology evoke memes of "kinship",[184] "co-existing",[185] and "collaboration".[186] More than this, as we explored in the previous chapter, recent ecological thinking has recognized far broader conceptual dynamics. Ecologies, like communities, are porous, diverse, fluid, and complex; communities, like ecologies, are "*not*" (so to speak) "contained within the skin"[187]; theorists of ecology conjure up images of an "aggregate of relationships",[188] and "the study of ourselves as part of the organic whole".[189] How we understand these suggestions in terms of both ecology *and* community is central to our study, and how we understand the synergies between their conceptual ideas is central to our thoughts.

Before embarking on our journey to explore these ecosystems and communities, we have one more concept to consider. In the next chapter, we'll explore the terrain of *performance* in the context of ecology.

Notes

1. Akira Miyawaki (1998), "Restoration of Urban Green Environments Based on the Theories of Vegetation Ecology", *Ecological Engineering*, 11, pp. 157–165, p. 159.
2. Lela Nargi (2019), "The Miyawaki Method: A Better Way to Build Forests?", *JSTOR* Daily, 24 July, https://daily.jstor.org/the-miyawaki-method-a-better-way-to-build-forests/. Accessed 15 March 2024.
3. Hannah Lewis (2022), *Mini-Forest Revolution: Using the Miyawaki Method to Rapidly Rewild the World*, White River Junction: Chelsea Green Publishing, p. 8.
4. Akira Miyawaki (1999), "Creative ecology restoration of native forests by native trees", *Plant Biotechnology*, 16.1, pp. 15–25, p. 24.
5. Lewis, *Mini-Forest*, p. 5.
6. Akira Miyawaki (2006), "Aiming for the Restoration of a Green Global Environment", *A better future for the planet Earth, vol.I–V (1992–2016): lectures, essays, and interviews of the winners of the Blue Planet Prize*, Tokyo: Asahi Glass Foundation, pp. 258–275, p. 271.
7. Anon. (2022), "Crowd Foresting: Miyawaki Model Afforestation", *CrowdForesting*, https://www.crowdforesting.org/. Accessed 8 March 2024.
8. Anon., "Crowd Foresting".
9. Lewis, *Mini-Forest*, p. 5.

10. Monika Egerer and Michael Suda (2023), "Designing 'Tiny Forests' as a lesson for transdisciplinary urban ecology learning", *Urban Ecosystems*, 26, pp. 1331–1339, p. 1334.
11. Akira Miyawaki (2006), "A Call to Plant Trees", *A better future for the planet Earth, vol.I–V (1992–2016): lectures, essays, and interviews of the winners of the Blue Planet Prize*, Tokyo: Asahi Glass Foundation, pp. 252–257, p. 256.
12. Kamala Thiagarajan (2020), "Could Akira Miyawaki's 50-Year Old Innovation Help Promote Biodiversity and Reduce the Risk of Climate Change?", *Ensia*, 11 June, https://ensia.com/features/miyawaki-trees-biodiversity-carbon-sequestration/. Accessed 8 March 2024.
13. Isabel Galwey (2020), "Taking Root: Failure and Success for the Great Forest Wall of Tōhoku", *Anthroposphere: The Oxford Climate Review*, 15 August, https://www.anthroposphere.co.uk/post/taking-root-failure-and-success-for-the-great-forest-wall-of-t%C5%8Dhoku. Accessed 8 March 2024.
14. Lewis, *Mini-Forest*, p. 19.
15. Miyawaki, "A Call", p. 254.
16. Miyawaki, "Creative ecology restoration", p. 16.
17. Akira Miyawaki and Frank B. Golley (1993), "Forest reconstruction as ecological engineering", *Ecological Engineering*, 2, pp. 333–345, p. 338.
18. Miyawaki and Golley, "Forest Reconstruction", p. 341.
19. Miyawaki and Golley, "Forest Reconstruction", p. 342.
20. Egerer and Suda, "Designing", p. 1334.
21. Miyawaki, "Creative ecology restoration", p. 16.
22. Egerer and Suda, "Designing", p. 1331.
23. Mike Charkow (2022), "Forests for the Future", *ARB Magazine*, 197, p. 30. https://www.avtree.co.uk/wp-content/uploads/2022/06/ARBMagazine197-pp29-33-Forests-of-the-future.pdf. Accessed 30 August 2023.
24. Bartolomeo Schirone, Antonello Salis, and Federico Vessella (2011), "Effectiveness of the Miyawaki method in Mediterranean forest restoration programs", *Landscape and Ecological Engineering*, 7, pp. 81–92, p. 91.

25. Fazal Rashid and Somil Daga (2023), "How Mr Miyawaki Broke My Heart", *Science: The Wire*, 25 April, https://science.thewire.in/environment/how-mr-miyawaki-broke-my-heart/. Accessed 9 March 2024.
26. Aike P. Rots (2021), "Trees of tension: re-making nature in post-disaster Tohoku", *Japan Forum*, 33:1, pp. 1–24; Annelise Giseburt (2023), "Miyawaki forests are a global sensation, but not everyone's sold on them", *Mongabay*, https://news.mongabay.com/2023/06/miyawaki-forests-are-a-global-sensation-but-not-everyones-sold-on-them/. Accessed 9 March 2024.
27. Lewis, *Mini-Forest*, p. 36.
28. Deepak Jaiswal et al. (2023), "Fast Growing Forests (FGF) to offset Greenhouse Gas (GHG) Emissions", *EGU General Assembly Conference Abstracts*.
29. Charkow, "Forests for the Future".
30. Anju Lis Kurian (2020), "Urban Heat Island Mitigation And Miyawaki Forests: An Analysis", *EM International*, pp. 186–191.
31. As the Royal Forestry Society reports, "high quality, diverse, green infrastructure in urban areas brings a wide range of benefits; from social cohesion, to health and wellbeing, as well as cognitive and educational improvement". Macarena L. Cárdenas et al. (2022), "Nature-based Solutions Building Urban Resilience for People and the Environment", *Quarterly Journal of Forestry*, 116:3, pp. 27–37, p. 27.
32. Geoffrey H. Donovan, Jeffrey P. Prestemon and Abigail R. Kaminski (2022), "The Natural Environment and Social Cohesion: Tree planting is associated with increased voter turnout in Portland, Oregon", *Trees, Forests and People*, 7:100215, p. 5.
33. Alain Daou, Mansour Saliba, and Antoine Kallab (2024), "A Review of the Miyawaki Method", *SSRN* 4,728,239, pp. 1–49; Donovan, Prestemon and Kaminski, "The Natural Environment". Other researchers such as Watkins et al. are more equivocal about the extent to which social cohesion is achieved through tree-planting initiatives. Shannon Lea Watkins et al. (2018), "Does collaborative tree planting between nonprofits and neighborhood groups improve neighborhood community capacity?", *Cities*, 74, pp. 83–99.

34. Watkins et al., "Does collaborative", p. 84.
35. Watkins et al., "Does collaborative", p. 91.
36. Donovan, Prestemon and Kaminski, "Natural Environment", p. 5.
37. Dana Fisher, Erika Svendsen, and James Connolly (2015), *Urban environmental stewardship and civic engagement: how planting trees strengthens the roots of democracy*, London and New York: Routledge, p. 41.
38. David Suzuki (2014), "Out of Darkness, the Light", *David Suzuki Foundation*, 9 January, https://davidsuzuki.org/story/out-of-darkness-the-light/. Accessed 8 March 2024.
39. Similar endeavours, such as the community practice of gathering together to refresh unused spaces, have been common for generations. Often referred to as "Guerilla Gardening", this originated from Liz Christy's "Green Guerillas" projects in New York in the 1970s. Communities take over unused, neglected or toxic land, and plant a few seeds, or take on more challenging projects such as reclaiming derelict and dangerous land. One example is the case of Smiling Hogshead Ranch in Queens, which was a project that took over a toxic railroad bed and transformed it into an "eco-oasis". Andrew Millison (2023), "Guerilla Gardeners turn Toxic Soil into Eco-Oasis", *YouTube*, https://www.youtube.com/watch?v=UqyK_9iybD8. Accessed 10 March 2025.
40. Anon. (2010), "David Cameron launches Tories' 'big society' plan", *BBC News*, 19 July. https://www.bbc.co.uk/news/uk-10680062. Accessed 19 March 2024.
41. Anon., "David Cameron".
42. "Community Empowerment" was one of three stated pillars of the policy, alongside "Opening up public services" and "Social Action". Rebecca Tunstall et al. (2011), "Building the Big Society", *Centre for Analysis of Social Exclusion (CASE)*, 67, White Rose University Consortium, p. 3, https://eprints.whiterose.ac.uk/75097/1/CASEreport67.pdf. Accessed 26 April 2024.
43. Ministry of Housing (2010), "Cameron and Clegg Launch the Big Society", *Gov.UK*, https://www.gov.uk/government/news/cameron-and-clegg-launch-the-big-society. Accessed 26 April 2024.

44. Simon Skidmore (2011), "Smaller Government: Bigger Society?", *Parliament UK*, https://publications.parliament.uk/pa/cm201012/cmselect/cmpubadm/writev/smaller/bs76.htm. Accessed 26 April 2024.
45. Alex Worrad-Andrews (2010), "'Big society' can give power back to communities", *The Guardian*, 20 July, https://www.theguardian.com/commentisfree/2010/jul/20/big-society-empower-communities. Accessed 26 April 2024.
46. Baljit Gill and Karen Cave (2011), "What do we know about the extent of community engagement in different communities?", in Tunstall et al., "Building", pp. 24–29.
47. Gill and Cave, "What do we know?".
48. Anon. (2022), "Connected Communities", *UKRI* website, https://www.ukri.org/what-we-offer/browse-our-areas-of-investment-and-support/connected-communities/. Accessed 11 January 2023.
49. Graham Crow and Alice Mah (2012), "Conceptualisations and meanings of 'community': the theory and operationalisation of a contested concept", *Connected Communities*, p. 6. https://www.community-methods.soton.ac.uk/resources/CC%20Final%20Report_30%20March%20GC.pdf. Accessed 26 April 2024.
50. Crow and Mah, "Conceptualisations", p. 9.
51. Megan Garber (2017), "What does 'community' mean?, *The Atlantic*, 3 July, https://amp.theatlantic.com/amp/article/532518/. Accessed 31 August 2020.
52. Fabian Pfortmüller (2017), "What does 'community' even mean? A definition attempt & conversation starter", *Together Institute*, 20 September, https://medium.com/together-institute/what-does-community-even-mean-a-definition-attempt-conversation-starter-9b443fc523d0. Accessed 31 August 2020.
53. Pfortmüller, "What does 'community'", n.p.
54. David W. McMillan and David M. Chavis (1986), "Sense of Community: A Definition and Theory", *Journal of Community Psychology*, 14, pp. 6–23, p. 9.
55. S.B. Sarason (1974), *The psychological sense of community: Perspectives for community psychology*. San Francisco: Jossey-Bass, p. 192.
56. Robert A. Nisbet (2004), *The Sociological Tradition* (sixth printing), New Brunswick: Transaction Publishers, p. 47.

57. Crow and Mah, "Conceptualisations", p. 6.
58. Crow and Mah, "Conceptualisations", p. 6.
59. Crow and Mah, "Research Report", p. 14.
60. "You are not alone" is the slogan of the official #MeToo movement; "Different together" is "a safe, supportive and understanding community for the partners of adults affected by Asperger's Syndrome". Anon. (n.d.), slogan, *#MeToo*, https://metoommvmt.org/. Accessed 1 September 2020; Anon. (n.d.), slogan, *Different Together*, https://different-together.co.uk/. Accessed 1 September 2020.
61. Peter Block (2018), *Community: the Structure of Belonging*, Oakland: Berrett-Koehler, p. xii.
62. Branden Harvey (2022), "37 Ways to Make a Difference in the World", *Good Good Good* website, 4 May. https://www.goodgoodgood.co/articles/how-to-make-a-difference. Accessed 11 January 2023.
63. Crow and Mah, "Conceptualisations", pp. 19–20.
64. David Suzuki (2007), *The Sacred Balance: Rediscovering our Place in Nature* (updated and expanded), Vancouver, Toronto, Berkeley: Greystone Books, p. 172.
65. Robert D. Putnam (2000), *Bowling Alone: The Collapse and Revival of American Community*, New York: Simon & Schuster, p. 23.
66. Crow and Mah, "Conceptualisations", p. 9; see also our discussion of these terms in Karen Savage and Dominic Symonds (2018), *Economies of Collaboration in Performance: More than the Sum of the Parts*, Cham: Palgrave Macmillan, pp. 179–180, and later in this book.
67. Crow and Mah, "Conceptualisations", p. 26.
68. Carla Pérez-Almendros, Luis Espinosa-Anke, and Steven Schockaert (2020), "Don't patronize me! An annotated dataset with patronizing and condescending language towards vulnerable communities", *arXiv preprint arXiv:2011.08320.*
69. Oliver Rudland (2023), "'Reaching Out' or Institutional Virtue-Signalling? The Role of Community Opera Projects in UK Opera Houses Today", *Tempo*, 77:304, pp. 62–72.
70. Marisol L. Meyer, Ceewin N. Louder, and Guerda Nicolas (2022), "Creating with, not for people: theory of change and

logic models for culturally responsive community-based intervention", *American Journal of Evaluation*, 43:3, pp. 378–393.
71. Claire Williams (2017), "Inclusive communities and robust causality: The constant struggle to balance access to the courts with protection for defendants", *Minnesota Law Review*, 102:969.
72. Michael K. Heiman (1996), "Race, waste, and class: New perspectives on environmental justice". *Antipode*, 28:2, pp. 111–121.
73. Mort Rosenblum and Doug Williamson (1990), *Squandering Eden: Africa at the Edge*, Boulder: Paladin Press.
74. Ted K. Bradshaw (2008), "The Post-Place Community: Contributions to the Debate about the Definition of Community", *Journal of Community Development*, 39: 1, pp. 5–16, p. 9.
75. Kathleen M. MacQueen et al (2001), "What is Community? An Evidence-Based Definition for Participatory Public Health", *American Journal of Public Health*, 91: 12, pp. 1929–1938, p. 1929.
76. Putnam, *Bowling Alone: The Collapse*, p. 273.
77. Nisbet, *The Sociological Tradition*, p. 47.
78. Vera L. Zolberg (2003), "Current Challenges for Cultural Policy: New Meanings of Community", *The Journal of Arts Management, Law, and Society*, 32:4, pp. 295–307, p. 299.
79. Felicia Wu Song (2009), *Virtual Communities: Bowling Alone, Online Together*, New York: Peter Lang.
80. Bradshaw, "The Post-Place Community", p. 5.
81. Ann Boyles (n.d.), "The Meaning of Community", *bahai.org*, https://www.bahai.org/documents/essays/boyles-ann/meaning-community. Accessed 31 August 2020.
82. Roberto Esposito (2009), "Community and nihilism", *Cosmos & History*, 5:1, pp. 24–36.
83. Crow and Mah, "Conceptualisations", p. 3.
84. Jeanette A. Bastian and Ben Alexander (2009), "Introduction: Communities and Archives—a Symbiotic Relationship", in Jeanette A. Bastian and Ben Alexander (2009), *Community Archives: The Shaping of Memory*, London: Facet Publishing, pp. xxi–xxiv, p. xxii.
85. Ferdinand Tönnies (2001), *Community and Civil Society* (ed. Jose Harris; trans. Jose Harris and Margaret Hollis), Cambridge: Cambridge University Press, p. 19.

86. Tönnies, *Community*, p. 22.
87. Tönnies, *Community*, p. 27.
88. Tönnies, *Community*, p. 28.
89. Tönnies, *Community*, p. 29.
90. Tönnies, *Community*, p. 52.
91. Tönnies, *Community*, p. 56.
92. Tönnies, *Community*, p. 63.
93. Louis Wirth (1926), "The Sociology of Ferdinand Tonnies", *American Journal of Sociology*, 32:3, pp. 412–422, p. 416.
94. Tönnies, *Community*, p. 186.
95. Émile Durkheim (2013), *The Division of Labour in Society* (ed. Steven Lukes; trans. W. D. Halls) (second edition), Houndmills, Palgrave Macmillan, p. 202.
96. Durkheim, *Division*, p. 202.
97. Durkheim, *Division*, p. 101.
98. Durkheim, *Division*, p. 141.
99. Durkheim, *Division*, p. 63.
100. Émile Durkheim (1976), *The Elementary Forms of the Religious Life* (intro. Robert Nisbet; trans. Joseph Ward Swain) (second edition), London: George Allen and Unwin Ltd, p. 387.
101. Vincent Descombes (2000), "The philosophy of collective representations" (trans. Anthony Cheal Pugh), *History of the Human Sciences*, 13:1, pp. 37–49, p. 44.
102. Durkheim, *Division*, p. 143.
103. Durkheim, *Division*, p. 177.
104. Durkheim, *Division*, p. 316.
105. Durkheim, *Division*, p. 204.
106. Paul Bohannan (1960), "Conscience Collective and Culture", in Kurt H. Wolff (ed.) (1960), Émile *Durkheim, 1858–1917: A Collection of Essays, with Translations and a Bibliography*, Columbus, Ohio State University Press, pp. 77–96, p. 81.
107. Émile Durkheim (1953), *Sociology and Philosophy* (trans. D.F. Pocock), Glencoe: The Free Press, pp. 11–12.
108. David Bloor (2000), "Collective Representations as Social Institutions", in William S. F. Pickering (2000), *Durkheim and Representations*, London: Routledge, pp. 157–166, p. 160.
109. Steven Brint (2001), "Gemeinschaft revisited: A critique and reconstruction of the community concept", *Sociological Theory*, 19:1, pp. 1–23, p. 2.

110. Brint, "Gemeinschaft revisited", p. 3.
111. Brint, "Gemeinschaft revisited", p. 1.
112. Alexis de Tocqueville (2004), *Democracy in America: the Complete and Unabridged Volumes I and II* (intro. By Joseph Epstein; trans. Henry Reeve), New York: Bantam Classics, p. 225.
113. De Tocqueville, *Democracy*, p. 329.
114. De Tocqueville, *Democracy*, pp. 628–632.
115. De Tocqueville, *Democracy*, pp. 636–639.
116. Pat Dudgeon et al. (2002), "Contemporary Aboriginal Perceptions of Community", in Adrian T. Fisher, Christopher C. Sonn, and Brian J. Bishop (eds) (2002), *Psychological Sense of Community: Research, Applications, and Implications*, New York: Kluwer Academic/Plenum Publishers, pp. 247–267, p. 265.
117. Mark Fettes (1998), "Indigenous education and the ecology of community", *Language Culture and Curriculum*, 11:3, pp. 250–271, p. 259.
118. Judy Taylor et al. (2012), "Towards a conceptual understanding of Aboriginal and Torres Strait Islander community and community functioning", *Community Development Journal*, 47:1, pp. 94–110, p. 99.
119. Fettes, "Indigenous", p. 65.
120. Dudgeon et al., "Contemporary Aboriginal", p. 249.
121. Barry Wellman (1979), "The Community Question: The Intimate Networks of East Yorkers", *American Journal of Sociology*, 84: 5, pp. 1201–31, p. 1201.
122. Nisbet, *Sociological Tradition*, p. 47.
123. Richard A. Brosio (1976), "Dewey's Concept Of Community: A Last Third Of The Twentieth Century Perspective", *The Journal of Educational Thought*, 10: 2, pp. 110–118, p. 116.
124. John Dewey (1946), *The Public and its Problems: an Essay in Political Enquiry*, Chicago: Gateway Books, p. 111.
125. Dewey, *Public*, pp. 126–127.
126. Dewey, *Public*, p. 127.
127. Dewey, *Public*, p. 148.
128. Timothy V. Kaufman-Osborn (1984), "John Dewey and the liberal science of community", *The Journal of Politics*, 46:4, pp. 1142–1165, p. 1147.
129. Dewey, *Public*, p. 151.

130. Dewey calls upon the dictionary for this definition: "the general sense, feeling, judgment of mankind or a community". John Dewey (1938), *Logic: The Theory of Inquiry*, New York: Henry Holt and Company, p. 62.
131. John Dewey (1916), *Democracy and Education: an Introduction to the Philosophy of Education*, New York: The Macmillan Company, p. 5.
132. Dewey, *Public*, p. 142.
133. Robert A. Nisbet (1953), *The Quest for Community: a Study in the Ethics of Order and Freedom*, New York: Oxford University Press, p. 7.
134. Nisbet, *Quest*, p. 9.
135. Timothy Carney (2020), *Alienated America: Why Some Places Thrive While Others Collapse*, New York: Harper Collins, pp. 13, 93, 147.
136. Nisbet, *Quest*, p. 49.
137. Carney, *Alienated*, p. 148.
138. Yuval Levin (2016), *The Fractured Republic: Renewing America's Social Contract in the Age of Individualism*, New York: Basic Books, p. 205.
139. Pfortmüller, "What does 'community'".
140. David W. McMillan and David M. Chavis (1986), "Sense of Community: A Definition and Theory", *Journal of Community Psychology*, 14, pp. 6–23, p. 9.
141. Barry Wellman and Barry Leighton (1979), "Networks, Neighborhoods, and Communities: Approaches to the Study of the Community Question", *Urban Affairs Quarterly*, 14:3, pp. 363–390.
142. Carney, *Alienated*, p. 248. Putnam uses the same metaphor and extends it: "Bonding social capital constitutes a kind of sociological superglue, whereas bridging social capital provides a sociological WD-40", Putnam, *Bowling Alone: The Collapse*, p. 23.
143. Yuval Levin (2017), "Conservatism in an Age of Alienation", *Modern Age*, 59:2, pp. 11–22, p. 19.
144. Carney, *Alienated*, pp. 98–99, citing Sandra Navidi (2017), "Why Hillbillies Don't' Do Networking", *Daily Beast*, 28 March, www.thedailybeast.com/why-hillbillies-dont-do-networking. Accessed 22 April 2024.

145. Robert D. Putnam (1995), "Bowling Alone: America's Declining Social Capital", *Journal of Democracy*, 6:1, pp. 65–78; Putnam, *Bowling Alone: The Collapse*; Robert D. Putnam and Lewis M. Feldstein with Don Cohen (2003), *Better Together: Restoring the American Community*, New York: Simon & Schuster.
146. Putnam, "Bowling Alone: America's Declining", p. 65.
147. Carney, *Alienated*, p. 160.
148. Putnam, "Bowling Alone: America's Declining", pp. 66–67.
149. Putnam, *Bowling Alone: The Collapse*, p. 136.
150. Carney, *Alienated*, p. 197.
151. Putnam, *Bowling Alone: The Collapse*, p. 189.
152. Putnam, *Bowling Alone: The Collapse*, p. 204.
153. Putnam, *Bowling Alone: The Collapse*, p. 216.
154. Putnam, *Bowling Alone: The Collapse*, p. 247.
155. Putnam, *Bowling Alone: The Collapse*, p. 277.
156. Putnam, *Bowling Alone: The Collapse*, p. 284.
157. Putnam, *Bowling Alone: The Collapse*, p. 295.
158. Putnam, *Bowling Alone: The Collapse*, pp. 404–412.
159. Barack Obama (2007), *The Audacity of Hope: Thoughts on Reclaiming the American Dream*, Edinburgh: Canongate Books, p. 9.
160. Obama, *Audacity*, p. 193.
161. Obama, *Audacity*, p. 193.
162. Obama, *Audacity*, p. 87.
163. Obama, *Audacity*, p. 55.
164. Obama, *Audacity*, p. 250.
165. Obama, *Audacity*, p. 263.
166. Obama, *Audacity*, p. 333.
167. Obama, *Audacity*, p. 356.
168. Donald Trump (2015), "Here's Donald Trump's Presidential Announcement Speech", *Time*, https://time.com/3923128/donald-trump-announcement-speech/. Accessed 19 March 2024.
169. Carney, *Alienated*, p. 46.
170. Levin, *Fractured*, p. 101.
171. Patrick J. Deneen (2018), *Why Liberalism Failed*, New Haven: Yale University Press, p. 192.
172. Nisbet, *Quest*, pp. 31–2.
173. Nisbet, *Quest*, p. 32.
174. Levin, *Fractured*, p. 205.

175. Levin, *Fractured*, p. 175.
176. Levin, *Fractured*, p. 181.
177. Deneen, *Why Liberalism*, p. 87, referencing Wendell Berry's ideas, especially as they emerge through fiction such as *The Way of Ignorance*.
178. Brint, "Gemeinschaft revisited", pp. 3–4.
179. Wellman, "Community Question", p. 1203.
180. Wellman, "Community Question", p. 1204.
181. Wellman, "Community Question", p. 1207.
182. "Their ties are not encapsulated in 'decoupled' little worlds [...] but are strands in the larger metropolitan web". Wellman, "Community Question", p. 1227.
183. Wellman, "Community Question", p. 1203.
184. Arne Naess (2001), *Ecology, Community and Lifestyle: Outline of an Ecosophy* (trans. David Rothenberg), Cambridge: Cambridge University Press, p. 168.
185. Timothy Morton (2016), *Dark Ecology: For a Logic of Coexistence*, New York: Columbia University Press, p. 5.
186. Timothy Morton (2012), "Art in the Age of Asymmetry: Hegel, Objects, Aesthetics", *Evental Aesthetics*, 1, pp. 121–142, p. 134.
187. Bateson, Gregory (1991), *A Sacred Unity: Further Steps to an Ecology of Mind*, New York: Cornelia and Michael Bessie Books, p. 261.
188. Bateson, *A Sacred*, pp. 304–305.
189. Devall, Bill, and George Sessions (1985), *Deep Ecology*, Layton: Gibbs Smith, p. 66.

Bibliography

Anon. (n.d.), slogan, *#MeToo*, https://metoomvmt.org/. Accessed 1 September 2020;

Anon. (n.d.), slogan, *Different Together*, https://different-together.co.uk/. Accessed 1 September 2020.

Anon. (2010), "David Cameron launches Tories' 'big society' plan", *BBC News*, 19 July. https://www.bbc.co.uk/news/uk-10680062. Accessed 19 March 2024.

Anon. (2022), "Connected Communities", *UKRI* website, https://www.ukri.org/what-we-offer/browse-our-areas-of-investment-and-support/connected-communities/. Accessed 11 January 2023.

Anon. (2022), "Crowd Foresting: Miyawaki Model Afforestation", *Crowd-Foresting*, https://www.crowdforesting.org/. Accessed 8 March 2024.

Bastian, Jeanette A. and Ben Alexander (2009), *Community Archives: The Shaping of Memory*, London: Facet Publishing.

Bateson, Gregory (1991), *A Sacred Unity: Further Steps to an Ecology of Mind*, New York: Cornelia and Michael Bessie Books, p. 261.

Block, Peter (2018), *Community: the Structure of Belonging*, Oakland: Berrett-Koehler.

Boyles, Ann (n.d.), "The Meaning of Community", *bahai.org*, https://www.bahai.org/documents/essays/boyles-ann/meaning-community. Accessed 31 August 2020.

Bradshaw, Ted K. (2008), "The Post-Place Community: Contributions to the Debate about the Definition of Community", *Journal of Community Development*, 39: 1, pp. 5–16.

Brint, Steven (2001), "Gemeinschaft revisited: A critique and reconstruction of the community concept", *Sociological Theory*, 19:1, pp. 1–23.

Brosio, Richard A. (1976), "Dewey's Concept Of Community: A Last Third Of The Twentieth Century Perspective", *The Journal of Educational Thought*, 10: 2, pp. 110–118.

Cárdenas, Macarena L., Bethany Pudifoot, Claire L. Narraway, Caroline Pilat, Victor Beumer, and Daniel B. Hayhow (2022), "Nature-based Solutions Building Urban Resilience for People and the Environment", *Quarterly Journal of Forestry*, 116:3, pp. 27–37.

Carney, Timothy (2020), *Alienated America: Why Some Places Thrive While Others Collapse*, New York: Harper Collins.

Charkow, Mike (2022), "Forests for the Future", *ARB Magazine*, 197, p. 30. https://www.avtree.co.uk/wp-content/uploads/2022/06/ARBMagazine197-pp29-33-Forests-of-the-future.pdf. Accessed 30 August 2023.

Crow, Graham, and Alice Mah (2012), "Conceptualisations and meanings of 'community': the theory and operationalisation of a contested concept", *Connected Communities*, https://www.community-methods.soton.ac.uk/resources/CC%20Final%20Report_30%20March%20GC.pdf. Accessed 26 April 2024.

Daou, Alain, Mansour Saliba and Antoine Kallab (2024), "A Review of the Miyawaki Method", *SSRN* 4728239, pp. 1–49.

De Tocqueville, Alexis (2004), *Democracy in America: the Complete and Unabridged Volumes I and II* (intro. By Joseph Epstein; trans. Henry Reeve), New York: Bantam Classics.

Deneen, Patrick J. (2018), *Why Liberalism Failed*, New Haven: Yale University Press.

Descombes, Vincent (2000), "The philosophy of collective representations" (trans. Anthony Cheal Pugh), *History of the Human Sciences*, 13:1, pp. 37–49.

Devall, Bill, and George Sessions (1985), *Deep Ecology*, Layton: Gibbs Smith.

Dewey, John (1916), *Democracy and Education: an Introduction to the Philosophy of Education*, New York: The Macmillan Company.

Dewey, John (1938), *Logic: The Theory of Inquiry*, New York: Henry Holt and Company.

Dewey, John (1946), *The Public and its Problems: an Essay in Political Enquiry*, Chicago: Gateway Books.

Donovan, Geoffrey H., Jeffrey P. Prestemon and Abigail R. Kaminski (2022), "The Natural Environment and Social Cohesion: Tree planting is associated with increased voter turnout in Portland, Oregon", *Trees, Forests and People*, 7:100215.

Durkheim, Émile (1953), *Sociology and Philosophy* (trans. D.F. Pocock), Glencoe: The Free Press.

Durkheim, Émile (1976), *The Elementary Forms of the Religious Life* (intro. Robert Nisbet; trans. Joseph Ward Swain) (second edition), London: George Allen and Unwin Ltd.

Durkheim, Émile (2013), *The Division of Labour in Society* (ed. Steven Lukes; trans. W. D. Halls) (second edition), Houndmills, Palgrave Macmillan.

Egerer, Monika, and Michael Suda (2023), "Designing 'Tiny Forests' as a lesson for transdisciplinary urban ecology learning", *Urban Ecosystems*, 26, pp. 1331–1339.

Esposito, Roberto (2009), "Community and nihilism", *Cosmos & History*, 5:1, pp. 24–36.

Fettes, Mark (1998), "Indigenous education and the ecology of community", *Language Culture and Curriculum*, 11:3, pp. 250–271.

Fisher, Adrian T., Christopher C. Sonn, and Brian J. Bishop (eds) (2002), *Psychological Sense of Community: Research, Applications, and Implications*, New York: Kluwer Academic/Plenum Publishers, pp. 247–267.

Fisher, Dana, Erika Svendsen, and James Connolly (2015), *Urban environmental stewardship and civic engagement: how planting trees strengthens the roots of democracy*, London and New York: Routledge.

Galwey, Isabel (2020), "Taking Root: Failure and Success for the Great Forest Wall of Tōhoku", *Anthroposphere: The Oxford Climate Review*, 15 August, https://www.anthroposphere.co.uk/post/taking-root-failure-and-success-for-the-great-forest-wall-of-t%C5%8Dhoku. Accessed 8 March 2024.

Garber, Megan (2017), "What does 'community' mean?, *The Atlantic*, 3 July, https://amp.theatlantic.com/amp/article/532518/. Accessed 31 August 2020.

Giseburt, Annelise (2023), "Miyawaki Forests are a Global Sensation, but not Everyone's Sold on Them", *Mongabay*, https://news.mongabay.com/2023/06/miyawaki-forests-are-a-global-sensation-but-not-everyones-sold-on-them/. Accessed 9 March 2024.

Harvey, Branden (2022), "37 Ways to Make a Difference in the World", *Good Good Good* website, 4 May. https://www.goodgoodgood.co/articles/how-to-make-a-difference. Accessed 11 January 2023.

Hayes, Nick, and Jon Moses (eds) (2024), *Wild Service: Why Nature Needs You*, London: Bloomsbury.

Heiman, Michael K. (1996), "Race, waste, and class: New perspectives on environmental justice." *Antipode*, 28:2, pp. 111–121.

Jaiswal, Deepak, , Sruthi Surendran, Merlin Lopus, Amit Kushwaha, Akhila K. Chandrabose, Anna Geveena, Shaji P. Saranga, Sethulakshmi Nair, and , Kalpuzha Ashtamoorthy Sreejith (2023), "Fast Growing Forests (FGF) to offset Greenhouse Gas (GHG) Emissions", *EGU General Assembly Conference Abstracts*.

Kaufman-Osborn, Timothy V. (1984), "John Dewey and the liberal science of community", *The Journal of Politics*, 46:4, pp. 1142–1165.

Kurian, Anju Lis (2020), "Urban Heat Island Mitigation And Miyawaki Forests: An Analysis", *EM International*, pp. 186–191.

Levin, Yuval (2016), *The Fractured Republic: Renewing America's Social Contract in the Age of Individualism*, New York: Basic Books

Levin, Yuval (2017), "Conservatism in an Age of Alienation", *Modern Age*, 59:2, pp. 11–22.

Lewis, Hannah (2022), *Mini-Forest Revolution: Using the Miyawaki Method to Rapidly Rewild the World*, White River Junction: Chelsea Green Publishing.

MacQueen, Kathleen M., Eleanor McLellan, David S. Metzger, Susan Kegeles, Ronald P. Strauss, Roseanne Scotti, Lynn Blanchard, and Robert T. Trotter (2001), "What is Community? An Evidence-Based Definition for Participatory Public Health", *American Journal of Public Health*, 91: 12, pp. 1929–1938.

McMillan, David W., and David M. Chavis (1986), "Sense of Community: A Definition and Theory", *Journal of Community Psychology*, 14, pp. 6–23.

Meyer, Marisol L., Ceewin N. Louder, and Guerda Nicolas (2022), "Creating with, not for people: theory of change and logic models for culturally responsive community-based intervention", *American Journal of Evaluation*, 43:3, pp. 378–393.

Millison, Andrew (2023), "Guerilla Gardeners turn Toxic Soil into Eco-Oasis", *YouTube*, https://www.youtube.com/watch?v=UqyK_9iybD8. Accessed 10 March 2025.

Ministry of Housing (2010), "Cameron and Clegg Launch the Big Society", *Gov.UK*, https://www.gov.uk/government/news/cameron-and-clegg-launch-the-big-society. Accessed 26 April 2024.

Miyawaki, Akira (1998), "Restoration of Urban Green Environments Based on the Theories of Vegetation Ecology", *Ecological Engineering*, 11, pp. 157–165.

Miyawaki, Akira (1999), "Creative ecology restoration of native forests by native trees", *Plant Biotechnology*, 16.1, pp. 15–25.

Miyawaki, Akira (2006), "A Call to Plant Trees", *A better future for the planet Earth, vol.I–V (1992–2016): lectures, essays, and interviews of the winners of the Blue Planet Prize*, Tokyo: Asahi Glass Foundation, pp. 252–257.

Miyawaki, Akira (2006), "Aiming for the Restoration of a Green Global Environment", *A better future for the planet Earth, vol.I–V (1992–2016): lectures, essays, and interviews of the winners of the Blue Planet Prize*, Tokyo: Asahi Glass Foundation, pp. 258–275.

Miyawaki, Akira, and Frank B. Golley (1993), "Forest reconstruction as ecological engineering", *Ecological Engineering*, 2, pp. 333.345.

Morton, Timothy (2012), "Art in the Age of Asymmetry: Hegel, Objects, Aesthetics", *Evental Aesthetics*, 1, pp. 121–142, p. 134.

Morton, Timothy (2016), *Dark Ecology: For a Logic of Coexistence*, New York: Columbia University Press, p. 5.

Naess, Arne (2001), *Ecology, Community and Lifestyle: Outline of an Ecosophy* (trans. David Rothenberg), Cambridge: Cambridge University Press.

Nargi, Lela (2019), "The Miyawaki Method: A Better Way to Build Forests?", *JSTOR* Daily, 24 July, https://daily.jstor.org/the-miyawaki-method-a-better-way-to-build-forests/. Accessed 15 March 2024.

Navidi, Sandra (2017), "Why Hillbillies Don't' Do Networking", *Daily Beast*, 28 March, www.thedailybeast.com/why-hillbillies-dont-do-networking. Accessed 22 April 2024.

Nisbet, Robert A. (1953), *The Quest for Community: a Study in the Ethics of Order and Freedom*, New York: Oxford University Press.

Nisbet, Robert A. (2004), *The Sociological Tradition* (sixth printing), New Brunswick: Transaction Publishers.

Obama, Barack (2007), *The Audacity of Hope: Thoughts on Reclaiming the American Dream*, Edinburgh: Canongate Books.

Pérez-Almendros, Carla, Luis Espinosa-Anke, and Steven Schockaert (2020), "Don't patronize me! an annotated dataset with patronizing and condescending language towards vulnerable communities", *arXiv preprint* arXiv: 2011.08320.

Pfortmüller, Fabian (2017), "What does 'community' even mean? A definition attempt & conversation starter", *Together Institute*, 20 September, https://medium.com/together-institute/what-does-community-even-mean-a-definition-attempt-conversation-starter-9b443fc523d0. Accessed 31 August 2020.

Pickering, William S. F. (2000), *Durkheim and Representations*, London: Routledge.

Putnam, Robert D. (1995), "Bowling Alone: America's Declining Social Capital", *Journal of Democracy*, 6:1, pp. 65–78.

Putnam, Robert D. (2000), *Bowling Alone: The Collapse and Revival of American Community*, New York: Simon & Schuster.

Putnam, Robert D., and Lewis M. Feldstein with Don Cohen (2003), *Better Together: Restoring the American Community*, New York: Simon & Schuster.

Rashid, Fazal, and Somil Daga (2023), "How Mr Miyawaki Broke My Heart", *Science: The Wire*, 25 April, https://science.thewire.in/environment/how-mr-miyawaki-broke-my-heart/. Accessed 9 March 2024.

Rosenblum, Mort, and Doug Williamson (1990), *Squandering Eden: Africa at the Edge*, Boulder: Paladin Press.

Rots, Aike P. (2021), "Trees of tension: re-making nature in post-disaster Tohoku", *Japan Forum*, 33:1, pp. 1–24.

Rudland, Oliver (2023), "'Reaching Out' or Institutional Virtue-Signalling? The Role of Community Opera Projects in UK Opera Houses Today", *Tempo*, 77:304, pp. 62–72.

Sarason, S.B. (1974), *The psychological sense of community: Perspectives for community psychology,* San Francisco: Jossey-Bass.

Savage, Karen, and Dominic Symonds (2018), *Economies of Collaboration in Performance: More than the Sum of the Parts*, Cham: Palgrave Macmillan.

Schirone Bartolomeo, Antonello Salis, and Federico Vessella (2011), "Effectiveness of the Miyawaki method in Mediterranean forest restoration programs", *Landscape and Ecological Engineering*, 7, pp. 81–92.

Skidmore, Simon (2011), "Smaller Government: Bigger Society?", *Parliament UK*, https://publications.parliament.uk/pa/cm201012/cmselect/cmpubadm/writev/smaller/bs76.htm. Accessed 26 April 2024.

Song, Felicia Wu (2009), *Virtual Communities: Bowling Alone, Online Together*, New York: Peter Lang.

Suzuki, David (2007), *The Sacred Balance: Rediscovering our Place in Nature* (updated and expanded), Vancouver, Toronto, Berkeley: Greystone Books.

Suzuki, David (2014), "Out of Darkness, the Light", *David Suzuki Foundation*, 9 January, https://davidsuzuki.org/story/out-of-darkness-the-light/. Accessed 8 March 2024.

Taylor, Judy, Jane Edwards, Sonia Champion, Simon Cheers, Alwin Chong, Rachel Cummins, and Brian Cheers, (2012), "Towards a conceptual understanding of Aboriginal and Torres Strait Islander community and community functioning", *Community Development Journal*, 47:1, pp. 94–110.

Thiagarajan, Kamala (2020), "Could Akira Miyawaki's 50-Year Old Innovation Help Promote Biodiversity and Reduce the Risk of Climate Change?", *Ensia*, 11 June, https://ensia.com/features/miyawaki-trees-biodiversity-carbon-sequestration/. Accessed 8 March 2024.

Tönnies, Ferdinand (2001), *Community and Civil Society* (ed. Jose Harris; trans. Jose Harris and Margaret Hollis), Cambridge: Cambridge University Press.

Trump, Donald (2015), "Here's Donald Trump's Presidential Announcement Speech", *Time*, https://time.com/3923128/donald-trump-announcement-speech/. Accessed 19 March 2024.

Tunstall, Rebecca, Ruth Lupton, Anne Power, and Liz Richardson (2011), "Building the Big Society", *Centre for Analysis of Social Exclusion (CASE)*, 67, White Rose University Consortium. https://eprints.whiterose.ac.uk/75097/1/CASEreport67.pdf. Accessed 26 April 2024.

Watkins, Shannon Lea, Jess Vogt, Sarah K. Mincey, Burnell C. Fischer, Rachael A. Bergmann, Sarah E. Widney, Lynne M. Westphal, and Sean Sweeney (2018), "Does collaborative tree planting between nonprofits and neighborhood groups improve neighborhood community capacity?", *Cities*, 74, pp. 83–99.

Wellman, Barry (1979), "The Community Question: The Intimate Networks of East Yorkers", *American Journal of Sociology*, 84: 5, pp. 1201–31.

Wellman, Barry, and Barry Leighton (1979), "Networks, Neighborhoods, and Communities: Approaches to the Study of the Community Question", *Urban Affairs Quarterly*, 14:3, pp. 363–390.

Williams, Claire (2017), "Inclusive communities and robust causality: The constant struggle to balance access to the courts with protection for defendants", *Minnesota Law Review*, 102:969.

Wirth, Louis (1926), "The Sociology of Ferdinand Tonnies", *American Journal of Sociology*, 32:3, pp. 412–422.

Wolff, Kurt H. (ed.) (1960), *Emile Durkheim, 1858–1917: A Collection of Essays, with Translations and a Bibliography*, Columbus, Ohio State University Press.

Worrad-Andrews, Alex (2010), "'Big society' can give power back to communities", *The Guardian*, 20 July, https://www.theguardian.com/commentisfree/2010/jul/20/big-society-empower-communities. Accessed 26 April 2024.

Zolberg, Vera L. (2003), "Current Challenges for Cultural Policy: New Meanings of Community", *The Journal of Arts Management, Law, and Society*, 32:4, pp. 295–307.

CHAPTER 4

Performance

The idea to plant a new woodland on the Swinton Estate in the Yorkshire Dales was inspired by a loss of trees, and of one tree in particular. "Like many areas of the countryside we have lost so many trees from larch blight", reported the estate owner Felicity Cunliffe-Lister, "so regenerating the area through this project is a perfect fit".[1] Chris Bradley takes up the tale: "In recent years, the area has lost swathes of trees to a fungal disease *phytophthora ramorum*", he explains; "The new woodland will regenerate part of the plantation".[2] The plan is to make the woodland a nature trail for visitors to the Swinton Estate, which already offers activities from wild swimming to forest bathing on its 20,000 acres.

However the more immediate impulse to plant a new woodland came as a response to the mysterious felling of a tree that was cherished by the whole nation. When she heard about the loss of the "Sycamore Gap tree" on 28 September 2023, North Yorkshire artist Lucy Pittaway, whose prints of the iconic tree have contributed to her winning the "UK's Most Popular Published Artist" award no less than three times,[3] pledged to plant a new tree for every print of the Sycamore Gap she sold[4]; the Swinton Estate offered its land to host the trees. There are now over 1000 trees in the woodland. Sycamore features heavily, of course, but so too do oak, cherry, and crab apple. The woodland trail is marked by arched bridges of larch logs, designed by another artist, Mark Wilkinson, and woven wicker "pods" by willow sculptor Jess Wilkinson hang above

K. Savage and D. Symonds, *Ecologies of Community in Performance*,
https://doi.org/10.1007/978-3-031-94343-0_4

the pathway to resemble birds' nests. "The log-pile arch will be packed with moss and other plant material, to create a sustainable wildlife haven for minibeasts", Bradley reports, while "hidden cameras will observe any wildlife that makes its home in these luxurious nesting pods".[5] The trail opened to the general public in the spring of 2025, and will stand as a tribute to the memory of the Sycamore Gap tree.

The Sycamore Gap tree has long been iconic in the North East, having stood in its characteristic dip in the ground for close to two hundred years. It was known as a "celebrity tree",[6] a "sacred tree",[7] and a "sentinel of time",[8] and it was awarded "English Tree of the Year" in the Woodland Trust Awards in 2016. More than this, it was part of the UNESCO World Heritage Site of Hadrian's Wall, which is a protected archaeological monument. The tree elicited powerful emotions from those who visited and from local communities—so much that it became a popular location for young couples getting engaged, or for ashes being scattered. Symbolically, too, the tree came to represent qualities of majesty, longevity, dignity, and strength, and was adopted as "a symbol of endurance, and even of the North East itself".[9] In 1991, Sycamore Gap achieved worldwide recognition when it featured in the Hollywood blockbuster *Robin Hood: Prince of Thieves*. Colloquially, it therefore also became known as the "Robin Hood tree". The Sycamore Gap tree, note Ben Davenport and Tanja Hoffman, "was for many THE tree—a perfect arboreal specimen located in a storied landscape".[10]

No one knew who committed the felling of the tree, though two men were later charged. Across the country millions expressed their outrage, and communities were left "devastated"[11] by the "act of meaningless vandalism".[12] In remarkably emotional terms, people speak of "personal tragedy",[13] of being "in mourning",[14] and of needing a "grieving process"[15] for the tree. In the wake of the loss, seeds have been planted in the hope of growing new saplings; cuttings have been grafted onto other rootstock to propagate hybrid trees; and new genetic methods are being tested to create clones of the original.[16] Meanwhile, the public showed an outpouring of commemorative responses: "People have been known to leave decorated memorial pebbles"[17]; poems and songs have been written and shared online[18]; "children's portraits of the tree line the walls of the swiftly-established 'celebration room' dedicated to the tree"[19]; and a section of the felled tree itself was put on display.[20] Northumberland printmaker Shona Branigan was commissioned to create a series of tree prints,[21] Poet Laureate Simon Armitage wrote his own homage,[22] North

East charity The Community Foundation pledged a considerable grant to create a Legacy Art Project,[23] and several crowdfunding sites were set up in partnership with the National Trust, to "celebrate the legacy of our tree but also benefit our community for generations to come".[24] One site even campaigned (unsuccessfully) to replace the tree with a full-size bronze replica.[25] These initiatives are "not just about preserving the memory of Sycamore Gap", Pittaway observes, "but also about fostering new growth and hope for the future".[26] That is very literally the case for the Swinton Estate woodland. "For one tree to inspire the planting of 1000 more is remarkable",[27] adds Cunliffe-Lister. The crowdfunding organizers, meanwhile, vouch that their project "will be a symbol of resilience and unity, just like the tree itself".[28]

From the perspective of conservationists, such extended benefits are opportune ways to curate the memory of the Sycamore Gap tree for the communities whom it has touched so deeply:

> While devastating, the loss of the Sycamore Gap tree opens an exciting new opportunity to decide—in collaboration with a tree—what heritage we choose to bring forward, including what we decide to remember, and what we choose to forget.[29]

In this, there is a sense of the community being able to "frame" the narrative of this event, to spin their storytelling for a particular purpose. Thus an act of meaningless vandalism becomes a story of new growth; the memory of a single tree comes to signify resilience and hope. A year after the felling, the National Trust announced that eight new shoots had emerged from the remains of the tree's stump, enabling the ancient forestry technique of coppicing to be used to revive at least a remnant of the tree. There will be a future for Sycamore Gap; we can frame its legacy as a story of rebirth.

These events clearly resonate in terms of the focus of this book, not only ecologically in terms of public attitudes to our environment, but also in terms of community as a tale of local tragedy leading to community cohesion. Rob Cowen is among a number of commentators to remark on the ecological perspectives surrounding the toppling of the tree. Writing about three weeks after it was felled, he laments a general apathy among the British public about nature and the environment, and notes the irony that the tree was cut down on the same day that the National Trust issued its "State of Nature" report, which spells out how "the UK is now one of the most nature-depleted countries on Earth".[30] That report's findings are stark, calling for a raft of coordinated conservation measures. But a point that Cowen makes is that this inexplicable act of hostility to nature is symptomatic of our alienated relationship with the land: we can't swim in the rivers because they're polluted; we can't walk through the countryside because we don't have extensive right to roam; we don't learn about nature as children because there isn't a syllabus in the mandated curriculum; and all of these circumstances have been brought about through government policy. "How can we ask people to protect something they have no sense of belonging in, or to?", asks Cowen, targeting the government for policies that privilege economic growth above other concerns. "If you're asking what I hope might rise from that poor stump, it is change", he concludes; "More than a new sculpture, a stone circle or commemorative bench, it is root-and-branch change".[31] We'll return to this notion of change later.

However, our purpose in using the example of the "Sycamore Gap" as the preface to this chapter is because for us it is an ideal focus for exploring *performance*, or at least, the *performative*. That's a concept that may require some clarification, but it is one that is central to our home discipline area of the performing arts. For us, performance concerns the ways in which people, objects, landscapes, and issues are framed to cast them in a certain light. We see this directly when a theatre production is staged, or a character is emphasized in a spotlight. But even beyond the trappings of theatre, we have a tendency to imbue things with certain qualities that make them performative. And we've already seen how performative this tree has been: it's been showcased as "majestic", as "resilient", and as "sacred"; it's even been capitalized as "THE tree". For us, this is a classic example of performativity: the tree literally stands out against the horizon, in its cultural context, and in our collective perceptions.

This chapter will consider contemporary understandings of performance in relation to the environment, noting that, as distinct from straightforward performances *in* or *about* nature, our relationship with the land is one that has become in itself *performative*—and as we've suggested, many of the acts that we see in the wake of the Sycamore Gap felling are examples of that performativity. We will start by considering some of the ways in which nature and our relationship to it has been reflected in cultural representations. Specifically, we will take a brief look at the classical idea of the *pastoral*, and the mediaeval representation of Robin Hood. Next, we will explore some of the tenets of performance theory introduced in the 1950s and 60 s by figures such as Richard Schechner and Erving Goffman. We will look at the notion of *performance ecology* as it has been developed in more recent years by Bonnie Marranca and Baz Kershaw, and we will argue that it is through performative dynamics that we represent the land with/in which we live and the relationship(s) we have with it.

Performing Nature: The Pastoral

Traditions of staging the land and our relationship with it have been strong not just in England but in many global contexts. The European pastoral tradition, for example, stems from the Golden Age of Greek mythology, whose region of Arcadia was characterized as a world of unspoiled tranquillity, presided over by the forest god Pan and his nymphs and dryads. The pastoral *bucolics* of the Greek poet Theocritus (third-century BCE) depicted lovelorn shepherds, toyed with by the gods but finding solace through song. Subsequently the imagery of this world was taken up by the Roman poet Virgil, whose *eclogues* (first-century BCE) featured shepherds lamenting unrequited love and trying to sustain their livelihoods despite being evicted from the land. The pastoral tradition was revived in the sixteenth century thanks to a fascination with classical culture. In translating its idioms into the world view of renaissance Europeans, writers and artists presented Arcadia as a version of both Eden and Utopia, suggesting themes of lost innocence and unreachable perfection. Thomas More's *Utopia* (1516) and John Milton's *Paradise Lost* (1667) are obvious references, but they straddle a period of prolific output across Europe, much of which evoked the imagery of the golden age. Among the features that Paul Alpers recognizes in this tradition are the portrayals of "idyllic landscape", "landscape as a setting for song", "an atmosphere of *otium* [self-realization]", and "a conscious attention to art and nature"; on the other hand, he also argues that the pastoral tradition is more concerned with "herdsmen and their lives, rather than landscape or idealized nature".[32] Hence the proliferation of nymphs and shepherds in pastoral imagery, from Theocritus and Virgil to Shakespeare.

In what Jeffrey Theis has christened "sylvan pastoral", the literature of sixteenth and seventeenth centuries England defines the way "in which English culture consciously and unconsciously navigates the complex spaces of forests and woodlands". He argues that the individual, nature, and society are "physically and symbolically shaped and reshaped by each other", noting that plays like *As You Like It* (1599) and *A Midsummernight's Dream* (1600) are among the most poignant examples, "creating and recreating English culture through its relation to the woods".[33] Charlotte Scott suggests that such reshaping is reflected in a focus on husbandry, a theme that is "at the forefront of their social and moral economies",[34] whether this be through farming the land, cultivating gardens, or herding livestock. In this, "the sixteenth century marked a decisive point in the shift from nature to culture", she remarks, "dependent upon an increasingly articulate investment in the language of cultivation".[35] Through cultivating the land we can control the way it serves our needs, whether those are basic survival, economic growth, or aesthetic enjoyment. This not only changes our relationship with the land, but also the human and social relations that we develop with/in it.

Raymond Williams calls this sort of sensibility a "structure of feeling",[36] and in his book *The Country and the City* (1973) he distinguishes between the twentieth-century emotional mindsets of rural and urban communities in the UK, how these emerged, and how they have shaped the development of society. He acknowledges a commonly-held view of the countryside offering "a natural way of life", reflecting "peace, innocence, and simple virtue", but prone to "backwardness, ignorance, limitation".[37] Such a view has entrenched an image of "Old England"[38] as "an earlier and happier rural England",[39] home to an "Organic Community".[40] At one point, he likens the image of this rural community to a wandering beggar (or, perhaps, a shepherd figure), whose "essential isolation and silence and loneliness have become the only carriers of nature and community".[41] It is the prevalent image of the "neo-classic pastoral",[42] and it conjures up the idea of a "Golden Age" or "paradise"[43] which has now been lost.[44]

Of course, Williams evokes a very particular sensibility from a very particular period in history (the mid-twentieth century), just as Charlotte Scott's identification of a particular relationship with the land is a reflection of a different point in time (the sixteenth century). After all, our relationships with the land and the way in which we reflect those relationships are hardly static, and changes rebalance our representations of nature and culture just as they impact more materially on our lives. If we use the land purely to feed our own family or clan, it becomes an articulation of our mechanical solidarity; if we begin to exchange our goods, it reflects a mercantile state; if we shape the land into an ornamental garden, it flaunts our status[45]; if we bring in cheap labour to enhance the land's production, it takes on the trappings of capitalism.[46] And our cultural representations of the land in art, music, or poetry reflect (or, as we will argue, *perform*) just such attitudes. In the Middle Ages, for example, the Enclosure Acts and the Charter of the Forest both had significant cultural impacts: Tom Williamson writes about the ways in which Enclosure led to "a radical reshaping of social geography. It changed patterns of human interaction and emphasised class divisions, while markedly reducing any notion of 'community'"[47]; meanwhile, Nicholas Robinson writes about how the Charter of the Forest caused the countryside to become "vital to the economy", with its "struggles [...] central to the creation of the rule of law".[48] These events both influenced the ways in which the land and our relationship with it was reflected.

A great deal of popular culture in poetry, ballads, and literature from the mediaeval period onwards has captured such reflections: "From the outlaw poems of Robin Hood, to the fantastical romances of the Round Table, to common complaint poems written in English and French, the medieval literature of England frequently features the forest as the setting of its composition, concerns, and conflicts", notes Chiykowski.[49] That's a broad range of references, but it points to the diversity of ways in which "Old England" and our relationship with it have been represented, from the wild woods of chivalric romance to the charm of the sylvan pastoral, and then to the social and economic spaces of the common forest which for communities reflected their sometimes fragile relationship with the land.[50]

For example, one of the most abiding ways in which the social transition of the Charter of the Forest (1217) has been imprinted on our imaginations is through the Robin Hood mythology with its themes of justice, community, and the land.[51] "Robin Hood reclaims the forest space and the symbols of its administration while establishing his own liveried court that stands in counterpoint to the royal ideology of arbitrary privilege", Chiykoski argues; thus "the ideology of the Robin Hood stories harkens back to a pre-Conquest tradition of land use and hunting law in which the common ownership of the land and its resources can only be asserted through their use".[52] Not least, this is demonstrated through Robin Hood's positioning, famously defending the rights of the common people, and stealing from the rich to give to the poor: "displaying total solidarity with the peasantry, he is the righter of wrongs, the restorer of justice, and the champion of social equity",[53] observes Pollard; "The whole point is that the outlaws take the law into their own hands and they defy in the name of the king his own officers, his corrupt ministers, and mock or kill them". And this sent a powerful message. It suggested that "direct action to make amends could be founded on the perception of an alternative 'popular' authority".[54] That the Robin Hood ballads did this during the fourteenth and fifteenth centuries through displacing the narrative into a previous era, and into the "strange and marginal world" of a wood[55] is at once a claim to historical precedent and an invocation of the force of the commons (people's access to the resource of the land). "Robin Hood's greenwood kingdom is thus a place where the impossible can happen; it is a paradigm of society as it should be", Pollard concludes.[56]

Representations of the land and our relationship with it therefore *perform* a certain way of seeing, and the popular dissemination of such mythologies gives ordinary people a licence to model their behaviours with/in the land according to the logic of their articulations. Thus a later ballad such as "The Lincolnshire Poacher" (c.1776), which is the unofficial anthem of our home county, reflects an ambivalence to the law, gives short shrift to the idea that the resources of the land are owned by anyone, and at the same time dismisses the servitude of "proto-capitalism" as, implicitly, a fool's game. Having served seven years as an apprentice, the narrator rejects the shackles of his indentures to take advantage of the more direct transactional economy of poaching. He bags a hare from poaching the land, and sells it to his neighbour: "Success to every poacher that wants to sell a hare", he concludes.[57] Handed down through the generations, popular folksongs such as this instil a particular viewpoint for communities, not only in this case about poaching as a means of survival, but also about attitudes towards the land and towards work. The song stages, as Stephen Ridgwell puts it, a "theatre of memory",[58] and our adoption of the song through singing it and thereby inhabiting its narrator figure is a performative act of solidarity with its views. Effectively, we side with the politics of the performance, and as the song becomes taken up in a more concerted way (as in this instance by its popular resonance to the county of Lincolnshire), the solidarity of this collective representation becomes a shared commitment. Even more recently, different invocations of our relationship with the land become the cultural norm, as we increasingly see ourselves as a part of the ecosystem. James Canton's *The Oak Papers* (2020), for example, is a study of our relationships with the oak tree, starting from a consideration of the Honywood oak, an 800-year-old tree on the Marks Hall Estate in north Essex, "a colossus of a tree"[59] which Canton's writing frames in similar terms to the Sycamore Gap tree. His style of "wild writing"—a first-person diary recounting his experiences in nature—somehow borrows the legacy of the tree's great age ("a mere sapling when the Magna Carta was signed")[60] to evoke some sort of ancient connection with the land and position him (and by extension, us) in a shared space with nature. His journey begins with quiet reflection, changing pace "to tune in to its slower time frame, to connect with the ecosystem that lives around it, inside it and beneath it"[61]:

> Then my hand reaches out to touch the skin of the old oak. I brush my fingertips, my palms against the rough bark. Something happens. I feel an ease that has not been felt within me for an age. My heart slows. My spirit stills.[62]

In Canton's evocative writing the spirit of nature is performed as a balm, tapping into contemporary resonances of natural wellbeing, spiritual encounter, and oneness with the earth. It's a different inflection to the pastorals, the Robin Hood mythologies and "The Lincolnshire Poacher" but it's no less a performative presentation of nature and our relationship with the natural world.

We could cite many other folksongs, stories, and images that turn to the land in both reflecting cultural identity and performing it to the world. As if conceived by way of example, the annual "Last Night of the Proms" concert parades a whole succession of such performances with its ritual rendition of "Fantasia on British Sea Songs", "Rule, Britannia!", "Pomp and Circumstance", "Land of Hope and Glory", "Jerusalem", "The National Anthem", and "Auld Lang Syne". By the time we gather round our TVs to join the community of promenaders at the Royal Albert Hall, all bobbing up and down, waving EU flags, and full-throatedly singing William Blake's paean about our "green and pleasant land",[63] we are well and truly lost to a complex cultural identity that both resists and absorbs the natural environment of the English landscape. In this—although there may be a hint of the transversal thinking Guattari invited in his *three ecologies* (social and mental and environmental), and no small degree of porosity in the way the ecosystemic boundaries of the Royal Albert Hall, cultural sensibility, and Englishness are all ritually exploded in a delightful display of multiple performativities—there is also evidence, as the scholar Baz Kershaw argues, of the centuries-long and ingrained "pathology" that has led to "the separation of humans from their environment"[64]: "culture versus nature, nature subservient to humans".[65] In an effort to address this, Kershaw has theorized a notion of theatre and performance ecology, which we will explore later in this chapter: "The ecologies of theatre and performance", he writes, "aim to refigure the relationship between 'culture' and 'nature' that all humans inevitably inherit".[66] But first, we need to step back to the 1950s to introduce the field of performance theory.

Performance Theory

The thinking behind performance theory initially comes from the work of linguistics scholar J.L. Austin, who reasoned that performative speech acts not only articulate a meaning, but also invoke a change in circumstances. Phrases such as "I do" in the context of a marriage ceremony, or "I bet" in the context of a wager are performative utterances that develop the social circumstances such that the couple becomes married, or the wager is placed. As the theory of performativity has developed, scholars have argued that a comparable transformation occurs in numerous social and cultural contexts, whether or not the performative utterance is explicitly stated. "People stage themselves and their relations, and in so doing produce culture", explains Christoph Wulf.[67]

This sort of perspective on performance—distinct from *drama* or *theatre*—has been a feature of our home disciplines in the performing arts since at least the Second World War, though it becomes very visible in the 1950s and 60s with the emergence of experimental theatre. Among the most prominent experimental theatre groups of the period was a collective of New York theorists and practitioners labelling itself "The Performance Group". Formed in 1967 by Richard Schechner, the group began to perform versions of the classics that experimented with the "rules" of theatre. He was particularly interested in the relationships between performers and audiences, in reconfiguring expectations of theatrical space, and in the use of ritual practices from other cultures. He termed his approach "Environmental Theater".

One of the first essays Schechner published was "Six Axioms for the Environmental Theater" (1967).[68] It should be said that his use of the term "environmental" is not influenced by an ecological agenda; nor is it an example of the "environmental thinking" we are tracking in this book; in the end, Schechner's term gave way to others such as "site-specific theatre" or "immersive theatre", and in his own writings, he turned to conceptualizing "performance". Still, many of the seeds of his later performance theories are evident in this early discussion. For example, Schechner conceptualizes "a continuum of theatrical events" that extend from the "impure" (public events, demonstrations), to the "pure" (orthodox theatre), which conceptually images a parallel to the open, porous ecology of the ecosystems we have been exploring. "It is because I wish to include this entire range in my definition of theater that traditional distinctions between art and life no longer apply", he writes,[69] articulating ideas redolent of the performance experiments epitomized by other artists such as composer John Cage ("Theatre Piece No.1", 1952) and conceptual artist Allan Kaprow ("Happenings", 1959), and by sociologists such as Erving Goffman (*The Presentation of Self in Everyday Life*, 1956).[70] All of these were questioning assumptions about what (a) "performance" should or could be, and in theorizing "environmental theatre", Schechner calls on those other writers and practitioners: "no one has had more effect than Cage and Kaprow", he acknowledges.[71] "One could view everyday life itself as theatre", stated Cage in 1965,[72] voicing what Goffman was explicitly writing about at around the same time. Like them, Schechner also understands the encounters of performance to be social and "relational",[73] recognizing that events establish meaningful interactions between all of those participating. As such, the study of performance encompasses all manner of scenarios including the day-to-day activities of spaces such as shopping malls, the work of athletes and sportspeople, the activities of Malibu Beach bodybuilders, and most significantly, the cultural rituals of other societies.

It's worth noting here how Schechner's work responded substantially to the work of anthropologist Victor Turner and his research into the ritual activities of other cultures.[74] Schechner, Turner, and others mounted three academic symposia on ritual and theatre.[75] "The goal of these meetings was to approach the genres of theatre, dance, music, sports, and ritual as a single, coherent group, *as performance*", wrote the organizers, "where it was hoped that the conferences would lay the groundwork for proposing general principles or 'universals of performance'".[76] Perhaps the most useful crystallization of those universals is found in Schechner's essay, "Toward a Poetics of Performance".[77] Here, he articulates the dynamic outcome of performance as a transformation, or as he occasionally terms it, "transformance":

> someone or some group begins to move to a new place in the social order; this move is acceded to or blocked; in either case a crisis occurs because any change in status involves a readjustment of the entire scheme; this readjustment is effected performatively—that is, by means of theater and ritual.[78]

Throughout his writing, Schechner turns to examples from other cultures of ritual practices that involve this sort of transformance, from the Orokolo of Papua New Guinea, to the Yoruban traditions of Nigeria, to the more prescribed performance forms such as Indian Kathakali. For Schechner, performance is genuinely intercultural and transformative.

Among Schechner's ideas is one articulating a conceptual nesting of "frames" which define the different layers of "rules" that constitute the organization of performance. He explains this initially by looking at the organization of ("pure") theatre: "the first frame concerns the physical stage or space", Schechner writes, "the second the conventions of [the performer's] epoch; the third the drama itself; and the fourth are the instructions given to the actress by her director".[79] A final frame, located within all of these, designates the performer's freedom—the way the actress may choose to play her part—though such freedom is only permitted once the other establishing rules have been taken into account. As Schechner sees it, other performance (the "impure" theatre of games, sports, play, or ritual) follows broadly the same principles. "Several basic qualities are shared by these activities", he writes: "1) a special ordering of time; 2) a special value attached to objects; 3) non-productivity in terms of goods; 4) rules".[80] Following this reasoning he expands the idea of performance to encompass a wide spectrum of activities.

The concept of a "frame" is also used by Erving Goffman, who explains it in a similar way in his book *Frame Analysis* (1974). Here, his focus is on the "social frameworks" that determine "guided doings"[81]—the actions that are permissible within the context of particular situations. "I shall consider the way in which the individual in ordinary work situations presents himself and his activity to others", Goffman explains; "The ways in which he guides and controls the impression they form of him, and the kinds of things he may or may not do while sustaining his performance before them".[82] Essentially, Goffman makes a series of observations about the ways in which human beings behave, and his premise is that we are always performing whenever we are in a social situation. He refers to this in terms of expression,[83] suggesting that our act/performance/behaviour is an attempt to communicate something about ourselves to other people.

Thus he defines performance as "all the activity of a given participant on a given occasion which serves to influence in any way any of the other participants".[84]

Theatre and Performance Ecology

The extended thinking about performance that figures such as Schechner, Cage, Kaprow, and Goffman initiated in the 1950s and 60 s has opened our eyes to performative dynamics well beyond the domain of theatre itself. Among these, there is a certain area that is of particular interest to our project. "Theatre and performance can take up different and diverse ways of thinking about our relationship to ecology",[85] writes Lisa Woynarski, introducing us to what she calls "ecodramaturgy",[86] and to what others have referred to as "environmental theatre"[87] or "eco-performance".[88]

In its simplest form, this deals with subjects in plays and performances that have an ecological agenda. Typically, that has meant either positioning characters in a natural setting ("landscape plays"),[89] or using the performance to mount an ecological statement ("ecocriticism").[90] Often, as might be expected, these manifestations of theatre/performance ecology go hand in hand: the play or performance will be set in nature partly because it seeks to magnify an ecological agenda. Not surprisingly, the Sycamore Gap tree inspired performances even before its tragic demise, and has done since: choreographer Eliot Smith's piece *LIFE* (2021), for instance, presented a "visual, physical, and auditory" dance piece inspired by the Sycamore Gap and reflecting on climate change[91]; after the felling, Italian composer Riccardo Pes performed his cello piece "Lament for the Tree" (2023) at the site[92]; and musicians Catriona Downie and Katy Lavinia Cooper include the vocal piece "Sycamore Gap" in their song cycle *Twelve Trees* (2024).[93] There is also, of course, drama to the narrative of the Sycamore Gap in and of itself: a dearly beloved icon is tragically cut down in their prime, leading to widespread outpourings of communal anguish and sorrow.

Ecologically-minded though such examples may be, scholars recognize that theatre and performance offer more complex relationships with ecology. "Theatre's ecological contribution is not found in any explicit eco-critical message it may purport to communicate", suggests Lavery, "but rather resides in the more oblique possibilities inherent in the theatrical medium itself".[94] Thus in contemporary understandings of performance, what becomes significant is the fact that the tree, this event, and our responses to it become *performative articulations*. Such a location articulates itself not just through the appearance, function, or navigability of its basic geography (a tree in a dip by a wall), but also through the traces of its histories, its mediation in images or stories, and the personal and emotional narratives it inspires. Thus Sycamore Gap's identity and our relationship(s) with it (even before the 2023 event) have been performed in multiple iterations through photographs, diary entries and scrapbooks, and latterly in a more public arena through a flow of memories and memes in selfies and social media. Add to this already rich performative identity the new saga of mystery and anguish that was caused by the tree's demise, and the identity of Sycamore Gap becomes abundant in what Schechner calls a "thickening" of significance. "Activities thicken" by virtue of their theatricality, he writes; explaining that they "get more complicated, dense, symbolic, contradictory, and multivocal—along a continuum of expanding consciousness". His inference is that within this "thickening" of significance, fundamental transformations can occur, underlining not only the symbolic power of conventional theatre but also the significance of the performative in ritual rites of passage within many societies (including our own). "By turning possibilities into action, into performances, whole worlds otherwise not lived are born",[95] he concludes.

The idea of culture being produced through that "thickening" of multiple performative utterances is something that not only reminds us of the way in which ecosystems are conceptually constituted, with various strands of their web intersecting and engaging in the messy tangle evoked by Bateson,[96] but also something that works in relation to context or environment. "Recent critical attention to space", observes David Crouch, "directs attention to its relational, dynamic, contingent character". He notes that "Space emerges from this as persistently 'in the making', through a complexity of forces, influences, practices".[97] That this is especially potent in the interrelations of human communities with the natural world confirms our performative relationship with/in the environment. "Isn't the human relationship with nature always already performed and performative?", asks Stephen Bottoms[98] in an essay that draws on Bateson and some of the perspectives we have been discussing. In this respect the Sycamore Gap tree becomes a personality, and the relationships we perform with and through it demonstrate a human sense of connection with the character of the landscape.

More than this, though, the personality of the focal point (in this case, the tree) also performs itself back along the strands of that systemic web, emanating beyond the confines of its spatial and conceptual boundaries to re-engage those whose photos, anecdotes, and memories have melded its character. And in that process, the individuals who are engaged become in turn imbued with a shared sense of experience, creating a cultural community. This is exactly what Wulf and other performance scholars describe when they write about the performativity of space:

> Cultural communities constitute themselves through verbal and non-verbal forms of interaction and communication. Many of the practices of intangible cultural heritage are, as it were, performed on "stages"; by means of staging and performing, forms of cohesiveness and intimacy, of communal solidarity and integration are produced.[99]

Here, the potential nuances—Lavery's "oblique possibilities"—are numerous, not least because of the multiple opportunities for signification that theatre and performance have at their disposal, which scholars have been debating for decades, but also because of the heightened dynamic of performance and its inclination towards transformance. "Green dramaturgy asks us to reconstitute the world", suggests May; "to re-conceive our notions of community in such a way that the very boundaries between nature and culture, self and other, begin to dissolve". She reminds us of performance's roots in ritual, and particularly in "the reciprocity between people and the natural world" that ritual incites, before concluding that "theater emerges not only as a means by which to investigate the long-standing humanist question 'who are we?' but also the urgent ecological question 'where are we?'".[100]

Bonnie Marranca is another theorist of theatre/performance and ecology who has written widely around this subject. Marranca was the founding editor of *Performing Arts Journal* in the 1970s, and her collected writings—an anthology titled *Ecologies of Theater* (1996)—include essays dating back to that period. Her topics in these essays cover a variety of themes from gardening to perestroika; but throughout them she retrospectively reads an obsession with ecology and the environment in her work.

If there is one element that binds Marranca's writings together, it is her focus on the twentieth century move away from literary or text-based dramatic forms (plays) and towards performance, and it is this she documents through writing about a cohort of significant avant-garde practitioners who resisted the literary conventions of nineteenth-century drama—figures such as Cage, but also Robert Wilson, Rachel Rosenthal, Heiner Müller, Maria Irene Fornes, and Meredith Monk. She is not so much interested in "the rhetoric of writings on nature or metamorphosis, or genres, like the eclogue or pastoral".[101] Instead, Marranca's view of theatre is that it should be interdisciplinary and transcultural/intercultural,[102] and it is this that encourages her use of metaphor: "a collective of texts, images, or sounds [is] an ecosystem", she muses, conceptualizing "the created work as an ecosystem/cultural system that includes its manner of production, reception, and funding, as well as institutions and interacting artistic communities".[103]

This is a characteristic employment of metaphor, though her use of ecological terminology is something that seems to develop over time. There is a sense that Marranca is gradually finding a vocabulary to talk about avant-garde performance, and as she grapples with the language, variations come to the fore: at first, it is "the idea of a theater culture" that she conceptualizes[104]; slightly later, she writes that "'performance culture' is a phrase I use to articulate the tendency in American life toward structuring all aspects of individual and social activity around the paradigm of theater"[105]; eventually, recognizing the shared etymology of "culture" and "cultivate", she begins "to think of the promise of culture as horticulture" and of herself "as a naturalist, so to speak".[106] Thus she hits on the metaphor of *ecology*. Even so, this is a use of the ecology metaphor that remains rather haphazard, as she refers variously to "the ecology of the image"[107]; "community ecology"[108]; "dramaturgy as an ecology"[109]; "the ecology of intertextuality"[110]; and even "ideology as an ecology",[111] all of which she groups together as "the ecologies of theater" (the title of her essay collection). Moreover, she uses other metaphors, which tend to blur the focus further still: Robert Wilson's work is reminiscent of an "anthology", a "database", a "menu", and an "archive", for instance, drawing together "fractal" or "dispersed" bits of text.[112] Here and elsewhere[113] it seems to be an association with chaos theory rather than ecology that stimulates her thoughts.

Baz Kershaw settles on the descriptors "theatre ecology" and "performance ecology",[114] and his use of the metaphor is more consistent, even if the opening gambit of his influential book *Theatre Ecology* (2009) makes the relationship between theatre/performance and ecology seem straightforward: "Theatre ecology is the ways theatres behave as ecosystems",[115] he writes, evoking a vision of the multiple goings-on in theatre buildings as audiences come and go, front-of-house personnel usher them into their seats, technicians manoeuvre around darkened backstage spaces, and performers bask in the spotlight. It's easy to describe such a scene in ecological terms ("a hive of activity"; "buzzing with people"). However, Kershaw's perspective becomes more sophisticated as the book proceeds: he calls attention to the common use of the term "seed funding" in the arts[116]; he talks of an erosion of "species" diversity in the theatre programming landscape[117]; he refers to audiences as an input of energy into the "theatre niche"[118]; and he brings into play the ecological terms "edge effect" and "ecotone" to discuss the actor-audience interaction of applause.[119] He also draws an analogy between the productivity push in UK cultural policy of the 1980s and a similar productivity push in the use of agricultural pesticides in the 1950s and 1960s, which was the target of Rachel Carson's seminal *Silent Spring* (1962).[120] We'll return to the implications of that approach to UK cultural policy in our final chapter; but for the time-being our focus is Kershaw's theatre ecology.

"'Ecology' fundamentally emphasizes the inseparable and reflexive interrelational and interdependent qualities of systems *as* systems", he writes, observing that "organisms—including humans—are both a part of and apart from their environments".[121] This qualification, in effect, enables his original statement ("Theatre ecology is the ways theatres behave as ecosystems") to serve as a double metaphor: the human (the actor, the technician, or the member of an audience) is called into play as an "organism" within the "environment" of the theatre building; and beyond this, his conceptualization extends the metaphor to evoke a "theatrical system", which might account not just for a particular instance of theatre (a production or performance event being a "theatrical system"), but for a whole systemic process (such as the development of a production from inception to realization), or even for a whole assemblage within the theatre industry (the mechanism of the funding landscape). In this conceptualization of theatrical systems, we can read influences from the theories of deep ecology: indeed, Kershaw acknowledges this himself—"Radical ecologists, such as Arne Naess, consider the organism-plus-environment formulation to be too biocentric", he notes,[122] invoking deep ecology's "rejection of the man-in-environment image in favour of *the relational, total-field image*"[123]; "organisms in interaction with each other and with the surroundings, organic as well as inorganic".[124]

Although his interest and commitment to theatre/performance and ecology demonstrates Kershaw's clear ecocritical stance (he would see himself as an environmental campaigner), his intellectual engagement with ecology reflects more the "oblique possibilities" of theatre. Indeed, Kershaw not only defines a conceptualization of "theatre ecology" (as we have seen), but also distinguishes between other associated engagements with the metaphor: "performance ecology", he suggests, explores "how performance is an integral part of global ecology and ecosystems"; "performance ecologies" cover "how particular performances operate ecologically as ecosystems"; and "ecologies of performance" consider "how humans might arrive at a better appreciation of performance processes in Earth's ecosystems through ecology".[125] These are subtly nuanced differences, to be sure, and arguably somewhat blurred with his invocation of "theatre ecology"; nevertheless, they establish an extended sense of the metaphor, which calls on the already-expanded understandings of performance that we have mentioned above. In short, as he puts it, "all human life is theatricalized and dramatized, including, crucially, its interactions with other species and the environment".[126] We could take this even further: insofar as any event or object can be *framed* and thereby made to *play a part* in a wider context, any constituent element of the world or our experience in it can be said to be *performative*, and any assemblage of those elements into a system can be understood as *ecological performance*.

In order to understand why this matters, it is useful to delve into Kershaw's extended discussion about "nature" and "culture", since theatre ecology "has to wrestle with the[ir] dominant meanings [...] as separate and oppositional entities".[127] As he points out—and as we have noted in chapter two—what we understand as natural ("nature") is in fact "a product of culture—a set of transformations through which we 'make sense' of the material qualities and features of the land and landscape through the operations of social narrative and myth".[128] Following that logic, the words "nature" and "culture" themselves have been "culturally scaped",[129] "constituted performatively",[130] and are "*always performed*".[131] Framing ourselves as the subject against the backdrop of a "natural" landscape demonstrates just how pervasively we have sought to use the world to our own advantage. Not only do we do that by exploiting the world's resources and causing immense environmental damage, but we also less obviously construct images that serve those purposes, establishing ourselves as the dominant species in our planetary biome, thereby perpetuating our global hold on the ecosystem (hence the advent of the "Anthropocene"). Thus there is a tension at the heart of performance when seen in ecological terms, for so long as performance is understood to be about *us* ("the self") and about the positioning of our claims within the world ("in everyday life"), and so long as the "many levels or modes of seeing, or experiencing"[132] remain the focus of our observations, we fail our role in the holistic ecological community of our biosphere.

Whether it has been through our material actions (our abuse of the earth's resources), or through our cultural representations (the way we present the world in relation to ourselves, using "theatrical and cinematic landscapes as units of human occupation"),[133] our view of nature has been "over-determined",[134] our understanding of ecology has been "conservative and anthropocentric",[135] and our relationship with our environment, according to Michael Serres, remains one of "narcissism"[136]: we have either made "a fetish out of wilderness or place" or "set out to place the so-called exceptionalism of the human subject in crisis".[137] And, we might add, we have very much maintained the same narcissism within our grasp of *performance.* Indeed, as Kershaw concedes, the use of such ecological metaphor can be challenged as no more than "a new relation of the human and the natural worlds, making the latter a privileged sign of the superiority of the former".[138] This is a particular concern of Una Chaudhuri, who critiques Marranca to make the point that "the metaphorical use of ecology can sometimes misrepresent the actual ecological issues at hand". "To use ecology as metaphor is to block the theater's approach to the deeply vexed problem of classification that lies at the heart of ecological philosophy", she contends, asking, "are we human beings—and our activities, such as theater—an integral part of nature, or are we somehow radically separate from it?".[139] She suggests that as our ecosophy becomes more sophisticated, theatre ecology "will call for a turn towards the literal, a programmatic resistance to the use of nature as metaphor".[140]

Still, it is difficult to see how any discussion of performance can escape from a nature/culture dichotomy, when the precepts of even an expanded view of performance are about its *framing*, *observation*, and *interpretation.* These seem all to be human concerns, emblematic of the human intellect, and redolent of our psychological and social play. To find a genuine departure from the human is to reach towards an altogether different dimension of performance, though bioperformativity—at least as Woynarski defends it—may be a step towards doing that. "*Bioperformativity*", she writes, "is about identifying non-anthropocentric ecodramaturgies through [...] the way the more-than-human performs as material and representation". She considers how trees "perform" "through their use as a recurrent symbol and popular image within ecological discourses",[141] something which also happens with animals and even assemblages such as the power grid.[142] Woynarski's thoughts here could be drawn directly from Timothy Morton's *dark ecology*, and particularly his writings on the "hyperobject": "All entities whatsoever are interconnected in an interobjective system", he writes[143]; "We coexist with human lifeforms, nonhuman lifeforms, and non-lifeforms, on the insides of a series of gigantic entities with whom we also coexist: the ecosystem, biosphere, climate, planet, Solar System". And to understand the "thickness" that implies, we need to remind ourselves of Morton's Jonah and the Whale analogy: "A multiple series of nested Russian dolls. Whales within whales within whales".[144] "The reality is that hyperobjects were already here, and slowly but surely we understood what they were already saying", writes Morton; "They contacted us".[145]

The value of such bioperformativity is that it "decentres the human as a singularity (the citizen, the white, liberal male subject) in favour of flora and fauna",[146] thereby "problematising human exceptionalism".[147] This takes us back to Richard Schechner: "I find nothing disturbing about relating the finest achievements of human art [...] to animal behavior", he writes; "because I detect no break between animal and human behavior. And especially in the realm of artistic-ritual behavior I find homologies, continuities, and analogies".[148] He goes on especially to compare us to primates and aquatic mammals, but ultimately notes that "All animals, including Homo sapiens, exist within the same ecological web".[149]

What can we draw from our discussion of contemporary performance scholarship as it relates to our ecological perspective? It's useful to remind ourselves of the terms of our project, which are that we are seeking to borrow discrete examples of processes in ecology, to apply their theories to activities in human communities, and thus to triangulate an understanding of these communities as human ecologies. Perhaps we can see this as looking backwards down the telescope: taking an ecological insight rather than one predicated on human concerns. But where exactly does performance play in to this? Here we make the assertion that it is in the dynamics of performance, within the "thickening" of significance, in its capacity for "transformance", and especially within a broad and ecological understanding of performance, that we can witness metaphorically the ecological processes of our case study examples. The instances we will encounter throughout the remaining chapters are of communities whose articulations and utterances, while rarely theatrical performances per se, are articulations heightened by artistic significance and thereby "thickened". In these moments of encounter, we hear, read, and feel the increased volume of these articulations, and explore the swelling of their presence.

Notes

1. Tom Wilkinson (2024), "Sycamore Gap: Artist creates woodland trail in memory of tree", *Forestry* Journal, 12 April, https://www.forestryjournal.co.uk/news/24245942.sycamore-gap-artist-creates-woodland-trail-memory-tree/. Accessed 27 December 2024.
2. Chris Bradley (2024), "Lucy's Vision to Create a Woodland Grows!", *Lucy Pittaway,* 8 April, https://www.lucypittaway.co.uk/blog/lucys-vision-to-create-a-woodland-grows. Accessed 27 December 2024.
3. Chris Bradley (2022), "Another win for the UK's Favourite Artist", *Lucy Pittaway,* 17 May, https://www.lucypittaway.co.uk/blog/another-win-for-the-uks-favourite-artist. Accessed 27 December 2024.
4. Emily Johnson (2024), "North Yorkshire artist plants woodland to honour Sycamore Gap tree", *BBC*, 13 April, https://www.bbc.com/news/uk-england-york-north-yorkshire-68770359. Accessed 27 August 2024.
5. Chris Bradley (2024), "Planting complete for Lucy Pittaway's woodland on the Swinton Estate", *Lucy Pittaway*, 20 December, https://www.lucypittaway.co.uk/blog/planting-complete-for-lucy-pittaways-woodland-on-the-swinton-estate. Accessed 27 December 2024.
6. Alli Templeton (2023), "Elegy for the Sycamore", *Mediaeval Wanderings blog*, https://medievalwanderings.com/2023/10/01/elegy-for-the-sycamore/. Accessed 27 August 2024.
7. Rory Stewart quoted in Anon. (2023), "The Observer view on the felling of Britain's famous Sycamore Gap tree", *The*

Observer, 1 October, https://www.theguardian.com/commentisfree/2023/oct/01/observer-view-sycamore-gap-tree-reminder-natures-precarity. Accessed 27 August 2024.
8. Si King, quoted in Natalie Finnigan (2023), "Hairy Bikers React to Felled Sycamore Gap Tree, Northumberland", *Hexham Courant*, 29 September, https://www.hexham-courant.co.uk/news/23820849.hairy-bikers-react-felled-sycamore-gap-tree-northumberland/. Accessed 27 August 2024.
9. Catriona Downie and Katy Lavinia Cooper (2024), "9. September: The Sycamore Gap", *Twelve Trees*, Sing for Pleasure Songbook 18, Bolton: Sing for Pleasure, p. 81.
10. Ben Davenport and Tanja Hoffman (2023), "The life and loss of the Sycamore Gap tree", *Heritage for Global Challenges Research Centre blog*, https://www.hgc.hosted.york.ac.uk/the-life-and-loss-of-the-sycamore-gap-tree/. Accessed 27 August 2024.
11. Anon. (2023), "Sycamore Gap: "We are devastated—it's so sad to see the tree gone'", *BBC*, 29 September, https://www.bbc.com/news/uk-66967101. Accessed 27 August 2024.
12. Anon., "The Observer", n.p.
13. Terry Deary quoted in Felicity Capon (2023), "Why did the Hadrian's Wall tree mean so much to us?", *The Week*, https://theweek.com/culture-life/why-did-the-hadrians-wall-tree-mean-so-much-to-us, Accessed 27 August 2024.
14. Daniel Hall (2023), "The last time I visited Sycamore Gap, I never imagined I would never see it again", *The Chronicle*, 28 September, https://www.chroniclelive.co.uk/news/north-east-news/sycamore-gap-northumberland-hadrians-wall-27806578. Accessed 27 August 2024.
15. Alex Barton (2024), "Sycamore Gap exhibition to offer 'grieving process' for felled historic tree", *The Telegraph*, 4 July, https://www.telegraph.co.uk/news/2024/07/04/sycamore-gap-felled-tree-sill-centre-exhibition-grieving/. Accessed 27 August 2024.
16. Anon. (2023), "Our response to the felling of the Sycamore Gap tree", *National Trust*, 29 September (updated 2 August 2024), https://www.nationaltrust.org.uk/who-we-are/news/our-statement-on-the-felling-of-the-sycamore-gap-tree. Accessed 26 August 2024.

17. Samantha Jagger (2023),"Sycamore Gap tree at Hadrian's Wall cut down by 'vandals'", *BBC*, https://www.bbc.com/news/uk-england-tyne-66947040. Accessed 27 August 2024.
18. See for example Lise Colas (2024), "Sycamore Gap", *Medium*, 5 February, https://lisecolas.medium.com/sycamore-gap-c988d1db45a2. Accessed 27 August 2024; Guy Fletcher (2023), "Sycamore Gap", *Friday Flash Fiction*, 13 October, https://www.fridayflashfiction.com/poetry/sycamore-gap-by-guy-fletcher. Accessed 27 August 2024; Samantha Teasdale (2023), "Hexham Courant readers' poems about Sycamore Gap", *Hexham Courant*, 29 September, https://www.hexham-courant.co.uk/news/23822947.hexham-courant-readers-poems-sycamore-gap/. Accessed 27 August 2024.
19. Rory Smith (2023), "A Tree Was Felled. No One Heard It. How Do You Find Out Who Did It?", *New York Times*, 6 October.
20. Ben Morris (2024), "Felled Sycamore Gap tree to go on public display in Northumberland", *The Observer*, 2 March, https://www.theguardian.com/uk-news/2024/mar/02/felled-sycamore-gap-tree-to-go-on-public-display-in-northumberland. Accessed 27 August 2024.
21. Anon., "Our Response", n.p.
22. News Desk (2023), "'One tree can't breathe for the whole world': Simon Armitage marks chainsaw felling at Sycamore Gap", *Write Out Loud*, https://www.writeoutloud.net/public/blogentry.php?blogentryid=132239. Accessed 27 August 2024.
23. Anon. (2024), "Community Foundation funds Sycamore gap legacy art project", *Community Foundation*, https://www.communityfoundation.org.uk/community-foundation-funds-sycamore-gap-legacy-art-project/. Accessed 26 August 2024.
24. Twice Brewed Inn (2023), "Celebrate the legacy of Sycamore Gap", *JustGiving*, https://www.justgiving.com/page/legacyofsycamoregap?utm_medium=fundraising&utm_content=page%2Flegacyofsycamoregap&utm_source=copyLink&utm_campaign=pfp-share. Accessed 26 August 2024; Gareth Carter (2023), "Alncom's fundraiser for National Trust", *JustGiving*, https://www.justgiving.com/page/alncom-1695915069468. Accessed 26 August 2024. A number of other non-official fundraisers have also been advertised: David Westwood (2023), "Help the National Trust restore Sycamore Gap", *GoFundMe*,

https://www.gofundme.com/f/vuss4-fund-a-full-size-replica-monument-at-sycamore-gap. Accessed 26 August 2024; Brian (2023), "Sycamore Gap Tree Replacement", *Crowdfunder*, https://www.crowdfunder.co.uk/p/sycamore-gap-tree-replacement. Accessed 26 August 2024.

25. Michael James (2023), "The Sycamore Gap Bronze Tree Sculpture", *Crowdfunder*, https://ba.crowdfunder.co.uk/p/the-sycamore-gap-tree-sculpture. Accessed 26 August 2024.
26. Chris Bradley (2024), "The Story Behind Lucy's Sycamore Gap Art and Trail", *Lucy Pittaway*, 13 June, https://www.lucypittaway.co.uk/blog/the-story-behind-lucys-sycamore-gap-art-and-trail. Accessed 26 August 2024.
27. Bradley, "Planting complete".
28. Twice Brewed Inn, "Celebrate the legacy".
29. Davenport and Hoffman, "The life", n.p.
30. Fiona Burns et al. (2023), *State of Nature 2023*, London: State of Nature Partnership, https://stateofnature.org.uk/wp-content/uploads/2023/09/TP25999-State-of-Nature-main-report_2023_FULL-DOC-v12.pdf. Accessed 27 August 2024.
31. Rob Cowen (2023), "I mourn the Sycamore Gap tree. But I also grieve for whoever chopped it down", *The Guardian*, 3 October. For similar perspectives, see Simon Rey (2023), "The outrageous felling of the Sycamore Gap tree", *The Spectator*, 29 September, https://www.spectator.co.uk/article/the-outrageous-felling-of-the-sycamore-gap-tree/. Accessed 27 August 2024; Andy Brown (2023), "The Sycamore Gap: when we brutalise nature we brutalise ourselves", *Yorkshire Bylines*, 1 October, https://yorkshirebylines.co.uk/politics/the-sycamore-gap-when-we-brutalise-nature-we-brutalise-ourselves/. Accessed 27 August 2024.
32. Paul Alpers (1997), *What is Pastoral?*, Chicago: University of Chicago Press, p. 22.
33. Jeffrey Theis (2002), "The wildwood and sylvan pastoral: Nature, history, and genre in early modern England", Ph.D. dissertation, University of Wisconsin-Madison, pp. 2–3.
34. Charlotte Scott (2014), *Shakespeare's Nature: From Cultivation to Culture*, Oxford: Oxford University Press, p. 6.
35. Scott, *Shakespeare's Nature*, p. 31.

36. Raymond Williams (2001), *The Long Revolution*, Peterborough: Broadview Press, p. 64.
37. Raymond Williams (2011), *The Country and the City*, Nottingham: Spokesman, p. 1.
38. Williams, *Country*, p. 9.
39. Williams, *Country*, p. 35.
40. Williams, *Country*, p. 9.
41. Williams, *Country*, p. 131.
42. Williams, *Country*, p. 13.
43. Williams, *Country*, p. 40.
44. Williams, *Country*, p. 56.
45. Tom Williamson (2000), "Understanding enclosure", *Landscapes*, 1:1, pp. 56–79, p. 71.
46. Jan Kott (1966), *Shakespeare, Our Contemporary* (trans. Boreslaw Taborski), New York: Anchor Books, p. 329.
47. Williamson, "Understanding enclosure", pp. 76–77.
48. Nicholas A. Robinson (2014), "The charter of the forest: Evolving human rights in nature", in Daniel Barstow Magraw, Andrea Martinez and Roy E. Brownell II (eds) (2014), *Magna Carta and the Rule of Law*, Chicago: American Bar Association, pp. 311–377, p. 316.
49. Chiykowski, "The Court", p. 11.
50. Jeremy Harte (1996), "Under the Greenwood Tree", *At the Edge: Exploring new interpretations of past and place in archaeology, folklore and mythology*, 1, pp. 15–19.
51. The character of Robin Hood emerged during the fourteenth and fifteenth centuries, thus 200 years after the Charter of the Forest, and therefore as Anthony Pollard notes, the early Robin Hood ballads would not have been performed to "an audience to whom the social and political tensions of the thirteenth and early fourteenth centuries were any longer relevant". Anthony J. Pollard (2016), "Political ideology in the early stories of Robin Hood", *Outlaws in medieval and early modern England*, London: Routledge, pp. 111–128, p. 113. Furthermore, as Jeremy Harte writes, "Between 1300 and 1325 the machinery of forest administration fell into disuse"; he goes on to explain that "visits of justices ceased and the administration of courts passed to local initiative. Forestry changed its emphasis from the production of venison to the growth of timber". Harte, "Under the Greenwood", p. 18.

Still, setting the Robin Hood ballads during the reign of King John places the landscape of the common land squarely within the tradition of the Charter and its relationship to communities. For after all, the political and social tensions surrounding that relationship were still very much in play.

52. Chiykowski, "The Court", pp. 3–4.
53. Pollard, "Political Ideology", p. 112.
54. Pollard, "Political Ideology", p. 127.
55. Harte, "Under the Greenwood", p. 17.
56. Pollard, "Political ideology", p. 126.
57. "Bold Poacher", or "My Delight on a Shiny Night", Printed by J. Cadman, Manchester, c.1850–55. From an image in Stephen John Ridgwell (2017), "'On a shiny night': The Representation of the English Poacher, c. 1830–1920", Ph.D. dissertation, University of Sussex, p. 48.
58. Ridgwell, "'On a shiny night'", p. 12.
59. James Canton (2020), "The Oak Papers", *James Canton website*, https://jamescanton.co.uk/books/the-oak-papers/. Accessed 10 March 2025.
60. James Canton (2020), *The Oak Papers*, Edinburgh: Canongate Books Ltd., p. 9.
61. Canton, "The Oak Papers".
62. Canton, *The Oak Papers*, p. 16.
63. William Blake (1804), "And Did Those Feet in Ancient Time", used as the setting for the hymn "Jerusalem" (1916), with music by Hubert Parry and a familiar arrangement by Edward Elgar (1922).
64. Kershaw, *Theatre Ecology*, p. 316.
65. Kershaw, *Theatre Ecology*, p. 266.
66. Kershaw, *Theatre Ecology*, p. 15.
67. Christoph Wulf (2011), "Performativity and dynamics of intangible cultural heritage", in Christiane Brosius and Karin M. Polit (eds) (2011), *Ritual, Heritage and Identity: The Politics of Culture and Performance in a Globalised World*, London: Routledge, pp. 76–94, p. 78.
68. Richard Schechner (1994), *Environmental Theater: An Expanded New Edition including "Six Axioms for Environmental Theater"*, New York: Applause Books.
69. Schechner, *Environmental Theater*, p. xix.

70. In the UK, comparable experimental theatre groups like The People Show, Action Space, and Welfare State International were being matched by establishment figures such as Peter Brook in the "Theatre of Blood" season at the Royal Shakespeare Company in 1964, the RSC's anti-Vietnam production *US* (1966), and subsequent work with the International Centre for Theatre Research (from 1971).
71. Schechner, *Environmental Theater*, pp. 60–61.
72. John Cage, Michael Kirby and Richard Schechner (1965), "An Interview with John Cage", *Tulane Drama Review*, 10:2, pp. 50–72, p. 50.
73. Schechner, *Environmental Theater*, p. xxiii.
74. Turner wrote the foreword to Schechner's *Between Theater and Anthropology* (1985), and Schechner reciprocated by writing the preface to Turner's posthumous *The Anthropology of Performance* (1986); meanwhile, Schechner and his collaborator Willa Appel dedicated *By Means of Performance* (1990) to the late anthropologist.
75. Schechner and his co-organizer Willa Appel document these symposia in their introduction to the proceedings that emerged. Richard Schechner and Willa Appel (1990), "Introduction", in Richard Schechner and Willa Appel (1990), *By Means of Performance: Intercultural Studies of Theatre and Ritual*, Cambridge: Cambridge University Press, pp. 1–7.
76. Anon. (1981), "Organizer's Statement", *Wenner-Gren Symposium #89–1*, https://wennergren.org/symposium-seminar/yaqui-ritual-and-performance/. Accessed 3 June 2024.
77. The material in this essay was originally delivered at a symposium at the University of Wisconsin-Milwaukee in 1975. Schechner, *Performance Theory*, p. xi.
78. Schechner, *Performance Theory*, p. 166.
79. Schechner, *Performance Theory*, p. 14.
80. Schechner, *Performance Theory*, p. 6.
81. Erving Goffman (1986), *Frame Analysis: An Essay on the Organization of Experience*, Boston: Northeastern University Press, p. 22.
82. Erving Goffman (1990), *The Presentation of Self in Everyday Life*, London: Penguin Books, p. 9.
83. Goffman, *Presentation*, p. 14.

84. Goffman, *Presentation*, p. 26.
85. Lisa Woynarski (2020), *Ecodramaturgies: Theatre, Performance and Climate Change*, Cham: Palgrave Macmillan, p. 19.
86. "Ecodramaturgy": "theatre and performance making that puts ecological reciprocity and community at the center of its theatrical and thematic intent". Woynarski, *Ecodramaturgies*, p. 9, quoting Theresa J. May.
87. May, "Greening", pp. 84–103.
88. Courtney B. Ryan (2023), *Eco-Performance, Art, and Spatial Justice in the US*, London: Routledge.
89. Such plays, their various representations of nature, and the theoretical concept of landscape in performance are explored extensively by Una Chaudhuri and others in eg. Una Chaudhuri (1997), *Staging Place: The Geography of Modern Drama*, Ann Arbor: University of Michigan Press, and Elinor Fuchs and Una Chaudhuri (eds) (2002), *Land/Scape/Theater*, Ann Arbor: University of Michigan Press. See also Bonnie Marranca (2012), *Ecologies of Theater*, New York: PAJ Publications.
90. Mohebat Ahmadi (2022), *Towards an Ecocritical Theatre: Playing the Anthropocene*, Abingdon: Routledge. The second of these types of performance, which May sums up as "environmentalist agit-prop", uses the play or performance's dramatic power to voice an environmental, or "ecocritical", position. May, "Greening", p. 94.
91. Eliot Smith (2021), "LIFE", *Eliot Smith Dance*, https://www.eliotsmithdance.com/life. Accessed 27 August 2024.
92. Anon. (2023), "Cellist mourns felled tree at Sycamore Gap with solo composition", *The Strad*, 11 October, https://www.thestrad.com/news/cellist-mourns-felled-tree-at-sycamore-gap-with-solo-composition/17126.article. Accessed 27 August 2024.
93. Anon. (2024), "What we do", *Sing for Pleasure*, https://singforpleasure.org.uk/what-we-do/. Accessed 27 August 2024.
94. Carl Lavery (2018), "Introduction: Performance and Ecology—What can Theatre Do?", in Carl Lavery (ed.) (2018), *Performance and Ecology: What can Theatre Do?*, London and New York: Routledge, pp. 1–22, p. 1.
95. Schechner, *Performance Theory*, p. 184.

96. Gregory Bateson (1991), *A Sacred Unity: Further Steps to an Ecology of Mind*, New York: Cornelia & Michael Bessie Books, p. 265.
97. David Crouch (2012), "Landscape, land and identity: a performative consideration", in Christine Berberich, Neil Campbell and Robert Hudson (eds) (2012), *Land and Identity: Theory, Memory, and Practice*, pp. 41–65, p. 45.
98. Stephen Bottoms (2007), "Introductions", in Stephen Bottoms and Matthew Goulish (eds) (2007), *Small Acts of Repair: Performance, Ecology, and Goat Island*, pp. 3–26, p. 19.
99. Wulf, "Performativity", p. 79.
100. May, "Greening", p. 100.
101. Marranca, *Ecologies*, p. xv.
102. Marranca, *Ecologies*, pp. 177, 214.
103. Marranca, *Ecologies*, p. xiv.
104. Marranca, *Ecologies*, p. 112.
105. Marranca, *Ecologies*, p. 89.
106. Marranca, *Ecologies*, p. 287.
107. Marranca, *Ecologies*, p. xv.
108. Marranca, *Ecologies*, p. xvii.
109. Marranca, *Ecologies*, p. 34.
110. Marranca, *Ecologies*, p. 47.
111. Marranca, *Ecologies*, p. 281.
112. Marranca, *Ecologies*, pp. 34–48, 233–241.
113. Marranca, *Ecologies*, pp. 24, 59–70, 75, 240, 281.
114. Baz Kershaw (2007), *Theatre Ecology: Environments and Performance Events*, Cambridge: Cambridge University Press. Other jargon has emerged depicting various sub-categories of environmental theatre, including Kershaw himself describing his project *Green Shade* as an "eco-installation", Dee Heddon's concept of "ecospectation", and Lynn Jacobson's evocation of an "eco-canon". Kershaw, *Theatre Ecology*, p. 6; Deirdre Heddon (2018), "Confounding Ecospectations: Disappointment and Hope in the Forest", in Lavery, *Performance and Ecology*, pp. 96; Lynn Jacobson (1992), "Green Theatre: Confessions of an Eco-reporter", *American Theatre*, 8:11, pp. 16–25, 55.
115. Kershaw, *Theatre Ecology*, p. 16.
116. Kershaw, *Theatre Ecology*, p. 158.
117. Kershaw, *Theatre Ecology*, p. 168.

118. Kershaw, *Theatre Ecology*, p. 172.
119. Kershaw, *Theatre Ecology*, pp. 179–205.
120. Kershaw, *Theatre Ecology*, p. 165. Where Carson documents the unintended but catastrophic consequences of spraying chemicals on fields to kill off insects and make the crop yields more productive, Kershaw points to the UK's neoliberal policies under the Thatcher government and the unintended consequences of its funding regime: "Money becomes a virus that debilitates the species it was supposed to benefit", he writes (p. 166). He doesn't quite hark back to the words of Rachel Carson, though he might as well have done: "By acquiescing in an act that can cause such suffering to a living creature, who among us is not diminished as a human being?". Carson, *Silent Spring*, p. 82.
121. Kershaw, *Theatre Ecology*, p. 16.
122. Kershaw, *Theatre Ecology*, p. 248.
123. Arne Naess (2001), *Ecology, Community and Lifestyle: Outline of an Ecosophy* (trans. David Rothenberg), Cambridge: Cambridge University Press, p. 28.
124. Naess, *Ecology*, p. 36.
125. Kershaw, *Theatre Ecology*, p. 14.
126. Kershaw, *Theatre Ecology*, p. 12.
127. Kershaw, *Theatre Ecology*, p. 17.
128. Adams, "Performative Locations", p. 86.
129. Gabriella Giannachi and Nigel Stewart (2005), "Introduction", in Gabriella Giannachi and Nigel Stewart (eds) (2005), *Performing Nature: Explorations in Ecology and the Arts*, Oxford: Peter Lang, pp. 19–62, p. 37. Giannachi and Stewart borrow this phrase from John Adams (2005), "Performative Locations: Wilderness Space and Place in Early Film", in Giannachi and Stewart, *Performing Nature*, pp. 85–101.
130. David Crouch (2003), "Performances and constitutions of natures: a consideration of the performance of lay geographies", in Bronislaw Szerszynski, Wallace Heim and Claire Waterton (eds) (2003), *Nature Performed: Environment, Culture and Performance*, Oxford: Blackwell Publishing, pp. 17–30, p. 28.
131. Giannachi and Stewart, "Introduction", p. 20.
132. Richard Schechner (1985), *Between Theater and Anthropology*, Philadelphia: University of Pennsylvania Press, p. 296.
133. Giannachi and Stewart, "Introduction", p. 41.

134. Carl Lavery and Claire Finburgh (2015), "Introduction: Greening the Absurd", in Carl Lavery and Claire Finburgh (eds) (2015), *Rethinking the Theatre of the Absurd: Ecology, the Environment and the Greening of the Modern Stage*, London and New York: Bloomsbury Methuen Drama, pp. 1–58, p. 18.
135. David Williams (2015), "Epilogue: 'The Ruins of Time (I've Forgotten this Before)'", in Lavery and Finburgh, *Rethinking*, pp. 241–249, p. 246.
136. Michel Serres (1995), *The Natural Contract* (trans. E. MacArthur and W. Paulson), Ann Arbor: University of Michigan Press, p. 33.
137. Lavery and Finburgh, "Introduction", p. 12.
138. Baz Kershaw (2005), "The Ecologies of Performance: On Biospheres and Theatres", in Giannachi and Stewart, *Performing Nature*, p. 74. Kershaw is referencing Una Chaudhuri's discussion of the glasshouse exhibits of the Great Exhibition in Chaudhuri, *Staging Place*, p. 77.
139. Una Chaudhuri (1994), "'There must be a lot of fish in that lake': Towards an ecological theatre", *Theatre*, 5, pp. 23–31, p. 27.
140. Chaudhuri, "There Must Be", p. 29. Like these authors, other writers have also found the ecological metaphors enticing. Even scholars whose focus is not explicitly theatre ecology have appropriated ecological language—in *Perform or Else*, for example, McKenzie toys with the idea of drilling for "core samples" from the "performance stratum", even doodling pseudo-geological images to illustrate that conceit. Jon McKenzie (2001), *Perform or Else: From Discipline to Performance*, London: Routledge, p. 179.
141. Woynarski, *Ecodramaturgies*, pp. 73–74.
142. Woynarski, *Ecodramaturgies*, pp. 80–81. We have seen this evidenced in the performance of the Sycamore Gap tree and its legacy, and will also see it in further "performances" of trees in each of our subsequent chapters.
143. Timothy Morton (2013), *Hyperobjects: Philosophy and Ecology After the End of the World*, Minneapolis, University of Minnesota Press, p. 83.
144. Morton, *Hyperobjects,* p. 128.
145. Morton, *Hyperobjects*, p. 201.
146. Woynarski, *Ecodramaturgies*, p. 95.

147. Woynarski, *Ecodramaturgies*, p. 85. Nevertheless, even bioperformance—an as-yet rather niche extension of "bioart" tends fundamentally to position the biotic from the perspective of the human concern: the bioperformance practice of someone like Catherine Fargher, for example, which brings "wet biology practice" into the science-art performance space of *BioHome*, still centralizes Fargher as a performer, her experiences of childbirth as a focus, and the bodily processes of gestation as the material of its performativity. Catherine Fargher and Terumi Narushima (2010), "Knit Two Together: Art/ Science Collaborations in *BioHome. The Chromosome Knitting Project*", in Anneke Smelik (ed.) (2010), *The Scientific Imaginary in Visual Culture*, Gottingen: V&R Unipress, pp. 181–194.
148. Schechner, *Performance Theory*, p. 184.
149. Schechner, *Performance Theory*, p. 215.

Bibliography

Ahmadi, Mohebat (2022), *Towards an Ecocritical Theatre: Playing the Anthropocene*, Abingdon: Routledge.

Alpers, Paul (1997), *What is Pastoral?*, Chicago: University of Chicago Press, p. 22.

Anon. (2023), "Cellist mourns felled tree at Sycamore Gap with solo composition", *The Strad*, 11 October, https://www.thestrad.com/news/cellist-mourns-felled-tree-at-sycamore-gap-with-solo-composition/17126.article. Accessed 27 August 2024.

Anon. (1981), "Organizer's Statement", *Wenner-Gren Symposium #89–1*, https://wennergren.org/symposium-seminar/yaqui-ritual-and-performance/. Accessed 3 June 2024.

Anon. (2023), "Our response to the felling of the Sycamore Gap tree", *National Trust*, 29 September (updated 2 August 2024), https://www.nationaltrust.org.uk/who-we-are/news/our-statement-on-the-felling-of-the-sycamore-gap-tree. Accessed 26 August 2024.

Anon. (2023), "Sycamore Gap: "We are devastated—it's so sad to see the tree gone"", *BBC*, 29 September, https://www.bbc.com/news/uk-66967101. Accessed 27 August 2024.

Anon. (2023), "The Observer view on the felling of Britain's famous Sycamore Gap tree", *The Observer*, 1 October, https://www.theguardian.com/commentisfree/2023/oct/01/observer-view-sycamore-gap-tree-reminder-natures-precarity. Accessed 27 August 2024.

Anon. (2024), "What we do", *Sing for Pleasure*, https://singforpleasure.org.uk/what-we-do/. Accessed 27 August 2024.

Anon. (2024), "Community Foundation funds Sycamore gap legacy art project", *Community Foundation*, https://www.communityfoundation.org.uk/community-foundation-funds-sycamore-gap-legacy-art-project/. Accessed 26 August 2024.

Bradley, Chris (2022), "Another win for the UK's Favourite Artist", *Lucy Pittaway*, 17 May, https://www.lucypittaway.co.uk/blog/another-win-for-the-uks-favourite-artist. Accessed 27 December 2024.

Bradley, Chris (2024), "Lucy's Vision to Create a Woodland Grows!", *Lucy Pittaway*, 8 April, https://www.lucypittaway.co.uk/blog/lucys-vision-to-create-a-woodland-grows. Accessed 27 December 2024.

Bradley, Chris (2024), "The Story Behind Lucy's Sycamore Gap Art and Trail", *Lucy Pittaway*, 13 June, https://www.lucypittaway.co.uk/blog/the-story-behind-lucys-sycamore-gap-art-and-trail. Accessed 26 August 2024.

Bradley, Chris (2024), "Planting complete for Lucy Pittaway's woodland on the Swinton Estate", *Lucy Pittaway*, 20 December, https://www.lucypittaway.co.uk/blog/planting-complete-for-lucy-pittaways-woodland-on-the-swinton-estate. Accessed 27 December 2024.

Barton, Alex (2024), "Sycamore Gap exhibition to offer 'grieving process' for felled historic tree", *The Telegraph*, 4 July, https://www.telegraph.co.uk/news/2024/07/04/sycamore-gap-felled-tree-sill-centre-exhibition-grieving/. Accessed 27 August 2024.

Bateson, Gregory (1991), *A Sacred Unity: Further Steps to an Ecology of Mind*, New York: Cornelia & Michael Bessie Books.

Berberich, Christine, Neil Campbell and Robert Hudson (eds) (2012), *Land and Identity: Theory, Memory, and Practice*, New York: Rodopi.

Bottoms, Stephen, and Matthew Goulish (eds) (2007), *Small Acts of Repair: Performance, Ecology, and Goat Island*

Brian (2023), "Sycamore Gap Tree Replacement", *Crowdfunder*, https://www.crowdfunder.co.uk/p/sycamore-gap-tree-replacement. Accessed 26 August 2024.

Brosius, Christiane, and Karin M. Polit (eds) (2011), *Ritual, Heritage and Identity: The Politics of Culture and Performance in a Globalised World*, London: Routledge.

Brown, Andy (2023), "The Sycamore Gap: when we brutalise nature we brutalise ourselves", *Yorkshire Bylines*, 1 October, https://yorkshirebylines.co.uk/politics/the-sycamore-gap-when-we-brutalise-nature-we-brutalise-ourselves/. Accessed 27 August 2024.

Burns, Fiona, S. Mordue, N. al Fulaij, P.H. Boersch-Supan, J. Boswell, R.J. Boyd, T. Bradfer-Lawrence, P. de Ornellas, A. de Palma, P. de Zylva, E.B. Dennis, S. Foster, G. Gilbert, L. Halliwell, K. Hawkins, K.A. Haysom, M.M.

Holland, J. Hughes, A.C. Jackson, F. Mancini, F. Mathews, A. McQuatters-Gollop, D.G. Noble, D. O'Brien, O.L. Pescott, A. Purvis, J. Simkin, A. Smith, A.J. Stanbury, J. Villemot, K.J. Walker, P. Walton, T.J. Webb, J. Williams, R. Wilson, R.D. Gregory (2023), *State of Nature 2023*, London: State of Nature Partnership, https://stateofnature.org.uk/wp-content/uploads/2023/09/TP25999-State-of-Nature-main-report_2023_FULL-DOC-v12.pdf. Accessed 27 August 2024.

Cage, John, Michael Kirby and Richard Schechner (1965), "An Interview with John Cage", *Tulane Drama Review*, 10:2, pp. 50-72.

Canton, James (2020), *The Oak Papers*, Edinburgh: Canongate Books Ltd.

Capon, Felicity (2023), "Why did the Hadrian's Wall tree mean so much to us?", *The Week*, https://theweek.com/culture-life/why-did-the-hadrians-wall-tree-mean-so-much-to-us, Accessed 27 August 2024.

Carson, Rachel (2000), *Silent Spring*, Dublin: Penguin Random House.

Carter, Gareth (2023), "Alncom's fundraiser for National Trust", *JustGiving*, https://www.justgiving.com/page/alncom-1695915069468. Accessed 26 August 2024.

Chaudhuri, Una (1994), "'There must be a lot of fish in that lake': Towards an ecological theatre", *Theatre*, 5, pp. 23–31.

Chaudhuri, Una (1997), *Staging Place: The Geography of Modern Drama*, Ann Arbor: University of Michigan Press.

Chiykowski, Peter (2011), "The Court of Beast and Bough: Contesting the Medieval English Forest in the Early Robin Hood Ballads", MA dissertation, Dalhousie University, Halifax, Nova Scotia.

Colas, Lise (2024), "Sycamore Gap", *Medium*, 5 February, https://lisecolas.medium.com/sycamore-gap-c988d1db45a2. Accessed 27 August 2024.

Cowen, Rob (2023), "I mourn the Sycamore Gap tree. But I also grieve for whoever chopped it down", *The Guardian*, 3 October.

Davenport, Ben and Tanja Hoffman (2023), "The life and loss of the Sycamore Gap tree", *Heritage for Global Challenges Research Centre blog*, https://www.hgc.hosted.york.ac.uk/the-life-and-loss-of-the-sycamore-gap-tree/. Accessed 27 August 2024.

David Westwood (2023), "Help the National Trust restore Sycamore Gap", *GoFundMe*, https://www.gofundme.com/f/vuss4-fund-a-full-size-replica-monument-at-sycamore-gap. Accessed 26 August 2024.

Downie, Catriona and Katy Lavinia Cooper (2024), "9. September: The Sycamore Gap", *Twelve Trees*, Sing for Pleasure Songbook 18, Bolton: Sing for Pleasure.

Durkheim, Émile (1976), *The Elementary Forms of the Religious Life* (intro. Robert Nisbet; trans. Joseph Ward Swain) (second edition), London: George Allen and Unwin Ltd.

Durkheim, Émile (2013), *The Division of Labour in Society* (ed. Steven Lukes; trans. W. D. Halls) (second edition), Houndmills, Palgrave Macmillan.

Dyer, Christopher (2006), "Conflict in the landscape: The enclosure movement in England, 1220–1349", *Landscape History*, 28.1, pp. 21–33.

Finnigan, Natalie (2023), "Hairy Bikers React to Felled Sycamore Gap Tree, Northumberland", *Hexham Courant*, 29 September, https://www.hexham-courant.co.uk/news/23820849.hairy-bikers-react-felled-sycamore-gap-tree-northumberland/. Accessed 27 August 2024.

Fletcher, Guy (2023), "Sycamore Gap", *Friday Flash Fiction*, 13 October, https://www.fridayflashfiction.com/poetry/sycamore-gap-by-guy-fletcher. Accessed 27 August 2024.

Fuchs, Elinor, and Una Chaudhuri (eds) (2002), *Land/Scape/Theater*, Ann Arbor: University of Michigan Press.

Giannachi, Gabriella, and Nigel Stewart (eds) (2005), *Performing Nature: Explorations in Ecology and the Arts*, Oxford: Peter Lang.

Goffman, Erving (1986), *Frame Analysis: An Essay on the Organization of Experience*, Boston: Northeastern University Press.

Goffman, Erving (1990), *The Presentation of Self in Everyday Life*, London: Penguin Books.

Hall, Daniel (2023), "The last time I visited Sycamore Gap, I never imagined I would never see it again", *The Chronicle*, 28 September, https://www.chroniclelive.co.uk/news/north-east-news/sycamore-gap-northumberland-hadrians-wall-27806578. Accessed 27 August 2024.

Harte, Jeremy (1996), "Under the Greenwood Tree", *At the Edge: Exploring new interpretations of past and place in archaeology, folklore and mythology*, 1, pp. 15–19.

Jacobson, Lynn (1992), "Green Theatre: Confessions of an Eco-reporter", *American Theatre*, 8:11, pp. 16–25, 55.

Jagger, Samantha (2023), "Sycamore Gap tree at Hadrian's Wall cut down by 'vandals'", *BBC*, https://www.bbc.com/news/uk-england-tyne-66947040. Accessed 27 August 2024.

James, Michael (2023), "The Sycamore Gap Bronze Tree Sculpture", *Crowdfunder*, https://ba.crowdfunder.co.uk/p/the-sycamore-gap-tree-sculpture. Accessed 26 August 2024.

Johnson, Emily (2024), "North Yorkshire artist plants woodland to honour Sycamore Gap tree", *BBC*, 13 April, https://www.bbc.com/news/uk-england-york-north-yorkshire-68770359. Accessed 27 August 2024.

Kershaw, Baz (2007), *Theatre Ecology: Environments and Performance Events*, Cambridge: Cambridge University Press.

Kott, Jan (1966), Shakespeare, Our Contemporary (trans. Boreslaw Taborski), New York: Anchor Books.

Lavery, Carl (ed.) (2018), *Performance and Ecology: What can Theatre Do?*, London and New York: Routledge.

Lavery, Carl, and Claire Finburgh (eds) (2015), *Rethinking the Theatre of the Absurd: Ecology, the Environment and the Greening of the Modern Stage*, London and New York: Bloomsbury Methuen Drama.

Magraw, Daniel Barstow, Andrea Martinez and Roy E. Brownell II (eds) (2014), *Magna Carta and the Rule of Law*, Chicago: American Bar Association.

Marranca, Bonnie (2012), *Ecologies of Theater*, New York: PAJ Publications.

May, Theresa J. (2005), "Greening the theater: Taking ecocriticism from page to stage", *Interdisciplinary Literary Studies*, 7:1, pp. 84–103.

May, Theresa J. (2007), "Beyond Bambi: Toward a Dangerous Ecocriticism in Theatre Studies", *Theatre Topics*, 17:2, pp. 95–110.

McKenzie, Jon (2001), *Perform or Else: From Discipline to Performance*, London: Routledge.

Morris, Ben (2024), "Felled Sycamore Gap tree to go on public display in Northumberland", *The Observer*, 2 March, https://www.theguardian.com/uk-news/2024/mar/02/felled-sycamore-gap-tree-to-go-on-public-display-in-northumberland. Accessed 27 August 2024.

Morton, Timothy (2013), *Hyperobjects: Philosophy and Ecology After the End of the World*, Minneapolis: University of Minnesota Press.

Naess, Arne (2001), *Ecology, Community and Lifestyle: Outline of an Ecosophy* (trans. David Rothenberg), Cambridge: Cambridge University Press, p. 28.

News Desk (2023), "'One tree can't breathe for the whole world': Simon Armitage marks chainsaw felling at Sycamore Gap", *Write Out Loud*, https://www.writeoutloud.net/public/blogentry.php?blogentryid=132239. Accessed 27 August 2024.

Pollard, Anthony J. (2016), "Political ideology in the early stories of Robin Hood", *Outlaws in medieval and early modern England*, London: Routledge, pp. 111–128.

Rey, Simon (2023), "The outrageous felling of the Sycamore Gap tree", *The Spectator*, 29 September, https://www.spectator.co.uk/article/the-outrageous-felling-of-the-sycamore-gap-tree/. Accessed 27 August 2024.

Ridgwell, Stephen John (2017), "'On a shiny night': The Representation of the English Poacher, c. 1830–1920", Ph.D. dissertation, University of Sussex.

Rosser, Gervase (2015), *The Art of Solidarity in the Middle Ages: Guilds in England 1250-1550*, Oxford: Oxford University Press.

Ryan, Courtney B. (2023), *Eco-Performance, Art, and Spatial Justice in the US*, London: Routledge.

Schechner, Richard (1985), *Between Theater and Anthropology*, Philadelphia: University of Pennsylvania Press.

Schechner, Richard (1994), *Environmental Theater: An Expanded New Edition including "Six Axioms for Environmental Theater"*, New York: Applause Books.

Schechner, Richard, and Willa Appel (1990), *By Means of Performance: Intercultural Studies of Theatre and Ritual*, Cambridge: Cambridge University Press.

Scott, Charlotte (2014), *Shakespeare's Nature: From Cultivation to Culture*, Oxford: Oxford University Press.

Serres, Michel (1995), *The Natural Contract* (trans. E. MacArthur and W. Paulson), Ann Arbor: University of Michigan Press.

Smelik, Anneke (ed.) (2010), *The Scientific Imaginary in Visual Culture*, Gottingen: V&R Unipress.

Smith, Eliot (2021), "LIFE", *Eliot Smith Dance*, https://www.eliotsmithdance.com/life. Accessed 27 August 2024.

Smith, Rory (2023), "A Tree Was Felled. No One Heard It. How Do You Find Out Who Did It?", *New York Times*, 6 October.

Szerszynski, Bronislaw, Wallace Heim and Claire Waterton (eds) (2003), *Nature Performed: Environment, Culture and Performance*, Oxford: Blackwell Publishing.

Teasdale, Samantha (2023), "Hexham Courant readers' poems about Sycamore Gap", *Hexham Courant*, 29 September, https://www.hexham-courant.co.uk/news/23822947.hexham-courant-readers-poems-sycamore-gap/. Accessed 27 August 2024.

Templeton, Alli (2023), "Elegy for the Sycamore", *Mediaeval Wanderings blog*, https://medievalwanderings.com/2023/10/01/elegy-for-the-sycamore/. Accessed 27 August 2024.

Theis, Jeffrey (2002), "The wildwood and sylvan pastoral: Nature, history, and genre in early modern England", Ph.D. dissertation, University of Wisconsin-Madison.

Twice Brewed Inn (2023), "Celebrate the legacy of Sycamore Gap", *JustGiving*, https://www.justgiving.com/page/legacyofsycamoregap?utm_medium=fundraising&utm_content=page%2Flegacyofsycamoregap&utm_source=copyLink&utm_campaign=pfp-share. Accessed 26 August 2024.

Wilkinson, Tom (2024), "Sycamore Gap: Artist creates woodland trail in memory of tree", *Forestry* Journal, 12 April, https://www.forestryjournal.co.uk/news/24245942.sycamore-gap-artist-creates-woodland-trail-memory-tree/. Accessed 27 December 2024.

Williams, Raymond (2001), *The Long Revolution*, Peterborough: Broadview Press.

Williams, Raymond (2011), *The Country and the City*, Nottingham: Spokesman.

Williamson, Tom (2000), "Understanding enclosure", *Landscapes*, 1:1, pp. 56–79.

Woynarski, Lisa (2020), *Ecodramaturgies: Theatre, Performance and Climate Change*, Cham: Palgrave Macmillan.

CHAPTER 5

Succession

Imagine not just planting a tree, but planting an entire living ecosystem. In the prologue to his book *Eden*, Tim Smit talks about the moment the Eden Project in Cornwall first opened its doors in 2001:

> As I watched the first visitor walk open-mouthed into our great green cathedral, and as small groups became hundreds and then thousands followed on behind, I thought to myself, the truly special thing about Eden is not what you see, although that is awesome enough; it is the spirit that brought so many ordinary people together, to add up to so much more than the sum of their parts. That was the real reason for hope. If we could do this, what could happen if even more were harnessed together. We'd build a magnificent Living Theatre, but the world is really the stage.[1]

It's interesting that Smit uses a theatrical metaphor to describe this vision.[2] In principle the living theatre that he built—Cornwall's celebrated Eden Project—could just be considered a tree-planting exercise on a massive scale, though he saw it in different terms. On the site of an abandoned clay pit in the heart of the UK's wildest county, Smit and his team set about building two colossal biomes within which to house an entire tropical rainforest and a dry, semi-arid chaparral. This was "a place like nothing anyone had ever seen before; a place that explored our place in nature, a place that demonstrated what could be done if people who wanted to make a difference got together".[3] With over a million plants

K. Savage and D. Symonds, *Ecologies of Community in Performance*,
https://doi.org/10.1007/978-3-031-94343-0_5

inside, The Eden Project has been called a "dramatic global garden",[4] "the eighth wonder of the world",[5] and "something of a miracle".[6] But for Smit, it was "a catalyst to make people think differently", and his explanation of this was rather urgent: "We are walking blindfold into an ecological catastrophe. I want Eden to help people and businesses to realise that the solution to nearly every big problem we face can be found in nature".[7] Elsewhere, he summed up that thought in a statement laden with provocation: "The underlying message of the project is that without plants there is no life on Earth".[8] Freighted with a message of that magnitude, the ambitions of The Eden Project were lofty in the extreme; but Eden wasn't the first audacious project Smit had found himself taking on.

Smit had escaped London in 1987 to live in Cornwall. His plan was to build a music studio so that he could write music without having to travel as a musician.[9] However, one fateful day brought him to an old estate that had fallen into ruin, where along with local builder John Nelson, he discovered what later became known as the Lost Gardens of Heligan: "hundreds of self-seeded sycamores and ash trees obscured the landscape. Ferns, mosses and lichens covered everything in this dank place, while here and there the eye was caught by exotic foliage and clumps of bamboo".[10] Smit was seized with an impulse to take on the restoration of the site, and on a later visit, Peter Thonday who had become well known as a Victorian garden expert from a TV series, *The Victorian Kitchen Garden*, teased him with a provocation: "Did we want to create a greatest hits record" by just cherry-picking aspects of the garden to restore, "or were we brave enough to perform the whole opera?"[11]—to bring the whole of the estate back to life and revive it "exactly as it would have been in the middle of the nineteenth century".[12]

This was an ambitious plan, but Smit and his colleagues set to work. They "sought the advice of every relevant specialist"[13]; "recovered the soil fertility" and "renovated all the structures"[14]; restocked the grounds with rare-breed cattle and sheep, "reared regeneratively to combat soil degradation and climate change"; and tamed the overgrowth "to reveal rare camellias and rhododendrons".[15] What they found was truly exciting:

> A group of Hooker rhododendrons, imported from the Himalayas in 1840, flourished behind the dense undergrowth. Trees included a 100ft New Zealand Yew and a group of 135ft sequoia trees from California. A Chinese ginkgo tree is from a species that dates to prehistoric times.[16]

The gardens began to open to the public in 1992 while work was still ongoing. Interest was high, and a six-part TV documentary about the restoration aired in 1997, enticing thousands to visit.

The success of the Lost Gardens and the vast knowledge learned had lit a fire in Smit, and it gave him the experience to be even more innovative when he conceived of the Eden Project. He took over an "exhausted, steep-sided clay pit" near the town of St Austell. It was "60 metres deep, with no soil, [and] 15 meters below the water table".[17] But he set about making it usable: "Some 1.8 million tonnes of dirt were shifted",[18] but "rather than ship in 90,000 tonnes of soil", he recalls, "we wanted to create it in a way that showed environmental regeneration was possible". As a result, he remembers, "we manufactured our soil from local minerals and compost in a pit nearby".[19] Then, they built a huge biome complex, which in itself was a feat of engineering. Designed by Nicholas Grimshaw and standing "55m high (11 double decker buses on top of each other), 100m long and 200m wide (2 football pitches)",[20] it is "tall enough to allow tropical trees such as teak and mahogany to grow to full height".[21] It consists principally of giant domes, "constructed from tubular steel and hexagonal cladding panels made of a special plastic".[22] Each of the panels inflates and is "large enough to contain a London taxi".[23] It's an awe-inspiring sight. "Nothing prepares you for your first glimpse",[24] remarks one reporter; "Sci-fi wasps' nests? Giant bubble wrap? Half-buried space ships?", queries another.[25] The building won an Institute for Civil Engineers Merit Award in 2002 as an outstanding example of engineering design and construction.

But of course, it is the life inside the Eden Project that is the main attraction. The largest biome simulates a rainforest, with over 1000 varieties of plants. Visitors can see rubber plants, cacao, banana trees, and even a full-size waterfall, as they weave their way through tropical undergrowth and past trees that tower above their heads. Meanwhile the Mediterranean biome encases another 1000 varieties from the mediterranean, California, South Africa, and Western Australia.[26] Smit explains that "the mediterranean-type areas of the world [...] are the richest in flora, but ironically the plants thrive on stress. Too much nutrient or water can kill them off".[27] Therefore this biome has different requirements to the rainforest biome, which is kept at a "constant 80 percent humidity".[28]

Since its inception, one of Eden's central aims has been to educate the public. Its message is that "Education is at the heart of everything we do", and its mission is to "connect people of all ages with the natural world, better understand their place within it and play their part in shaping the future".[29] A short walk from the biomes sits The Core, packed with learning materials to engage audiences from pre-school to degree level and beyond. Recently, state-of-the-art 360° cameras and augmented reality have been introduced at The Core and on the website, in collaboration with art collective Marshmallow Laser Feast. These experiences reveal the "unseen perspectives of the rainforest"[30]: through immersive 360° technology, users can interact with the Eden Project, and thereby with the natural world. One of the installations, *The Weather Maker*, reveals what happens above and below ground, exploring the journey of rainwater, nutrients, and sugar from the leaves, into the plant, and down to the roots and the soil. Another, *The Living Lens*, asks: "if we could view the world the way other animals do [...] would we better understand the vital connections between plants and animals? Would we be more driven to protect them?".[31] Revealing these processes through experiential technology enables visitors to understand the whole living ecosystem and its interactions.

The Eden Project soon became "One of Britain's top ten attractions",[32] exceeding expectations from the start. "Original estimates put the projected number of visitors at 750,000 for the year", reported Steve Keenan after its first Easter-weekend rush; "but within six weeks of opening, it has already welcomed 200,000 people".[33] By 2004, Peter Hardy was reporting "five million visitors to date who have tailgated their way off the A30",[34] adding the Eden Project to a bucket list of Cornwall sights and activities they want to tick off: the clifftops and beaches; the icecreams and pasties; the surfing in Newquay; the Lost Gardens of Heligan; Land's End; Rick Stein's; the Minack.

The tourists start arriving in the Spring, perhaps as early as Easter, but certainly by the long bank holiday weekends of May Day and Whitsuntide. From then until late September there'll be a steady stream of caravans and roof racks nose-to-tail down the A30 and over Bodmin Moor until they turn off to their destination coves. Nowadays, the county welcomes around 5 million visitors annually,[35] heading to the gardens, the beaches, the landmarks, or the attractions—to places like Cornwall's Eden Project. And a good number of them make their way further down the county to the picturesque destination of this chapter—the seaside resort of St Ives.

Our hypothesis in this book is that we can productively turn to the language and concepts of ecology in order to understand the dynamics of community, especially when they relate to the arts. Through the next few chapters, we are going to explore various communities in the arts—sometimes communities that are depicted; sometimes communities of practitioners; and sometimes communities whose characteristics are articulated through the arts—and in each chapter we're going to do this using the prism of an ecological concept. In this, our first case study chapter, we turn to the celebrated "artist colony" of St Ives in Cornwall. This became a magnet for artists during the late nineteenth century, then a magnet for tourists through the late twentieth and early twenty-first. We'll trace its history as an artist colony, viewing the artist as its *keystone species*. We'll consider the *ecological succession* that the town went through as it became a tourist hotspot, and we'll conceptualize the tourist as an *invasive species* within its isolated ecosystem. As its fortunes fluctuated throughout these changes, the ecosystem became comparable to an *imbalanced biome*. Finally we'll see how an approach of *ecosystem stewardship* ultimately helped it to build sustainability and resilience.

The Artist Colony of St Ives

St Ives nestles in a bay on the Cornish peninsula's northern coast, just twenty miles from Land's End. Out here at the extremity of the UK mainland, Cornwall is on the one hand exposed to the gales of the Atlantic and the Irish Sea, and on the other, insulated by the warm waters of the Gulf Stream. Consequently, it has developed a characteristically rugged and windswept terrain while enjoying some of the warmest temperatures in the UK, qualities that make Cornwall one of the country's favoured domestic holiday spots.

According to accounts, it was a feature of this microclimate—the idiosyncratic light—that first attracted artists in the eighteenth century. The celebrated landscape painter J.M.W. Turner visited as early as 1811, and once rail connections to London were established, others like James Whistler and Walter Sickert visited in the early 1880s. By 1885 St Ives was becoming known as an appealing location, and more artists started to arrive as images from the area gained attention at the Royal Academy and the Paris Salon.

The tendency for artists to gravitate towards a location was popular in the late nineteenth century; Nina Lübbren documents eighteen artist colonies across Europe during this period, including Concarneau and Pont-Aven in France, and St Ives and nearby Newlyn in the UK.[36] Although St Ives is not the largest colony she describes, Lübbren records

106 artists in the town between 1886 and 1907 ("16 per cent women. 51 per cent British, 24 per cent American, 10 per cent Australian, 5 other nationalities").[37] Among the pioneers were painters such as Louis Grier and Julius Olsson, who established the seascapes of the St Ives School as distinct from the nearby Newlyn School with its depictions of the fishing community.

The first wave of artists to put down roots in St Ives established a recognizable community even before the turn of the century, and the formation of the St Ives School in 1889 and St Ives Arts Club in 1890 consolidated their activities. But what made St Ives globally renowned is the fact that subsequent generations of artists continued to migrate to the town: by the end of the First World War, notes David Tovey, the number of studios had grown to "over 200".[38] That figure would fluctuate, but the sheer scale of the artistic community in St Ives, relative to its size, had a profound impact.

A new generation began to arrive during the late 1920s. This was the point at which a prominent mythology emerged, when two visiting artists from London, Christopher Wood and Ben Nicholson, chanced upon an open door—the home of retired sailor and one-time scrap merchant Alfred Wallis. Wallis had taken up painting aged 67 as a hobby following his wife's death. "He produced paintings on odd scraps of paper and torn card, remembering and painting scenes of his experiences of boats, the sea, harbour and Godrevy lighthouse",[39] writes Marion Whybrow. Wallis's painting style was distinctive—the images, flattened onto the surface of the material, are representational but primitive, lacking elegance but oozing life: "Sailing Ships and Two Steamers, Newlyn Harbour" (c.1928–1933) shows a murky sea on which trawlers are coming in and out of the harbour (see Fig. 5.1); "The Hold House Port Mear Square Island Port Mear Beach" (1932) depicts the town in the foreground, with its "Island" promontory jutting into the sea, and in the distance a precarious looking fishing vessel being tossed about by the waves. "He had no knowledge of perspective or proportion", remarks Cross, "but a great feeling for the motion, isolation and danger of the sea".[40]

It was around this time that some of the colony's most celebrated exponents of different art forms began to arrive. "The colony now attracted retired writers, playwrights, actors and musicians, so that as much interest was shown in literature, drama and music, as in art", writes Tovey,[41] and as the pursuits of the colony widened, so too did the subject matter of their work. The potter Bernard Leach, whose work was influenced by

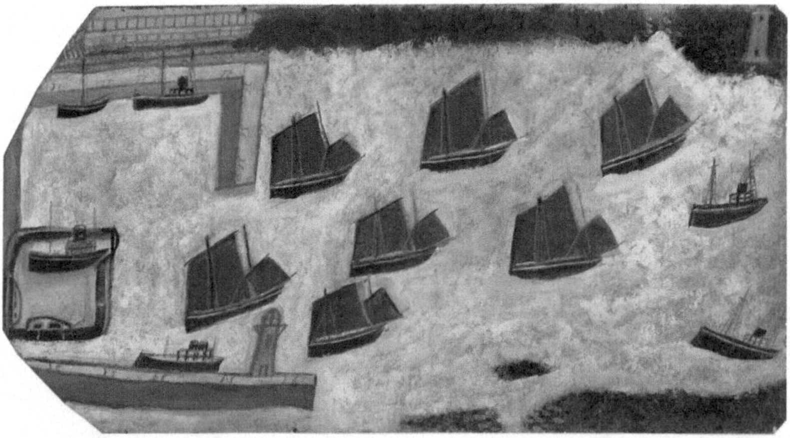

Fig. 5.1 Alfred Wallis, "Sailing Ships and Two Steamers, Newlyn Harbour" (c.1928–1933). Image credit: Kettle's Yard, University of Cambridge

visits to Japan, set up his pottery in 1920; Nicholson himself—already known in London for his geometrical wood reliefs—moved to St Ives with the sculptor Barbara Hepworth in 1939; and shortly afterwards they were joined by the Russian Realist sculptor, Naum Gabo. All four worked against the backdrop of a colony whose output was becoming increasingly abstract, though still deeply touched by the particularities of the place.

Nicholson and Hepworth led the way, and to earn a living with more commercially viable work, Nicholson turned from the austere abstractions of his London days to capture hints of the world around him. At first, these were straightforward depictions of St Ives, often framed through a window with the paraphernalia of his studio in the foreground (see Fig. 5.2).[42] Increasingly, that paraphernalia became more abstracted,[43] flattened into shapes (see Fig. 5.3), and eventually deflecting any representational quality as the image moved towards full abstraction.[44] This was to be Nicholson's signature.

Throughout the war, a wave of younger artists made their way to St Ives: Sven Berlin arrived in 1938; Wilhelmina Barns-Graham in 1942; Patrick Heron in 1944; and Terry Frost in 1946. Peter Lanyon, born in St Ives, would return from training in London to join the new wave. It was a new phase in the fortunes of the colony, not just in terms of its popularity with a younger generation, but also because of the emerging

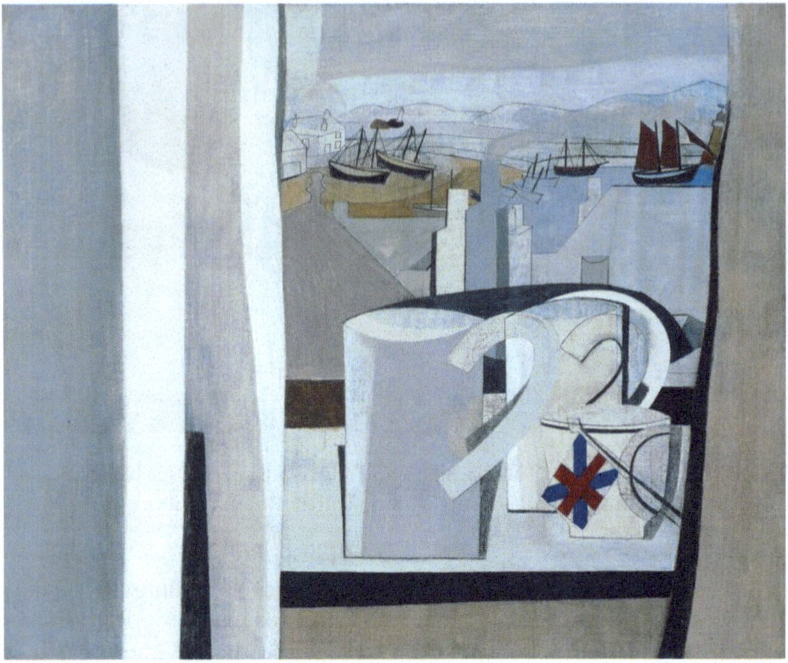

Fig. 5.2 Ben Nicholson, "1943–45 (St Ives, Cornwall)" (1943–45), © Angela Verren Taunt. All rights reserved, DACS 2025. Image credit: Tate

style that became associated with their work. Together, these would represent the St Ives School of the post-war period, breaking away from the more traditional St Ives Society, and under the influence of Nicholson and Hepworth, forming the Penwith Society of Arts in 1948. Here, they explored their creativity not only on canvas, but also through screen-print, glass, ceramics, and writing.

As their work became increasingly abstract, it became more difficult to see the reflection of place in the paintings of these younger artists. Nevertheless, what still marked the St Ives painters of the 1950s was a clever ability to both represent and abstract their subject matter.[45] Here, Peter Lanyon is an interesting figure, and one who clearly felt a deep affinity with the town. "His concern was not to make pure shape or colour on a surface", writes Whybrow, "but to charge every work with information from the world in which he lived".[46] David Crouch and Mark Toogood

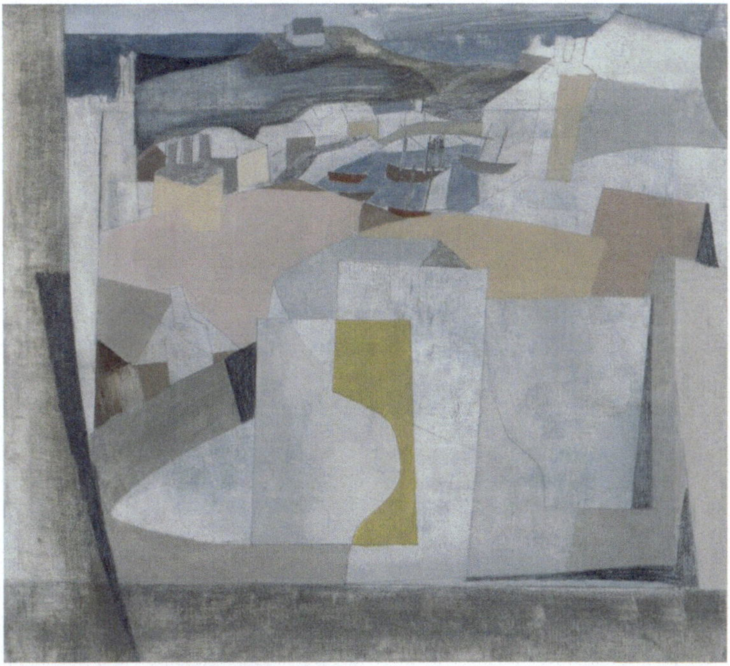

Fig. 5.3 Ben Nicholson, "July 15 1949 (St Ives Harbour)" (1949), © Angela Verren Taunt. All rights reserved, DACS 2025. Image credit: Yale Center for British Art. Paul Mellon Collection, B2006.14.7

consider Lanyon's work as "social, spatial narrative", suggesting that his paintings reflect "different levels of relation to the land as well as his strong self-identification as Cornish".[47] Lanyon's techniques were immersive, discovering the environment through walking, driving, climbing, and even gliding. "In a painterly version of Method Acting", writes Michael Bird, "he vaulted over Bronze Age walls, clung to vertical cliffs and disappeared down mine shafts" to research his paintings.[48] "For Lanyon", recalled his contemporary Patrick Heron, "art and nature corroborate one another: each seeks its own identity in the other" (see Fig. 5.4).[49]

For a time—at least according to Heron—"a case could well be made for considering St Ives the most influential centre of Western painting".[50] This was just before the New York abstractionists like Jackson Pollock and Mark Rothko burst onto the scene. The prominence was short-lived, to

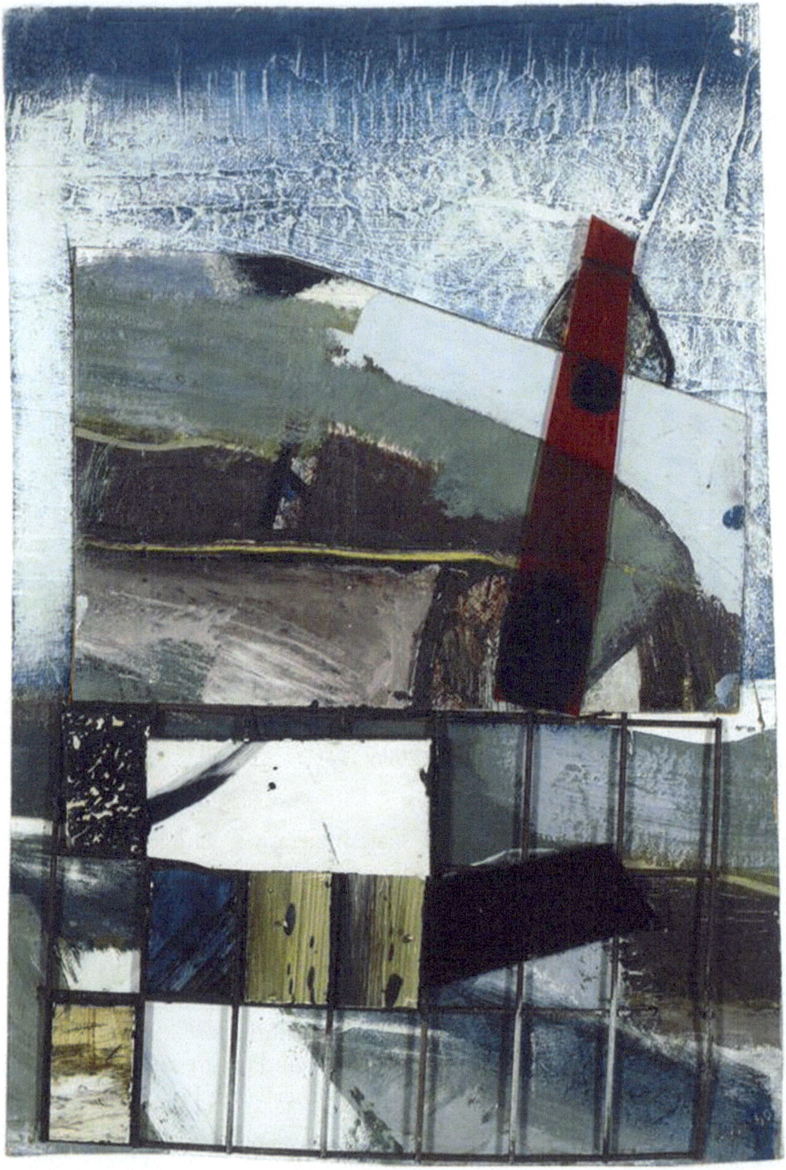

Fig. 5.4 Peter Lanyon, "Built Up Coast" (1960), © Estate of Peter Lanyon. All Rights Reserved, DACS 2025. Image credit: Manchester Art Gallery

be sure, and the claim to world dominance might well be contested. Still, the fact that such a claim has been made at all is another part of the myth-making that surrounds this artist colony: “the sense of art-in-its-place that St Ives has come to embody”.[51] In 1968, Nicholson, Hepworth, and Leach were given Freedom of the Borough: a tangible indication of how much art meant to the town.

Nicholson moved away from St Ives in 1958, Lanyon died following a tragic glider crash in 1964, and Hepworth—along with two other prominent contemporaries, Bryan Wynter and Roger Hilton, died in 1975, bringing to a “symbolic end” the era of the St Ives School.[52] “No group could survive the three deaths”, suggests Cross. Still, with a legacy spanning fully one hundred years, “there was definitely a recognizable ‘St Ives’ character”, he argues; “one that brought abstraction in free play with landscape elements”.[53]

An Ecosystem and Its Keystone Species

It’s clear that St Ives has been bonded by a firm sense of community since at least the 1880s, and that the solidarity of the artists was affirmed by the constitution of formal associations like the St Ives Society and the Penwith Society. But in order to understand this colony as an organic community, or even a community in an ecological sense, we need to consider how the townsfolk, the artists, and other “species” shared the environment; we need to understand in Durkheim’s terms the “intra-social relationships”,[54] or in Bateson’s language the “cooperating tangle” of this mountainside.[55]

In her analysis of artists’ colonies, Nina Lübbren considers the wider make-up of the towns. They “needed to provide an infrastructure to serve the artistic population”, she notes, listing the demand for “hotels, pubs, studios, exhibition venues and shops selling artists’ supplies”.[56] Tovey adds other ways in which the infrastructure developed, citing “a variety of different employment opportunities, ranging from full-time domestic servants and cooks, through to casual helpers and errand-boys, and, of course, models”.[57] Alongside these, artists had need of “various ancillary helpers, such as procurers of paints, artists’ equipment, costumes and props, lessors of property and animal models, and finally, children to carry artists’ bulky outdoor sketching apparatus and to busy themselves as brush washers, boot cleaners, and, intriguingly, as ‘model holders’”.[58] In other words, an entire community of tradespeople, businesses, and peripheral services grew up around the artists; and "this community", writes Wilkinson, "would be incomplete without its critics, writers, editors and photographers, the media by which a painter's reputation is made or broken".[59]

A picture emerges of St Ives beginning to function around the practices and needs of the growing artist population, and this also seems to have affected the town socially. "As early as 1891", writes Tovey, "a local correspondent paid tribute to the artists for 'the marked influence they have brought to bear on the artistic education of the residents'":

> They also improved, beyond measure, the business of many local shopkeepers, and, although this raised prices, which did lead to resentment from those unable to benefit from the artists' presence, there would have been considerably greater hardship, particularly during the winter months, without them.[60]

As they became further integrated, "artists felt motivated to become more involved in the affairs of the community by standing for places on the local Council or by becoming Borough Magistrates or by taking part in local politics". They were deemed to have brought "a beneficial moral influence", "enlargement of the mental vision of the townspeople", "more liberal opinions", "broader views of the modern condition of social life", and further influence on "cultural activities, such as music, drama, and sport", "support for worthy causes", and "involvement in campaigns on such matters as women's rights and animal cruelty".[61] By integrating into the social and cultural fabric of St Ives in multi-faceted ways, artists became not only meaningful community members, but also dominant influences—*keystone species*—within the ecosystem of the town.

Put simply, because of their high "community importance values" and their role as "ecosystem engineers", some species "matter more than others",[62] as ecologists put it, and these are typically referred to as "keystone species", a term coined by Robert T. Paine in 1969.[63] Examples in the natural world vary: beavers are notable ecosystem engineers in wetland areas; coral is the keystone species of a reef ecosystem; and wolves play a fundamental role in balancing their habitats. Although scientists recognize that the term is used somewhat indiscriminately to refer to a variety of different qualities,[64] the concept of *keystone species* has stabilized to refer to species that have a "disproportionate importance in their community"[65]; or, in another way of putting it, species whose "removal would produce a dramatic effect"[66] on the remaining ecosystem.

Such could be said of the artist in St Ives, around which so much of the town's infrastructure was organized, and in relation to which so many of the community's dynamics orbited, at least during the colony era. The artist supported countless other members of the community through using services, generating trade, employing other professionals, and attracting visitors. Metaphorically, these are equivalent network relations to those found in other ecosystems; the artist can be considered a keystone species because removing them from the ecosystem "would produce a dramatic effect".

By way of honouring that legacy, London's Tate gallery held a retrospective of the St Ives artists in 1985: "St Ives 1939–1964: Twenty-five Years of Painting, Sculpture and Pottery". This exhibition was to be the catalyst for a more permanent memorial to the artists, a newly-built gallery called Tate St Ives. The initiative responded to a local government report: "Although traditionally Cornwall has attracted artists, writers, craftworkers—and its tourist reputation partly rests on this", concluded *The Survey of the Arts in the South West* (1984), "actual facilities for the arts are very scarce". A new Gallery, it suggested, "could be a popular tourist asset, reinforcing the historic importance of the west of England as a force in the visual arts".[67] Two explicit drivers, then: to provide facilities for the arts, and to attract tourists.

The gallery was opened in 1993 on the site of a former gasworks on the seafront, and thus "not only provided a solution to an 'eyesore'", but also "formed part of regenerating the town of St Ives itself".[68] It had an instant effect in attracting tourists and enabling the knock-on development of associated attractions. "A relatively wealthy group of new tourists visiting the gallery has had an immediate impact on retailing and restaurants in the town",[69] reported one study, with the initiative precipitating "year round activity and employment, new restaurants and cafes, a growing number of galleries, craft and jewellery shops and rising house prices".[70] Local craftworkers and gallery owners reported an 18% surge in business, while general trade rose by at least 5% and in some cases as much as 20%.[71] "Commercial galleries capitalize on the 'Tate Effect' and proliferate quickly", suggests another report; "Many of these sell tourist kitsch, but serious works by new St Ives artists are also on view".[72] All in all, Tate St Ives has been "a net contributor to the local economy".[73]

Indeed, as a report for the Department of Culture, Media and Sport makes clear, "within two years [...] people whose main reason for visiting St Ives was to visit the gallery were contributing £16 million per annum to the local economy".[74] Thanks to the keystone species of the artist, the ecosystem of St Ives has thrived, attracting other species into the community at different conceptual positions of the food web: the supporting trades who are reliant on the artists in one respect, and the tourists who are reliant on them in another.

A Town Through Time

St Ives nowadays is indeed a magnet for tourists drawn by the picturesque town, pristine beaches, natural beauty of the rugged coastline, and world-class art gallery. However, that was by no means always the case: go back 300 years and the town was home to a community making a living from the sea. Over time, the ecosystem of this community has changed, in a process that is not dissimilar to the concept of *ecological succession*. To understand this, let's take a look at the various stages St Ives has gone through.

Tracking back, we find the ecosystem of St Ives in what Cyril Noall calls its "classical" state.[75] It was a town that relied on natural resources—a sea brimming with fish stock, especially pilchards and mackerel; and "a relatively sheltered position [...] shielded from the prevailing winds"[76]—something that made St Ives "the first good harbour on the north Cornish coast", especially for the fishing method known as "seining".[77] Indeed, so much did the fishing industry flourish in St Ives that throughout the mid-nineteenth century, "four hundred or so [fishing boats] made up one of the largest fleets in Europe".[78] This was big business: "in 1844 a calculated catch of 75,000,000 fish was made in a single day and something like £60,000 flowed into the coffers". With the catch salted and pressed, it was taken back to the harbour and shipped to Europe. "In one year", we're told, "as many as 60,000,000 pilchards were exported from St Ives"[79]; and the export business was to benefit further "by the opening of through railway communication between Penzance and London in 1859, which enabled vast new markets to be reached".[80]

To sustain an industry of that size, of course, the town needed an infrastructure of associated services: "Fishing gave rise to a number of important subsidiary industries", notes Noall; "the vast quantities of herrings landed at St Ives towards the close of last century led to a curing industry being established for making kippers, bloaters and red herrings".[81] The fishing trade was further supported by "ship and boat builders, rope makers, blacksmiths and carpenters".[82] And the town itself would have thrived from the additional day-to-day services that such a workforce needed: hostelries for food and drink; market stalls for essentials; tradespeople like waggoners, coopers, and weavers; and professionals such as bankers, doctors, or lawyers. In this period, St Ives was fundamentally different from the later artist colony, held together by a different keystone trade and woven from a different food web of interactions. But the community still worked like an ecosystem, with its make-up characterized by the dynamics of the "population-community" and its "process-functions" driven by the fishing trade.[83]

According to accounts, St Ives was a dirty place. "Poor drainage and sanitation, when coupled with the odours of the fishing industry, made a vile combination", explains Tovey; "the fishermen also gutted fish on the harbour beach and this offal was left for scavenging gulls and, unsurprisingly, attracted rats".[84] Reports from the time bring such unpleasantness vividly into relief: "'The stench arising from the stores, and from the putrid *rejectamenta* lying about the town [was] to strangers almost intolerable'", wrote one observer; "and its reputation as a 'dirty fishing town' continued through the nineteenth century",[85] when it was seen as nothing but a "rancid little fishing port".[86]

As it happens, though, St Ives in this period wasn't exclusively characterized by its fishing; a second industry also prevailed in the "hinterland" behind the town,[87] where it benefited from rich seams of underground metals and minerals like tin, copper, and radium. "Cornwall was the cradle of the world's industry", claims Bryan Harvey[88]; or as Kennedy and Kingcome put it, "a powerhouse of technical innovation"[89]: it was the mining industry that precipitated the development of Watt's steam engine, in order to drain the mines of water, and it was also the mining industry that led to the development of gas lighting, with nearby Redruth having gas streetlights before London.[90] This too was big business: "by the early 1860s, mining and quarrying directly employed 30% of adult men and 10% of waged women".[91] Like other industries, mining also relied on and provided for a variety of other associated activities: St Ives was a centre for smelting the extracted ore; it was also one of the ports used to bring in coal for the "fire engines" (the shaft-mine water pumps).[92]

In short, until close to the end of the nineteenth century, there was no indication at all that St Ives would soon become known as a thriving artist colony, or even a tourist hotspot: Cornwall was known for "fish, tin, and copper", "the three most important commodities that have moulded the county's history".[93] As Bernard Deacon reports, "reminiscences from locally born contemporaries are striking for their lack of mention of the artists", who are "almost totally absent from accounts of local life".[94] The significance of the artist was to emerge later, and only because of considerable changes that altered the status quo.

The discovery of tin lodes in Australia and Tasmania caused the price of tin to plummet. "The effect in Cornwall was catastrophic", writes H.V. Williams[95]; and throughout the 1870s, dozens of mines shut down. Temporarily, some of the miners—"glad of the employment"—may have found work on the construction of the St Ives branch railway line, from 1874 to 1877.[96] But "by 1880 it was [also] clear to all that copper mining in Cornwall was at an end".[97] Accounts speak of "slow-onset change", which was "measured in decades and generations",[98] as the town recalibrated its identity, its connection to local resources, and its survival. And this is a change that can be viewed from an ecological perspective, because such a shift in the town's identity over time is reminiscent of the process of *ecological succession*.

St Ives and Ecological Succession

According to scientists, ecological succession is "the sequential replacement of species following a disturbance". It "can encompass change in communities [...], ecosystems [...], and structural change".[99] It is also a concept that has been applied to human societies[100]; in fact, the concept of succession has been used for decades to track demographic changes in sociological contexts, particularly in respect of changes involving different racial communities in America.[101]

Eugene P. Odum proposed three main parameters for ecological succession, suggesting that "it is an orderly process of community development"; "It results from modification of the physical environment by the community"; and "it culminates in a stabilized ecosystem". He goes on to explain it in more detail, starting at the point at which a young ecosystem forms. Early on in the development of the ecosystem, production (growth) exceeds respiration (carbon emission): "quantity production characterizes the young ecosystem", writes Odum, "so that biomass accumulates". Sooner or later, though, a more nuanced dynamic emerges as the ecosystem reaches a mature stage of "quality production and feedback control". Finally, the ecosystem reaches its climax state (Odum pinpoints this stage at 60–80% of the way through its lifecycle), when quantities stabilize: "autotrophic and heterotrophic metabolism [are] balanced". In most cases, Odum writes, "population regulation" of a climax ecosystem will necessarily involve "adaptation": species within the community will change their behaviour, or in some cases, even their DNA, in order to find an equilibrium—"overall homeostasis". He suggests that such stability can last for anywhere from a hundred days to a hundred years. Moreover, the stability will often be strengthened if a diverse array of species has emerged during the formation of the ecosystem in more-or-less balanced ways. "The variety of species [...] tends to increase during the early stages of community development", he writes, leading to "reduced dominance by any one species". "The overall strategy", he reasons, "is [...] directed toward achieving as large and diverse an organic structure as is possible within the limits set by the available energy input and the prevailing physical conditions". Odum concludes with a speculative hypothesis, that "after a long period of relative stability or 'adulthood', [...] ecosystems again develop unbalanced metabolism and become more vulnerable to diseases and other perturbations"; a new dominant species emerges, growth of that newcomer increases, and the next iteration of the process—successional change—begins.[102]

This is an analogue to the process that St Ives experienced as it developed from a fishing and mining port (its "classical" mature ecosystem) to an artists' colony (its next successional iteration), and it was rather an abrupt ecological disturbance that made such a change happen: "The disaster hit Cornwall hard", writes Noall of the discovery of cheap Tasmanian tin[103]; "children were sent out into the fields to work, but their earnings were so low that they could not keep themselves, let alone bring in any money to their distressed parents".[104] And if the demise of the mining industry was not bad enough, St Ives also faced a parallel slump in fish stocks, a second decline "commonly acknowledged to have gained momentum around 1880".[105] Hitting the seining industry hard, the end of the nineteenth century saw a virtual collapse in production: "For several years this industry, once so lucrative, especially at St Ives, Newquay, Mullion and Cadgwith, has been gradually dwindling", reported the *West Briton* in 1891, "but this is probably the first year in which not a single seine has been put into the water".[106] "The commercial failure of that season sounded the death knell for seine adventuring", writes Tony Pawlyn; "By 1892 there were only 142 seines left at St Ives, and of these only 70 were fitted out for the season".[107] By the 1900s, writes Paul Thompson, St Ives "presented a scene of unmitigated melancholy, its men 'despondent and disheartened', their abandoned luggers 'moored up'".[108] "The decline continued remorselessly until by the mid-1920s virtually all had disappeared",[109] writes Pawlyn. But into the gap, as we have seen, came the artists, initiating the process of successional change. Gradually, St Ives' identity would transform from one ecosystem (mining and fishing port) into another (artist colony), finding equilibrium with a new dominant species.[110]

To some, Odum's explanation of succession is rather simplistic.[111] A more recent account of the processes of successional change is provided by Karel Prach and Lawrence R. Walker (2011). They discuss three main components of succession, noting the *mechanisms* through which it operates, the *stages* through which it goes, and the *trajectories* which define its direction of travel.[112] *Mechanisms* such as the "dispersal and establishment of species" can clearly be seen in the demise of the former industries, the departure of the miners and fishermen, and the arrival of the artists, between around 1870 and 1920; "competition" is established by the arrival of that different species (the artists), which becomes increasingly dominant; and "facilitation" is evident in the emergence of a subsidiary infrastructure providing accommodation, refreshment, materials, and services. *Stages* determine the maturity of the succession: at first come pioneers like Louis Grier and Julius Olsson during the "initial" stage of establishment; following that, developments through the "early", "mid", and "late" stages of the colony (from the St Ives School of 1889 to the St Ives Society of Artists in 1927 to the Penwith Society of 1948) introduce nuances in the styles of art (from seascapes to abstractions) and in the types of art that are practiced (from fine art to sculpture to pottery). *Trajectories* can be varied: "convergent; divergent; parallel; circular; or complex". Thus, a "parallel" trajectory of mining and fishing in the earlier St Ives succeeds to a more "singular" trajectory of a dedicated artist colony with its associated needs. Subsequently, the ecosystem succeeds again through a much more "complex" trajectory whose "overall homeostasis" is fashioned as much by the historical resonance as it is by the modern town, as tourism and heritage and their associated services commodify tin, fish, and art into brand St Ives.

Using ecological succession as a way of understanding St Ives is a helpful analogue; but we should not forget the warning that even a stable ecosystem in its successional climax may "develop unbalanced metabolism and become more vulnerable to diseases and other perturbations". Recognizing that threat also means acknowledging a responsibility to steward the ecosystem's future. And St Ives may be particularly at risk thanks to its vulnerable isolation.

An Isolated Ecosystem

"Cornwall is remote by definition", remark Kennedy and Kingcome, a truism that has generated a "dominant myth of isolated peripherality and insularity".[113] To be sure, part of its appeal has been its "remoteness from the centre", something that not only distances this far end of Cornwall from the rest of the country, but perhaps also bonds its people within a "sheltered community".[114] It's a remoteness that is romanticized, whether in the forceful claims of the Cornish people to an independent identity, or in the rugged mystique of narratives like *Poldark*, *Rebecca*, or *Jamaica Inn*. According to Lanyon, Cornwall "has for centuries been almost an island"[115]; or as Kennedy and Kingcome put it, "a fantasy island at the end of Great Britain, where nothing has been touched by urban industries".[116] Frequently referred to as "insular people"[117] (even as "fierce and insular people"),[118] the Cornish have their own flag, their own language, and a strong sense of nationalism. Meanwhile, Bernard Deacon describes Cornish fishing ports as "spatially adrift; [...] temporally unmoored",[119] with one 1868 travelogue noting that "St Ives is a queer-looking town, with all the seeming of having been drifted in here whole, from some portion of an olden world".[120]

Appropriately, in terms of an environmental perspective, this compares with the idiosyncratic qualities of islands, which "by virtue of their isolation have relatively simple ecosystems made up of the few vagrant species that succeeded in colonizing them".[121] Islands are often habitats in which certain "endemic" species are uniquely represented: the lemurs of Madagascar, for instance, or the koalas of Australia. Because of their isolation, and sometimes their inaccessibility, islands can be biodiversity hotspots; UNESCO recognizes, for example, that Madagascar is "home to 15,000 native species of land plants—eighty-five per cent of which are endemic".[122] On the other hand, the concentration of an entire species population on a single island can make that species vulnerable, rendering the ecosystem fragile, especially when new species are introduced. "Invasive mammals are particularly devastating", write Spatz and

Holmes; "predators such as cats and mongooses prey upon native species, while herbivores, such as goats and pigs, alter landscapes and contribute to habitat loss".[123] As the infiltrators flourish, the stability of ecosystems is threatened, and even if this doesn't lead to extinction (as it often does), the status quo can be radically altered as new species overwhelm the old. "In the Galapagos, introduced goats outcompete tortoises for food", writes Ben Garrod, "while snakes accidentally taken to Guam, where the birds had never seen a snake before, are destroying the fine balance of island ecosystems there".[124]

Tourists as an Invasive Species

Tourists began arriving in St Ives as soon as the railway line was cut. "The Victorian railway brought artists and holiday-makers", writes Robert McCrum; "the sail lofts became studios, and the cottages of Downalong became guest-houses".[125] This was a development that the town welcomed; after all, as Michael Bird points out, "scenic beauty was a raw material whose exploitation might go some way towards replacing the disastrous loss of pilchards and tin".[126] For a long time the town thrived, in part because of the keystone species of the artist. "The artists helped significantly to convert St Ives from a failing industrial town into a prime tourist destination", Tovey reflects.[127] But as the ecosystem passed its climax point, the balance was visibly seen to shift. "In the late 1950s and 1960s the town of St Ives began to attract summer visitors as never before", reports Cross; "Townsfolk and artists alike felt the pressure, as from June to September the town was overwhelmed by cars and visitors".[128] By the last throes of the colony era, and the symbolic deaths of Hepworth, Wynter, and Hilton, the ecosystem was over-run. When the artist Terry Frost moved back to the area in 1974, St Ives "struck him as having changed for the worse", reports Bird; "The tourist crowds oppressed him". More than that, the inspirational art had given way: "There were just as many artists", writes Bird, but "the vacuous mantra 'inspired by the Cornish landscape' was mysteriously expected to explain and excuse every grey-green slap of paint, every scrap of blue washing hung out on a horizon line".[129] Sure enough, the tourist crowds thronging around St Ives even today encounter little of the former colony, despite the countless prints for sale: childlike houses a lame nod to the primitive authenticity of Alfred Wallis; one-dimensional terraces stacked in a faint gesture to abstraction; dinghies bobbing on a pastel sea in a ubiquitous signal of the seaside. "A stroll down St Andrews Street or Fore Street past the commercial galleries that crowd the narrow streets illustrates the kind of art the general public is interested in", observes McCrum; "tasteful watercolours, sunny landscapes, and smiling realistic portraits. This is the world of pictures we can all understand: place-mat art".[130]

In September 1994 Tate St Ives played host to the conference "Imagining Cornwall: English Modernism, Tourism, and the Pleasure Periphery". The keynote address was given by the renowned tourism scholar Dean MacCannell, who spoke about "the narratives and pleasures of St Ives as tourist town, artists' colony and beatnik centre".[131] That description is very much brand St Ives, and an identity that the town still claims to this day. However, in his book *Empty Meeting Grounds* from around the same period (1992), MacCannell invokes the idea of *positive involution*, a process that takes place within a tourist site whereby "the community turns in on itself, transforming its qualities and characteristics into symbolic resources for its own future development". Such a transformation can potentially work to the community's benefit, he suggests; but if it is an initiative driven from the top (an effort "on the part of governments and other agencies to implement economic and social change [...] in the name of 'progress' and modernization"), the results can be counter-productive:

> Rather than making the community a better place to live for all of its actual and potential residents, the effect of the restoration of poor neighborhoods is to drive out the people whose homes have been improved, forcing them to the margins, eventually into homelessness. In a society that needs to think of itself as 'humane', this sort of thing cannot be explicit or avowed. It could only be accomplished by policymakers who claim to be doing the opposite, who have mastered the art of disguising narrow class interests in a rhetoric of 'diversity', 'preservation' and the like.

Various performative mechanisms are brought into this process, MacCannell observes, which he refers to as "staged authenticity" and "reconstructed pseudo-gemeinschaft"[132]—the site, in other words, begins to *perform* its attractions to the visitors in order to maximize its tourist appeal.

Initially, the construction of an international art institution led to regeneration and brought unexpectedly high levels of visitors to the town:

Tate St Ives was recognized "as a social and economic success", as a report acknowledged in 1996. Positivity was high, as we have seen, and "the benefits have spread over the whole of Cornwall".[133] But underneath that headline news were somewhat different reports. "The Tate does not appeal to the average tourist", suggested McCrum as early as 1995, and nor did it seem to appeal to the locals: "The local fishermen call it 'the Ice-Cream Factory'", he wrote; "disaffected local artists describe it as 'a theme park'".[134] By 2004, as another government report acknowledges, "critics from neighbouring districts [maintained] that the trickle down effects have not benefited them".[135]

When the Tate decided to expand in 2005 its plans were met with a wave of local resistance. "More than 2,000 local people have signed a petition against the expansion plans", reported Steven Morris, "many of them believing the gallery is losing touch with the town". In a town of only around 10,000 inhabitants, this was a sizable proportion of the population; and it included "members of the artistic community and the architects of the original building, [who] argue Tate St Ives has lost sight of what it was supposed to be—an exciting but relatively modest gallery in a small town".[136] Remember that one of the original drivers for the Tate was that "actual facilities for the arts are very scarce"; yet as far as locals were concerned, the gallery had pandered exclusively to the other driver—"a popular tourist asset".[137] More than this, it was gaining a reputation for ignoring the needs of local people; that "the gallery is determined to drive its proposals through no matter what local people think of them".[138]

The cachet of an international institution brought in a clutch of commercial London gallery owners, (the New Millennium; Belgrade St Ives), but with the four- and five-figure prices they listed, all but the most affluent of local buyers were left out of the market. Fore Street and Wharf Road were peppered with upmarket restaurants to service clientele who would sweep into town for a long weekend, while being out of reach to the locals. Worse than this, the warnings from scholars like MacConnell seemed to be coming into effect. Unlike the organic succession of St Ives from fishing port to artist colony, its transition into a tourist hotspot was engineered by committee.

St Ives as an Imbalanced Biome

Much has been written about how tourism can have an adverse effect on a place, perverting the very character that visitors have set out to see. “Tourism destroys unique place identities”, argues Moya Kneafsey[139]; it serves to “trivialise a unique way of life that cannot be replaced”, suggests Gary Mason[140]; meanwhile, Davydd Greenwood talks of a “collapse of cultural meaning” in destinations that are targeted by excessive visitors.[141] These reports consider tourism in rural Ireland, Hawai'i and the Basque country, though the comments echo others made about St Ives. “You feel you have come somewhere made especially for the tourists”, remarked one visitor to a reporter, who agreed: “It is polished, empty, immaculate: an ideal of a fishing village that looks nothing like a real fishing village”.[142] Others make similar complaints, noting also the social impact: “the sense of community in the town has been lost as the economy moves away from fishing to tourism”, writes Milo Boyd, adding that for some residents “the town is synonymous with rude tourists, empty second homes, ‘plain greedy’ landlords and poverty”.[143] The flip side of the summer tourist boom brings a seasonal economy, a creaking infrastructure, and an unregulated accommodation system that leaves hundreds of second homes empty in the winter while locals can't afford to get on the housing ladder.[144] And social consequences can be profound, as Huw Thomas has explored: “A slow-onset loss of community houses to second homes and holiday homes led to another experiential change that included a decline in core population and a decline in social networks”, he explains.[145] “Why is somebody's holiday worth the death of the culture?”, asks one resident; “If you apply colonial theory to Cornwall, you'd call it ethnocide”.[146]

Rather than colonial theory, we are applying an ecological perspective, and complaints such as these resonate with ecological warnings about the impact of invasive species on vulnerable ecosystems, resulting in what ecologists call an *imbalanced biome*. Remember Odum's warning? (“after a long period of relative stability [...] ecosystems again develop unbalanced metabolism and become more vulnerable to diseases and other perturbations”)[147]; this is just another route to successional change. “The enormous problem still is to manage, control, and where necessary alter the pattern of food-chains in the world, without upsetting the balance of their populations”, as Charles Elton puts it; in his view, we need to “find out how the broken balance can be restored and protected”.[148]

Ecosystem Stewardship

Against a background of local animosity, Tate St Ives CEO Mark Osterfield "went out and met with possibly every single community group in St Ives, attended all the town council meetings and created a working group with the most vociferous voices against the Tate expansion".[149] The new development was approved, and opened in 2016, but significantly, it also involved a number of cultural and community initiatives such as the "St Ives Legacy Project", the "Look Group Project" and "Tate Collective". These have engaged with local communities, and especially with young people; they've reached out more widely around Cornwall to enable the whole county to feel ownership of the gallery; they've improved accessibility for the visually impaired and hard of hearing; they've instituted a "Locals Pass" to encourage residents to keep visiting; and they've built associations with national events in order to strengthen a spirit of involvement. "I think there is a big shift in the sector moving away from the idea that you have the expert curator who puts on a show and people come and respond", says Osterfield; "people have their own interests, skills and knowledge and it's about creating opportunities for them to make their own programme, which is really beneficial".[150] Subsequent Town Deal Funding has also been secured "to support a new integrated, low carbon transport system, develop new skills, businesses and jobs, as well as provide transformed housing and community assets"—a move away from simply attracting tourists.[151] Crucially, art (more specifically: the arts) features prominently in this redevelopment: among the nine projects in the Town Deal are "the refurbishment of the Palais de Danse [formerly Barbara Hepworth's studio] to become a new, digitally enabled, training, learning and enterprise hub", "the refurbishment of St Ives Theatre and The Guildhall [to] enable the extension of the performance season across the year and provide more opportunities for local children and young people", and "a major expansion of the Leach Pottery [which] will generate additional opportunities for residents whilst creating year-round jobs".[152]

Thus the most recent wave of activity by the authorities and the major players in the town (like Tate St Ives) has shown a far more considered approach to handling its ecosystem: more of a collaboration with the community. We see a more balanced awareness of the different species (residents as well as tourists), coupled with an acknowledgment of the dominant drivers (the cultural legacy; the visitor economy); emerging populations (children); vulnerable groups (the visually impaired or hard of hearing); and dynamics beyond the immediate biome (in the wider county and country). Rather than leaving St Ives prone to future disturbance—whether those events are natural (the deaths of Hepworth, Wynter and Hilton), cultural (the sudden slump of fishing and mining), or man-made (the original construction of the Tate)—the approach now is a sort of stewardship aimed at making the ecosystem sustainable and the community resilient.

Theorists of ecosystem stewardship note "a gradual shift in resource management paradigms"[153] away from "steady-state resource management", which attempts "to reduce variability and prevent change", and towards a strategy that "embraces change as a basic feature of the way the world works".[154] The ecosystem stewardship model involves three principles: "reducing vulnerability to expected changes", "fostering resilience to sustain desirable conditions", and "transforming from undesirable trajectories"[155]; and develops through three phases: "from [1] an intensive resource exploitation phase to [2] a steady-state resource management paradigm aimed at maximum or optimum sustained yield of a single resource, such as fish or trees, and subsequently to [3] ecosystem management to sustain a broader suite of interdependent ecosystem services in their historic condition".[156] We could substitute the "fish or trees" of this explanation with "tourists" to recognize a similar transition in St Ives: first (c.1870-c.1930), the town exploits its attractions of "an old fishing village and a picturesque art colony"[157] to build a tourist economy; next (c.1930-c.2005), the delights of the Cornish Riviera and the artistic reputation of the town are pushed to maximize tourist yield ("'Your mines may close but you can go on mining your heritage for ever'")[158]; finally (c.2005–), a "broader suite of ecosystem services" is instituted ("support[ing] a new integrated, low carbon transport system, develop[ing] new skills, businesses and jobs, [...] provid[ing] transformed housing and community assets").[159] In the world of conservation it's a solution that aims "to identify pragmatic strategies that increase the likelihood of socially beneficial outcomes and reduce the risk of bad outcomes".[160] For anyone committed to social regeneration in St Ives, these would seem to be desirable pursuits.

Nowadays the focus for St Ives is holistic, geared towards the whole community rather than just a section of it. Part of the community continues to be its artists, and there are still one or two holdouts for their work: the Penwith (Penwith Society of Artists), the Mariners (St

Ives Society of Artists), the Porthminster Gallery, and the New Craftsman support local artists and continue the colony traditions. Meanwhile, some of the hallmark names of the colony era are represented through the Leach Pottery and the Barbara Hepworth Museum. And recent successful artists keep the town's artist colony legacy alive. The presence of the artist remains strong in St Ives, though perhaps the broader vision for the town will secure a sustainable and resilient future for this community.

Notes

1. Tim Smit (2001), *Eden,* London: Bantam Press, p. 13.
2. Elsewhere, he called it "a blend of science and theatre, set in an epic location"; and a "living theatre of plants and people". See Marcus Binney (2000), "Now Here's a Dome for the Millennium", *The Times*, 29 February, Arts and Sports, n.p.; Richard Morrison (2001), "Why I Created Eden", *The Times*, 12 March, p. 2.
3. Eden Project (n.d.), "Our Origins", *Eden Project*, https://www.edenproject.com/mission/origins. Accessed 27 December 2024.
4. Anon. (n.d.) "Eden Project near St Austell", *Great British Gardens*, https://www.greatbritishgardens.co.uk/cornwall/item/eden-project-near-st-austell.html. Accessed 30 December 2024.
5. This is usually attributed to *The Times* as if in a review, though Morrison actually reports the phrase as "the epithet favoured west of the Tamar". Morrison, "Why I Created".
6. Susan Allen Toth (2001), "More Than One Eden in Cornwall", *New York Times*, 22 July, Section 5, p. 8.
7. Morrison, "Why I", p. 2.
8. Deborah King (2001), "Genesis of the Garden", *The Times*, 14 March, p. 3.
9. Smit had been working as a writer, arranger, and producer in the record industry for artists such as Twiggy, Louise Tucker, and Barry Manilow.
10. Tim Smit (1997), *The Lost Gardens of Heligan*, London: Victor Gollancz, p. 144.
11. Smit, *Eden*, p. 20.
12. Smit, *Eden*, p. 20.
13. Smit, *The Lost*, p. 26.

14. Smit, *Eden*, p. 21.
15. Clare Hargreaves (2022), "Cornwall's sleeping beauty: the tale of Heligan's lost gardens", *The Observer*, 17 April, https://www.theguardian.com/lifeandstyle/2022/apr/17/the-tale-of-heligan-lost-gardens-cornwall-treasure-rediscovered-30-years-ago. Accessed 30 December 2024.
16. Anon. (1993), "Secret garden found on ruined estate", *The Times*, 19 August, p. 16.
17. Eden Project, "Our Origins".
18. Christopher Warman (2001), "Paradise Found", *The Times*, 10 November, Section 3, p. 13.
19. Tim Smit in Dave Simpson (2021), "Inspired by a sink full of soap bubbles: how we made the Eden Project", *The Guardian*, 15 March, https://www.theguardian.com/artanddesign/2021/mar/15/how-we-made-eden-project-cornish-lost-civilisation-soap-bubbles. Accessed 29 December 2024.
20. Institute of Civil Engineers (2024), "Eden Project", *Institute of Civil Engineers*, https://www.ice.org.uk/what-is-civil-engineering/infrastructure-projects/eden-project. Accessed 27 December 2024.
21. Binney, "Now Here's".
22. Institute of Civil Engineers (2024), "Eden Project".
23. Binney, "Now Here's".
24. Peter Hardy (2004), "Dome Sweet Dome", *The Daily Mail*, 11 August, p. 59.
25. Toth, "More Than One".
26. Eden Project (n.d.), "Mediterranean Biome", *Eden Project*, https://www.edenproject.com/visit/things-to-do/mediterranean-biome. Accessed 27 December 2024.
27. Smit, *Eden*, p. 263.
28. Smit, *Eden*, p. 282.
29. Eden Project (n.d.), "Learn with Eden", *Eden Project*, https://www.edenproject.com/learn. Accessed 27 December 2024.
30. Eden Project (n.d.), "The Invisible Rainforest", *Eden Project*, https://www.edenproject.com/learn/eden-at-home/the-invisible-rainforest. Accessed 27 December 2024.
31. Eden Project, "The Invisible".
32. Ben Webster and Helen Rumbelow (2002), "Newquay ready for the weekend jet set", *The Times*, 1 February, p. 3.

33. Steve Keenan (2001), "How long is the queue for Eden?", *The Times*, 28 April, Sect. 5:2, p. 7.
34. Hardy, "Dome".
35. Natasha Carthew (2023), "Welcome to Cornwall! Please don't ruin it for us local people", *The Guardian*, 29 May, https://www.theguardian.com/commentisfree/2023/may/29/welcome-to-cornwall-please-dont-ruin-it-for-us-local-people#:~:text=Cornwall%20is%20one%20of%20the,meant%20unaffordable%20housing%20for%20residents. Accessed 1 July 2023.
36. Nina Lübbren (2001), *Rural Artists' Colonies in Europe, 1870–1910*, Manchester: Manchester University Press.
37. Lübbren, *Rural*, p. 173.
38. David Tovey, (2009), *St Ives (1860–1930): The Artists and the Community: A Social History*, Tewkesbury: Wilson Books, p. 117.
39. Marion Whybrow (1994), *St Ives 1883–1993: Portrait of an Art Colony*, Woodbridge: The Antique Collectors' Club, p. 120.
40. Tom Cross (1995), *Painting the Warmth of the Sun: St Ives Artists 1939–1975*, Tiverton: Westcountry Books, p. 20.
41. Tovey, *The Artists*, p. 26.
42. See, for example, "1943–45 (St Ives, Cornwall)" (1943–5); "1946 (Window in Cornwall)" (1946); "Mousehole, November 11, 1947" (1947); or "15 July, 1949 (St Ives Harbour)" (1949).
43. "December 1951 (St Ives—Oval and Steeple)" (1951).
44. "February 1953 (Contrapuntal)" (1953).
45. As can be seen, for example, in Terry Frost's "Walk Along the Quay" (1950), William Scott's "Cornish Harbour" (1951); Wilhelmina Barns-Graham's "Rock Forms" series (1953).
46. Whybrow, *Portrait*, p. 118.
47. David Crouch and Mark Toogood (1999), "Everyday Abstraction: Geographical Knowledge in the Art of Peter Lanyon", *Ecumene*, 6:1, pp. 72–89, pp. 73, 76.
48. Michael Bird (2016), *The St Ives Artists: A Biography of Place and Time* (second edition), London: Lund Humphries, p. 97.
49. Patrick Heron, "Peter Lanyon", in Anon. (1992), *Peter Lanyon: Air, Land and Sea*, London: South Bank Centre, p. 6.
50. Patrick Heron, quoted in Bird, *The St Ives Artists*, p. 13.
51. David Wilkinson (2005), "St Ives: The Legendary Quality of Light?", in Laura Newton (ed.) (2005), *Painting at the Edge:*

British Coastal Art Colonies, 1880–1930, Bristol: Sansom & Co., pp. 65–77, p. 76.

52. Anon. (n.d.), "1945-1960s: The St Ives School", *St Ives School of Painting website*, https://www.schoolofpainting.co.uk/about-us/st-ives-art-scene/the-st-ives-school/. Accessed 30 March 2023.
53. Cross, *Painting*, p. 196.
54. Émile Durkheim (2013), *The Division of Labour in Society* (ed. Steven Lukes; trans. W. D. Halls) (second edition), Houndmills, Palgrave Macmillan, p. 204.
55. Gregory Bateson (1991), *A Sacred Unity: Further Steps to an Ecology of Mind*, New York: Cornelia & Michael Bessie Books, p. 265.
56. Lübbren, *Rural*, p. 64.
57. Tovey, *The Artists*, p. 395.
58. Lübbren, *Rural*, p. 64.
59. Wilkinson, "The Legendary", p. 66.
60. Tovey, *The Artists*, pp. 287, 395.
61. Tovey, *The Artists*, pp. 287, 315.
62. Shahid Naeem and F. Stuart Chapin III (1999), "Biodiversity and Ecosystem Functioning: Maintaining Natural Life Support Processes", *Issues in Ecology*, 4:11, p. 6.
63. L. Scott Mills, Michael E Soulé, and Daniel F. Doak (1993), "The Keystone-Species Concept in Ecology and Conservation", *BioScience*, 43:4, pp. 219–224, p. 219.
64. Mills, Soulé, and Doak, "The Keystone-Species".
65. Nicolas Mouquet et al. (2013), "Extending the concept of keystone species to communities and Ecosystems", *Ecology Letters*, 16:1, pp. 1–8, p. 1.
66. Robert H. MacArthur (1972), "Strong, or weak, interactions", *Transactions of the Connecticut Academy of Arts and Sciences*, 44: 177, p. 88.
67. Janet Axten (1995), *Gasworks to Gallery: The Story of Tate St Ives*, St Ives: Janet Axten and Colin Orchard, p. 53. See also Sir Richard Carew Pole (2002), "The Planning and Development of the Tate Gallery St Ives, 1988–93", in Anne Douglas (ed.) (2002), "On the Edge: Culture and the arts in remote and

rural locations" (conference), Aberdeen: Robert Gordon University, pp. 130–7. https://ontheedgeresearch.files.wordpress.com/2011/12/ontheedge.pdf. Accessed 9 July 2023.
68. Charles Landry et al. (1996), *The Art of Regeneration: Urban Renewal through Cultural Activity*, Stroud: Comedia, p. 70.
69. Landry et al., *The Art*, p. 71.
70. Graeme Evans and Phyllida Shaw (2004), *The Contribution of Culture to Regeneration in the UK: A Review of Evidence*, London: Department for Culture, Media, and Sport, p. 36.
71. Landry et al., *The Art*, pp. 70–71.
72. Patricia Huang (2005), "Where Art, Architecture, and Landscape Meet", *National Museum of Natural Science*, 19, pp. 7–22, p. 16.
73. Landry et al., *The Art*, p. 71.
74. Evans and Shaw, "The Contribution", p. 24.
75. Cyril Noall (1977), *The Book of St Ives*, Chesham: Barracuda Books Ltd, p. 47.
76. Noall, *The Book*, p. 13.
77. Elizabeth Sparrow and Michael Williams (1976), *About St Ives*, Bodmin: Bossiney Books, p. 4.
78. Reg Watkiss (2005), *Reflections of Old Cornwall: A Further Selection of Historic Photographs*, Tiverton: Halsgrove, p. 51.
79. Sparrow and Williams, *About*, p. 9.
80. Noall, *The Book*, p. 77.
81. Noall, *The Book*, pp. 76–77.
82. Sparrow and Williams, *About*, p. 8.
83. Robert V. O'Neill (1986), *A hierarchical concept of ecosystems*, Princeton: Princeton University Press, p. 1.
84. David Tovey (2008), *St Ives Art pre-1890: The Dawn of the Colony*, Tewkesbury: Wilson Books, p. 10.
85. Tovey, *St Ives Art pre-1890*, p. 10.
86. Wilkinson, "The Legendary", p. 65.
87. Beryl James (1988), *Tales of the Tinners' Way*, Redruth: Dyllansow Truran, p. 58.
88. Bryan Harvey (1970), "Who started it all?", *British Journal of Industrial Medicine*, 27:1, pp. 81–83, p. 81.
89. Neil Kennedy and Nigel Kingcome (1998), "Disneyfication of Cornwall—developing a Poldark heritage complex", *International Journal of Heritage Studies*, 4:1, pp. 45–59, p. 47.
90. Harvey, "Who?", pp. 81–82.

91. Kennedy and Kingcome, "Disneyfication", p. 47.
92. A.K. Hamilton-Jenkin (1962), *The Cornish Miner: An Account of his Life Above and Underground from Early Times*, London: George Allen and Unwin, p. 120.
93. Roger David Penhallurick (2017), *Tin in Antiquity: its mining and trade throughout the ancient world with particular reference to Cornwall*, London: Routledge, p. 148.
94. Bernard Deacon (2001), "Imagining the Fishing: Artists and Fishermen in Late Nineteenth Century Cornwall", *Rural History*, 12:2, pp. 159–178, p. 169. Deacon cites William H. Paynter (1999), *Old St Ives: The reminiscences of William Paynter*, Penzance: Oakmagic Publications, and Eddie Murt (1994), *Downlong Days: A St Ives Miscellany*, St Ives: St Ives Printing and Publishing Co.
95. H.V. Williams (1969), *Cornwall's Old Mines*, Truro: Tor Mark Press, p. 18.
96. Richard C. Long (2022), *The St Ives Branch Line: A History*, Barnsley: Pen and Sword Books Ltd, n.p.
97. Williams, *Cornwall's Old Mines*, p. 10.
98. Huw Thomas (2017), "Social Resilience in Cornish Fishing Communities", Ph.D. thesis, Plymouth: Plymouth University, p. 5–172.
99. Karel Prach and Lawrence R. Walker (2011), "Four opportunities for studies of ecological succession", *Trends in Ecology and Evolution*, 26:3, pp. 119–123, p. 120.
100. Eugene P. Odum (1969), "The Strategy of Ecosystem Development: An understanding of ecological succession provides a basis for resolving man's conflict with nature", *Science*, 164:3877, pp. 262–270, p. 262.
101. See Howard Aldrich (1975), "Ecological succession in racially changing neighborhoods", *Urban Affairs Quarterly*, 10:3, pp. 327–348; Howard Aldrich and A. J. Reiss Jr (1976), "Continuities in the study of ecological succession: Changes in the race composition of neighborhoods and their businesses", *American Journal of Sociology*, 81:4, pp. 846–866; Paul Frederick Cressey (1938), "Population succession in Chicago: 1898–1930", *American Journal of Sociology*, 44:1, pp. 59–69.
102. Odum, "The Strategy", pp. 264–266. Odum's theories have been tested and in some cases challenged since the publication

of his 1969 "Strategy of Ecosystem Development", and there are nuanced understandings of succession modelled at individual, population, community, and ecosystem levels. Nevertheless, his "classical" perspective is still cited as the baseline for successional theory.

103. Cyril Noall (1958), *Beloved St Ives*, St Ives: W. & J. Jacobs, p. 60.
104. Noall, *Beloved*, p. 60.
105. Thomas, "Social Resilience", pp. 4–143.
106. Anon. (1891), *West Briton*, 8 January.
107. Tony Pawlyn (2014), "The Cornish Sea Fisheries in the Nineteenth Century", in Philip Payton, Alston Kennerley, and Helen Doe (eds) (2014), *The Maritime History of Cornwall*, Exeter: University of Exeter Press, pp. 263–275, p. 268.
108. Paul Thompson (1983), *Living the Fishing*, London: Routledge and Kegan Paul, p. 26. Even so, fishing did not completely die out in St Ives; indeed, even in 1931, the town was still rated among the top three fishing ports, along with Newlyn and Porthleven. See Bennett, "A Century", p. 222.
109. Pawlyn, "The Cornish", p. 268.
110. Fikret Berkes uses a very relevant example when he discusses "transformative change". "Such changes involve shifts in the nature of the system, such as when a household adopts a new way of making a living, or when a coastal region may move from a fishing economy to a tourism-based economy". Fikret Berkes (2017), "Environmental governance for the anthropocene? Social-ecological systems, resilience, and collaborative learning", *Sustainability*, 9:7, 1232, p. 5.
111. See Robert P. McIntosh (1981), "Succession and Ecological Theory", in Darrell C. West, Herman H. Shugart, Daniel B. Botkin (eds) (2012), *Forest Succession: Springer Advanced Texts in Life Sciences*, New York: Springer, pp. 10–23.
112. Prach and Walker, "Four opportunities", p. 120.
113. Kennedy and Kingcome, "Disneyfication", p. 56.
114. Cross, *Painting*, pp. 198, 38.
115. Lanyon, "The Face".
116. Kennedy and Kingcome, "Disneyfication", p. 55.
117. Lanyon, "The Face".
118. Robert McCrum (1995), "St Ives: Rogue Gallery", *The Guardian*, 29 July, p. T36.

119. Deacon, "Imagining", p. 163.
120. Elihu Burritt (1868), "A Walk from London to Land's End and Back", London: Sampson Low, Son, and Marsden, p. 239.
121. Paul Jepson and Cain Blythe (2020), *Rewilding: The Radical Science of Ecological Recovery*, London: Icon Books, p. 60.
122. Dena R. Spatz and Nick D. Holmes (2021), "Islands: Fragile showcases of biodiversity", *UNESCO*, 16 June, https://www.unesco.org/en/articles/islands-fragile-showcases-biodiversity-0#:~:text=Islands%20represent%20only%20a%20small,with%20extinction%20by%20invasive%20species. Accessed 1 July 2023.
123. Spatz and Holmes, "Islands".
124. Ben Garrod (2016), "A Philippines island has the world's greatest concentration of unique mammals – here's why", *The Conversation*, 15 July, https://theconversation.com/a-philippines-island-has-the-worlds-greatest-concentration-of-unique-mammals-heres-why-62523. Accessed 1 July 2023.
125. McCrum, "St Ives", p. T36.
126. Bird, *The St Ives Artists*, p. 37.
127. Tovey, *The Artists*, p. 395.
128. Cross, *Painting*, p. 195.
129. Bird, *The St Ives Artists*, p. 272.
130. McCrum, "St Ives", p. T36.
131. Margaret Garlake (1994), "Artnotes: Imagining Cornwall", *Art Monthly* 180, p. 26.
132. Dean MacCannell (1992), *Empty Meeting Grounds: the Tourist Papers*, London: Routledge, pp. 303–306.
133. Landry et al., *The Art*, pp. 70–71.
134. McCrum, "St Ives", p. T36.
135. Evans and Shaw, "The Contribution", p. 36.
136. Steven Morris (2005), "Tate Expansion Makes Waves in St Ives", *The Guardian*, 8 October, https://www.theguardian.com/uk/2005/oct/08/arts.artsnews. Accessed 7 July 2023.
137. Axten, *Gasworks*, p. 53. See also Pole, "The Planning and Development", pp. 130–137.
138. Morris, "Tate Expansion".
139. Moya Kneafsey (1998), "Tourism and Place Identity: A case-study in rural Ireland", *Irish Geography*, 31:2, pp. 111–123, p. 113.
140. Gary Mason (1996) "Manufactured myths: packaging the exotic for visitor consumption", in Mike Robinson, Nigel Evans, and

Paul Callaghan (eds), *Tourism and Culture: Image, Identity and Marketing*, Northumberland: Centre for Travel and Tourism, pp. 121–135, pp. 130–131.
141. Davydd J. Greenwood (1978), "Culture by the pound: an anthropological perspective on tourism as cultural commoditization", in Valene L. Smith (ed.), *Hosts and Guests*, Oxford: Blackwell, 129–138, p. 135.
142. Tanya Gold (2022), "A Story of Two St Ives: 'It is Cursed and Confused by its Raging Beauty", *The Guardian*, 25 July, https://www.theguardian.com/travel/2022/jul/25/a-story-of-two-st-ives-it-is-cursed-and-confused-by-its-raging-beauty. Accessed 7 July 2023.
143. Milo Boyd (2023), "'I Live in the UK's Best Seaside Town—People are Clueless about its Dark Side'", *The Mirror*, 27 March, https://www.mirror.co.uk/travel/uk-ireland/i-live-uks-best-seaside-29559061. Accessed 7 July 2023.
144. Carthew, "Welcome".
145. Thomas, "Social Resilience", pp. 8–431.
146. Gold, "A Story", n.p.
147. Odum, "The Strategy", pp. 264–266.
148. Elton, *The ecology of invasions*, pp. 112, 31, 110, 130, 93, 151. As with many of the "classic" ecological theories, Elton's has in more recent years been queried, contested, and advanced, though theorists acknowledge that the basic premise of his ideas still stands. See Jason D. Fridley and Dov F. Sax (2014), "The imbalance of nature: revisiting a Darwinian framework for invasion biology", *Global Ecology and Biogeography*, 23:11, pp. 1157–1166.
149. Adrian Murphy (2017), "Tate St Ives—winning over the community to deliver a £20m extension", *Museums and Heritage*, 20 December, https://advisor.museumsandheritage.com/features/tate-st-ives-winning-community-deliver-20m-extension/. Accessed 2 July 2023.
150. Osterfield in Murphy, "Tate St Ives", n.p.
151. Anon. (2021), "St Ives secures £19.9 million investment from Government's Towns Fund", press release, *St Ives Town Deal*, 15 June, https://www.stivestowndeal.org.uk/media-release-19-9-million-for-st-ives/. Accessed 2 July 2023.
152. Anon., "St Ives secures", n.p.

153. F. Stuart Chapin III et al. (2009), "Resilience-Based Stewardship: Strategies for Navigating Sustainable Pathways in a Changing World", in F. Stuart Chapin III, Gary P. Kofinas and Carl Folke (eds) (2009), *Principles of Ecosystem Stewardship: Resilience-Based Natural Resource Management for a Changing World*, New York: Springer, pp. 319–337, p. 319.
154. F. Stuart Chapin III, Carl Folke and Gary P. Kofinas (2009), "A Framework for Understanding Change", in Chapin III, Kofinas, and Folke, *Principles*, pp. 3–28, p. 5.
155. F. Stuart Chapin III, et al. (2010), "Ecosystem stewardship: sustainability strategies for a rapidly changing planet", *Trends in ecology & evolution* 25:4, pp. 241–249, p. 242.
156. Chapin III et al., "Resilience-Based", p. 319.
157. James S. Martell (1927), "St Ives", *The Dalhousie Review*, 17:2, pp. 199–202, p. 199.
158. Kennedy and Kingcome, "Disneyfication", p. 54.
159. Anon. (2021), "St Ives secures", n.p.
160. Chapin III et al., "Ecosystem stewardship", p. 242.

Bibliography

Aldrich, Howard (1975), "Ecological succession in racially changing neighborhoods", *Urban Affairs Quarterly*, 10:3, pp. 327–348.

Aldrich, Howard, and A. J. Reiss Jr (1976), "Continuities in the study of ecological succession: Changes in the race composition of neighborhoods and their businesses", *American Journal of Sociology*, 81:4, pp. 846–866.

Anon. (1992), *Peter Lanyon: Air, Land and Sea*, London: South Bank Centre, p. 6.

Anon. (1993), "Secret garden found on ruined estate", *The Times*, 19 August, p. 16.

Anon. (2021), "St Ives secures £19.9 million investment from Government's Towns Fund", press release, *St Ives Town Deal*, 15 June, https://www.stivestowndeal.org.uk/media-release-19-9-million-for-st-ives/. Accessed 2 July 2023.

Anon. (n.d.) "Eden Project near St Austell", *Great British Gardens*, https://www.greatbritishgardens.co.uk/cornwall/item/eden-project-near-st-austell.html. Accessed 30 December 2024.

Anon. (n.d.), "1945–1960s: The St Ives School", *St Ives School of Painting website*, https://www.schoolofpainting.co.uk/about-us/st-ives-art-scene/the-st-ives-school/. Accessed 30 March 2023.

Axten, Janet (1995), *Gasworks to Gallery: The Story of Tate St Ives*, St Ives: Janet Axten and Colin Orchard.

Bateson, Gregory (1991), *A Sacred Unity: Further Steps to an Ecology of Mind*, New York: Cornelia & Michael Bessie Books.

Berkes, Fikret (2017), "Environmental governance for the anthropocene? Social-ecological systems, resilience, and collaborative learning", *Sustainability*, 9:7, 1232.

Binney, Marcus (2000), "Now Here's a Dome for the Millennium", *The Times*, 29 February, Arts and Sports.

Bird, Michael (2016), *The St Ives Artists: A Biography of Place and Time* (second edition), London: Lund Humphries.

Boyd, Milo (2023), "'I Live in the UK's Best Seaside Town—People are Clueless about its Dark Side'", *The Mirror*, 27 March, https://www.mirror.co.uk/travel/uk-ireland/i-live-uks-best-seaside-29559061. Accessed 7 July 2023.

Burritt, Elihu (1868), "A Walk from London to Land's End and Back", London: Sampson Low, Son, and Marsden.

Chapin III, F. Stuart, Gary P. Kofinas and Carl Folke (eds) (2009), *Principles of Ecosystem Stewardship: Resilience-Based Natural Resource Management for a Changing World*, New York: Springer.

Chapin III, F. Stuart, Stephen R. Carpenter, Gary P. Kofinas, Carl Folke, Nick Abel, William C. Clark, Per Olsson, D. Mark Stafford Smith, Brian Walker, Oran R. Young, Fikret Berkes, Reinette Biggs, J. Morgan Grove, Rosamond L. Naylor, Evelyn Pinkerton, Will Steffen, and Frederick J. Swanson (2010), "Ecosystem stewardship: sustainability strategies for a rapidly changing planet", *Trends in ecology & evolution* 25:4, pp. 241–249.

Cressey, Paul Frederick (1938), "Population succession in Chicago: 1898-1930", *American Journal of Sociology*, 44:1, pp. 59–69.

Cross, Tom (1995), *Painting the Warmth of the Sun: St Ives Artists 1939-1975*, Tiverton: Westcountry Books.

Crouch, David and Mark Toogood (1999), "Everyday Abstraction: Geographical Knowledge in the Art of Peter Lanyon", *Ecumene*, 6:1, pp. 72–89.

Deacon, Bernard (2001), "Imagining the Fishing: Artists and Fishermen in Late Nineteenth Century Cornwall", *Rural History*, 12:2, pp. 159–178.

Douglas, Anne (ed.) (2002), "On the Edge: Culture and the arts in remote and rural locations" (conference), Aberdeen: Robert Gordon University. https://ontheedgeresearch.files.wordpress.com/2011/12/ontheedge.pdf. Accessed 9 July 2023.

Durkheim, Émile (2013), *The Division of Labour in Society* (ed. Steven Lukes; trans. W. D. Halls) (second edition), Houndmills, Palgrave Macmillan.

Eden Project (n.d.), "Learn with Eden", *Eden Project*, https://www.edenproject.com/learn. Accessed 27 December 2024.

Eden Project (n.d.), "Mediterranean Biome", *Eden Project*, https://www.edenproject.com/visit/things-to-do/mediterranean-biome. Accessed 27 December 2024.

Eden Project (n.d.), "Our Origins", *Eden Project*, https://www.edenproject.com/mission/origins. Accessed 27 December 2024.

Eden Project (n.d.), "The Invisible Rainforest", *Eden Project*, https://www.edenproject.com/learn/eden-at-home/the-invisible-rainforest. Accessed 27 December 2024.

Elton, Charles S. (1969), *The ecology of invasions by animals and plants*, Trowbridge and London: Redwood Press.

Evans, Graeme, and Phyllida Shaw (2004), *The Contribution of Culture to Regeneration in the UK: A Review of Evidence*, London: Department for Culture, Media, and Sport.

Fridley, Jason D., and Dov F. Sax (2014), "The imbalance of nature: revisiting a Darwinian framework for invasion biology", *Global Ecology and Biogeography*, 23:11, pp. 1157–1166.

Garlake, Margaret (1994), "Artnotes: Imagining Cornwall", *Art Monthly* 180, p. 26.

Garrod, Ben (2016), "A Philippines island has the world's greatest concentration of unique mammals – here's why", *The Conversation*, 15 July, https://theconversation.com/a-philippines-island-has-the-worlds-greatest-concentration-of-unique-mammals-heres-why-62523. Accessed 1 July 2023.

Gold, Tanya (2022), "A Story of Two St Ives: 'It is Cursed and Confused by its Raging Beauty", *The Guardian*, 25 July, https://www.theguardian.com/travel/2022/jul/25/a-story-of-two-st-ives-it-is-cursed-and-confused-by-its-raging-beauty. Accessed 7 July 2023.

Hamilton-Jenkin, A.K. (1962), *The Cornish Miner: An Account of his Life Above and Underground from Early Times*, London: George Allen and Unwin.

Hardy, Peter (2004), "Dome Sweet Dome", *The Daily Mail*, 11 August, p. 59.

Hargreaves, Clare (2022), "Cornwall's sleeping beauty: the tale of Heligan's lost gardens", *The Observer*, 17 April, https://www.theguardian.com/lifeandstyle/2022/apr/17/the-tale-of-heligan-lost-gardens-cornwall-treasure-rediscovered-30-years-ago. Accessed 30 December 2024.

Harvey, Bryan (1970), "Who started it all?", *British Journal of Industrial Medicine*, 27:1, pp. 81–83.

Huang, Patricia (2005), "Where Art, Architecture, and Landscape Meet", *National Museum of Natural Science*, 19, pp. 7–22.

Institute of Civil Engineers (2024), "Eden Project", *Institute of Civil Engineers*, https://www.ice.org.uk/what-is-civil-engineering/infrastructure-projects/eden-project. Accessed 27 December 2024.

James, Beryl (1988), *Tales of the Tinners' Way*, Redruth: Dyllansow Truran.

Jepson, Paul and Cain Blythe (2020), *Rewilding: The Radical Science of Ecological Recovery*, London: Icon Books.

Keenan, Steve (2001), "How long is the queue for Eden?", *The Times*, 28 April, Section 2, p. 7.

Kennedy, Neil, and Nigel Kingcome (1998), "Disneyfication of Cornwall—developing a Poldark heritage complex", *International Journal of Heritage Studies*, 4:1, pp. 45–59.

King, Deborah (2001), "Genesis of the Garden", *The Times*, 14 March, p. 3.

Kneafsey, Moya (1998), "Tourism and Place Identity: A case-study in rural Ireland", *Irish Geography*, 31:2, pp. 111–123.

Landry, Charles, Lesley Greene, Francois Matarasso, and Franco Bianchini (1996), *The Art of Regeneration: Urban Renewal through Cultural Activity*, Stroud: Comedia.

Long, Richard C. (2022), *The St Ives Branch Line: A History*, Barnsley: Pen and Sword Books Ltd, n.p.

Lübbren, Nina (2001), *Rural Artists' Colonies in Europe, 1870-1910*, Manchester: Manchester University Press.

MacArthur, Robert H. (1972), "Strong, or weak, interactions", *Transactions of the Connecticut Academy of Arts and Sciences*, 44: 177, p. 88.

MacCannell, Dean (1992), *Empty Meeting Grounds: the Tourist Papers*, London: Routledge.

Martell, James S. (1927), "St Ives", *The Dalhousie Review*, 17:2, pp. 199–202.

McCrum, Robert (1995), "St Ives: Rogue Gallery", *The Guardian*, 29 July, p. T36.

Mills, L. Scott, Michael E Soulé, and Daniel F. Doak (1993), "The Keystone-Species Concept in Ecology and Conservation", *BioScience*, 43:4, pp. 219–224.

Morris, Steven (2005), "Tate Expansion Makes Waves in St Ives", *The Guardian*, 8 October, https://www.theguardian.com/uk/2005/oct/08/arts.artsnews. Accessed 7 July 2023.

Morrison, Richard (2001), "Why I Created Eden", *The Times*, 12 March, p. 2.

Mouquet, Nicolas, Dominique Gravel, Francois Massol, and Vincent Calcagno (2013), "Extending the concept of keystone species to communities and Ecosystems", *Ecology Letters*, 16:1, pp. 1–8.

Murphy, Adrian (2017), "Tate St Ives – winning over the community to deliver a £20m extension", *Museums and Heritage*, 20 December, https://advisor.museumsandheritage.com/features/tate-st-ives-winning-community-deliver-20m-extension/. Accessed 2 July 2023.

Murt, Eddie (1994), *Downlong Days: A St Ives Miscellany*, St Ives: St Ives Printing and Publishing Co.

Naeem, Shahid, and F. Stuart Chapin III (1999), "Biodiversity and Ecosystem Functioning: Maintaining Natural Life Support Processes", *Issues in Ecology*, 4:11.

Noall, Cyril (1958), *Beloved St Ives*, St Ives: W. & J. Jacobs.

Noall, Cyril (1977), *The Book of St Ives*, Chesham: Barracuda Books Ltd.

O'Neill, Robert V. (1986), *A hierarchical concept of ecosystems*, Princeton: Princeton University Press.

Odum, Eugene P. (1969), "The Strategy of Ecosystem Development: An understanding of ecological succession provides a basis for resolving man's conflict with nature", *Science*, 164:3877, pp. 262–270.

Paynter, William H. (1999), *Old St Ives: The reminiscences of William Paynter*, Penzance: Oakmagic Publications.

Payton, Philip, Alston Kennerley, and Helen Doe (eds) (2014), *The Maritime History of Cornwall*, Exeter: University of Exeter Press.

Penhallurick, Roger David (2017), *Tin in Antiquity: its mining and trade throughout the ancient world with particular reference to Cornwall*, London: Routledge.

Prach, Karel, and Lawrence R. Walker (2011), "Four opportunities for studies of ecological succession", *Trends in Ecology and Evolution*, 26:3, pp. 119–123.

Robinson, Mike, Nigel Evans, and Paul Callaghan (eds), *Tourism and Culture: Image, Identity and Marketing*, Northumberland: Centre for Travel and Tourism.

Simpson, Dave (2021), "Inspired by a sink full of soap bubbles: how we made the Eden Project", *The Guardian*, 15 March, https://www.theguardian.com/artanddesign/2021/mar/15/how-we-made-eden-project-cornish-lost-civilisation-soap-bubbles. Accessed 29 December 2024.

Smit, Tim (1997), *The Lost Gardens of Heligan*, London: Victor Gollancz.

Smit, Tim (2001), *Eden*, London: Bantam Press.

Smith, Valene L. (ed.), *Hosts and Guests*, Oxford: Blackwell.

Sparrow, Elizabeth, and Michael Williams (1976), *About St Ives*, Bodmin: Bossiney Books.

Spatz, Dena R. and Nick D. Holmes (2021), "Islands: Fragile showcases of biodiversity", *UNESCO*, 16 June, https://www.unesco.org/en/articles/islands-fragile-showcases-biodiversity-0#:~:text=Islands%20represent%20only%20a%20small,with%20extinction%20by%20invasive%20species. Accessed 1 July 2023.

Thomas, Huw (2017), "Social Resilience in Cornish Fishing Communities", Ph.D. thesis, Plymouth: Plymouth University.

Thompson, Paul (1983), *Living the Fishing*, London: Routledge and Kegan Paul.

Toth, Susan Allen (2001), "More Than One Eden in Cornwall", *New York Times*, 22 July, Section 5, p. 8.
Tovey, David (2003), *Creating a Splash: The St Ives Society of Artists—The First 25 Years (1927-1952)*, Tewkesbury: Wilson Books.
Tovey, David (2008), *St Ives Art pre-1890: The Dawn of the Colony*, Tewkesbury: Wilson Books.
Tovey, David (2009), *St Ives (1860-1930): The Artists and the Community: A Social History*, Tewkesbury: Wilson Books.
Warman, Christopher (2001), "Paradise Found", *The Times*, 10 November, Section 3, p. 13.
Watkiss, Reg (2005), *Reflections of Old Cornwall: A Further Selection of Historic Photographs*, Tiverton: Halsgrove.
Webster, Ben, and Helen Rumbelow (2002), "Newquay ready for the weekend jet set", *The Times*, 1 February, p. 3.
West, Darrell C., Herman H. Shugart, Daniel B. Botkin (eds) (2012), *Forest Succession: Springer Advanced Texts in Life Sciences*, New York: Springer.
Whybrow, Marion (1994), *St Ives 1883–1993: Portrait of an Art Colony*, Woodbridge: The Antique Collectors' Club.
Wilkinson, David (2005), "St Ives: The Legendary Quality of Light?", in Laura Newton (ed.) (2005), *Painting at the Edge: British Coastal Art Colonies, 1880-1930*, Bristol: Sansom & Co., pp. 65–77.
Williams, H.V. (1969), *Cornwall's Old Mines*, Truro: Tor Mark Press.

CHAPTER 6

Charismatic Species

When "The Woman Who Planted Trees"[1] was awarded the Nobel Peace Prize in 2004, she received it with a characteristically passionate response:

> The tree for me is hope, it is the future. When I look at the tree I see promise. It starts from a seed. Eventually it becomes a huge tree. It becomes an ecosystem in itself. It is a home for other species. It is a symbol of many things. A tree is everything to me.[2]

Wangari Maathai led her first community tree-planting event on 5 June 1977, World Environment Day, planting seven trees that she had propagated in her Kenyan back yard. In the same year, she founded the Green Belt Movement (GBM), built on a platform of environmental and community ideals. It was set up "to strive for better environmental management, community empowerment, and livelihood improvement, using tree-planting as an entry point", explains its current executive director, Aisha Karanja.[3] Since it started, GBM has overseen the planting of more than 51 million trees, and helped over 30,000 women develop trades in forestry and associated conservation skills.[4]

The Green Belt Movement defines a green belt as a newly-planted "woodlot" of at least 1000 trees on public land, and 100 trees on private land. These are overseen by community-based "tree-nursery" groups, so that "a large section of the community (or stakeholders) is mobilized", and "operations do not heavily rely on a small minority". Maathai

K. Savage and D. Symonds, *Ecologies of Community in Performance*,
https://doi.org/10.1007/978-3-031-94343-0_6

encouraged groups to "form freely—without coercion", for democratic processes to establish group leadership, and for regular communication between group members. Once a group has been formed, members look for an appropriate piece of land, and prepare the site, which includes "digging trenches, making terraces, laying seedbeds and building fences", "searching for seeds and cuttings, transplanting seedlings, supervising work [...] and carrying out follow-ups".

Maathai continues: "When tree-nursery groups have sufficient seedlings that are ready for issue to the public, they make announcements in public gatherings such as church services and schools, in barazas (or local administration meetings) and also through house visits". Those interested are asked to "dig holes with diameters and depths of about two feet", and "after the holes are inspected and approved by a member of a given tree nursery, the seedlings are issued". Once planted, there are two follow-up visits. The first, after a month, is to check that seedlings have been planted properly; the second, after 3–6 months, is to ensure they have become established and are likely to survive. If the growths are considered acceptable, GBM will purchase them, enabling a grassroots income from the project and sustainable continuity for the enterprise.[5]

GBM's tree planting has not always been without incident: Maathai was attacked violently on several occasions, especially before her Nobel recognition gained her international attention. "Why was tree planting so dangerous?", asks Stuart Jeffries; "The main reason was that it flew in the face of then president Moi's policy of development". This highlights the political aspect of Maathai's initiative, both in developing and empowering communities, but also in standing against the corruption or environmental complacency of the state. On one occasion, GBM initiated a green belt planting on common land that had been earmarked for a new development; this was effectively "Nairobi's only remaining forest", yet it was being turned over to luxury housing. "As she raised her hoe to dig a hole for the sapling, she and her friends were attacked by 200 guards armed with machetes, whips, bows and arrows, and swords", reports Jeffries.[6] Though standing in the way was clearly a great risk, Maathai over the years managed to thwart dozens of environmentally-damaging projects while rallying local communities to her cause.

As Marc Michaelson notes, the use of "village-based self-help work is a longstanding tradition in much of the South, as local people have pooled their collective abilities to meet community needs for centuries". To do this through planting trees is a relatively straightforward way of engaging communities: “Tree planting is a low technology undertaking”, he writes; “the necessary skills and training can be executed locally without outside ‘experts’”. More than this, “it is also inexpensive, for seeds are locally available and nursery establishment is labor intensive, not resource intensive”. Maathai’s oversight of the projects ensured that local community agency—particularly that of women—was at its core.[7] But the significance of GBM’s work is that it has effectively upscaled existing village self-help initiatives to enable “movements on a regional, national, or international scale”, and thereby to bring about “society-wide social change”.[8] Today, the organization claims to have seeded over 4000 community groups, 70% of which are focused on women.[9] As Heather Ball notes, “Wangari’s idea gave confidence to women who did not realize they had the ability to bring about changes in their communities. They are now empowered to take charge of their lives and the lives of their families”.[10] As she says, “Wangari has proved that one person can make a difference, that tiny seeds can become forests, and great ideas can become reality”.[11]

One of the scheme’s motivations has been to encourage communities to replace tree stock.[12] Fewer than 7% of households have access to electricity in Kenya, which means that 90% are reliant on firewood, and as a result, “the closed canopy forest has been reduced to less than 2%”.[13] As Eileen E. Schell notes, this is a particular concern, because “deforestation and forest degradation are the second leading cause of global warming, responsible for about 15% of global greenhouse gas emissions”.[14] Accordingly, whenever Wangari Maathai hosted a tree-planting ceremony, she would insist on participants reciting a ritual promise in which they resolved to save their land from desertification by planting trees.[15]

But Maathai’s initiatives were not only to ensure a sustainable source of fuel; they also had a more ambitious community impulse, “to generate income for rural women”; “to promote environmental consciousness among the youth”; “to empower people at the grassroots”; “to demonstrate the capacity of women in development”; “to curb soil erosion”; and “to disseminate information on environment conservation”.[16] In addition, she sought to educate communities about food security, sustainable farming, and water harvesting[17]; to raise civic awareness of issues such as human and environmental rights[18]; and to lead advocacy against forest destruction and illegal development by corporations or the state.[19] “I guess we [are] planting people simultaneously as we plant trees”, one of her admirers comments.[20]

Given Maathai's work, it is not surprizing that the tree has emerged as both a literal and rhetorical symbol of the Green Belt Movement, "perhaps second only to the rhetorical presence of Wangari Maathai herself".[21] In the branding of the movement, the huge, green canopy of an African Umbrella Tree takes pride of place, silhouetted against the African continent. It's not that Maathai held a particular affinity for the Umbrella Tree; it's just that, like other examples that have become rhetorically or symbolically significant—the Tree of Life; Ygdrassil; the Giant Redwood; the English Oak; the Indian Bodhi; the Japanese Bonsai—the tree, for Maathai, became in ecological terms *charismatic*.

In this chapter, the idea of the *charismatic species* is going to offer our perspective. We'll introduce the concept, noting how a focus on the "cute" or the "cuddly" can divert attention away from other endangered species. In terms of applying this to the arts, we'll discuss how the concerns of privileging one thing at the expense of another might be brought to the UK arts funding landscape, and to our cultural fascination with Shakespeare. Next we'll observe how global conservation efforts have often been led by well-known personalities who we consider as *charismatic leaders*. Here we'll explore the differing levels of profile enjoyed by celebrity environmentalists like Greta Thunberg, and lesser-known climate activists such as Vanessa Nakate in the global South. We'll consider how by privileging certain narratives over others, we risk losing out on valuable conservation initiatives. Finally we'll explore how the idea of the charismatic species—in some ways a *single species* perspective—can be viewed in terms of community. We'll link this idea, through an observation of Thunberg's huge global following, with the idea of an *ecosystem approach* to conservation.

Charismatic Species

The idea of the "charismatic species" emerged in the 1980s to describe animals or plants that had become "symbols and rallying points for major conservation initiatives".[22] Animals are considered charismatic or "flagship" species because they are "beautiful" or "cute"; "dangerous" or "impressive"; "rare" or "endangered".[23] Perhaps the most obvious example is the giant panda, which became the symbol of the World Wildlife Fund in 1961, and which is still used on the organization's logo. We're enchanted by these over-sized teddy-bears; we coo over them in zoos; we celebrate the birth of offspring; and like other charismatic animals, we marvel over them in nature documentaries.

Pandas are not the only charismatic animals—we're also drawn to lions and tigers, koalas, and llamas, even otters and chipmunks. There's not a definitive list—after all, charisma is a subjective quality. Nor are these species charismatic in the way that humans are: in human terms, the notion of charisma is a unique trait which epitomizes someone's appeal as an individual; in the natural world, "charisma lies not in a mysterious and intriguing individual but in the relationship that emerges between an entire species and human observers".[24] Still, we use the term to imply charisma in every elephant, every dugong, and every squirrel. Nor is it that these species possess particular traits that make them charismatic; they just appeal to us—or at least, they are *framed* in a way to appeal. Perhaps it's because we project human-like emotions onto them; perhaps it's because we see them anthropomorphized in films and books; perhaps it's because they are represented by evocative imagery and marketing[25]; perhaps it's because they are introduced to us by celebrity environmentalists.

Clayton Fordahl recognizes that "charismatic species are socially constructed".[26] This happens, he suggests, because of the ways in which they are presented—or performed—to us by experts, and it is perpetuated because we reiterate those same performances in our representations of the species. "The charisma of charismatic megafauna is not innate", he writes, "but requires both emotional connection and reputation management on the part of reputational entrepreneurs".[27] The cultural performances of certain species are particularly compelling, he argues, because of "emotional resonance": "Some species simply hold more emotional or cultural sway over humans than others", he notes, adding that "the coincidence of certain variables (size, diet, range, behaviour, etc.) is strongly associated with charisma".[28] This helps us understand "the general appeal of species that are phylogenetically or physiognomically closer to humans".[29] As one study shows, "Characteristics such as large bodies, exquisite features, distinctive behaviours, striking colours, and pleasant sounds or fragrances are thought to be broadly appealing to humans and influence our perception of species".[30]

Jamie Lorimer, in his study of "nonhuman charisma", also understands it to be a quality emerging from our connection to a species, rather than something that animals or plants possess themselves.[31] He argues that our particular human characteristics and habits—the fact that we are hominids of a certain size who tend to live on land and to be active during daylight—predispose us to feel connected to certain species that intersect with these qualities. This explains why we relate to gorillas and chimpanzees, and constitutes what he terms their *ecological charisma*. He goes on to suggest that the close connections experienced by conservationists or naturalists are thanks to them "tuning in", "learning to be affected", or even (in Deleuzian terms) "becoming-animal" with the species. "Ecological charisma would be very different if we had gills or night vision", he reasons; because we are the way we are, "some animals [...] are easier to become than others".[32] The second element Lorimer points to is *aesthetic charisma*—our engagement with a species because of its looks, or perhaps the way it sounds. This works in tandem with ecological charisma, and Lorimer distinguishes between the *cuddly charisma* we have noted, that spawns close connections based on cuteness, and a *feral charisma*, that "trigger[s] strong and visceral feelings of disgust and even panic".[33] The final element he discusses is *corporeal charisma*, a combination of the "epiphany" that triggered someone's connection to an animal in the first place, and the "jouissance" that comes later from classifying, collecting, or identifying examples. "Together", he writes, "ecological, aesthetic, and corporeal charisma chart the facets of nonhuman charisma in UK biodiversity conservation and thus sketch an alternative taxonomy to the nonhuman realm".[34] Crucially, Lorimer emphasizes that his thinking "does not elevate nonhuman organisms to the humanist status of subjects. Neither does it reduce them to the instrumental domain of the objects of natural science".[35] Instead, he argues, it allows us to understand our relationships with nonhuman species to be part of a continuum within the ecosystem of the more-than-human world. He refers to charismatic species as "boundary objects", making the point that they allow "the acceptable entry of emotion and vernacular understandings of nature into 'objective' conservation biology".[36]

All of these examples highlight the connection between what we might call "cuddlability" and conservation. "Cuddlability [...] makes children and seals so much more successful than amoebas and ozone layers as 'securitisable' goods", remark Jeroen Warner and Richard Meissner.[37] That's a connection that the conservation industry exploits: supporters who "adopt" a giant panda through the WWF today, from as little as £3 a month, receive "a cuddly toy", "a welcome pack", "regular updates", and "extra goodies",[38] suggesting that the organization is pandering to the charismatic image of this creature, and the inference that the public can easily bring a panda into the family. Conservation organizations and charities are thus able to "draw financial support more easily" for charismatic species than they are for other animals,[39] note researchers. As Vernon H. Heywood remarks, "It is often easier to mobilize public opinion in support of such species than for more subtle ecosystem conservation priorities".[40] Thus the use of charismatic species in conservation has its merits. Other conservationists agree: "From a practical stand-point", write Nigel Leader-Williams and Holly T. Dublin, "the image helps to focus attention on a single species, rather than on the fuzzy concept of species richness, or fuzzier concept of genetic diversity".[41] What's more, by protecting the flagship species, there can be a trickle-down effect: "Many other species and entire ecosystems are preserved as a secondary benefit under the 'umbrella' of these high-visibility, so-called 'charismatic' vertebrates", confirm Meffe and Carroll.[42] "The campaigns using them as symbols are excellent examples of the way in which key groups of animals can be used to sell the whole issue of conservation, both in tropical countries and in the developed world", write Leader-Williams and Dublin; in short, "there is a strategic, rather than an ecological or biological, sense to the original use of flagship species".[43]

Charismatic Leaders

In the wider environmental context, we also find such issues championed by another group of charismatic individuals: celebrity environmentalists. These are the "reputational entrepreneurs" noted by Fordahl. In the most high-profile instances they can command a great deal of media attention, thereby adding promotional weight to environmental concerns.

One example is former vice-president Al Gore, who since the 1980s has used his profile to promote environmental awareness. More recently, he has written publications on global warming and the environment, the most well-known of which is his book and documentary film *An Inconvenient Truth* (2006). It focuses on the actions individuals and corporations can take to reduce emissions. The film in particular has been generally well received, and has arguably "set the precedent for how we talk about climate change".[44] In 2007 Gore was also awarded the Nobel Peace Prize. Nowadays he promotes The Climate Reality Project, an awareness initiative to upskill climate ambassadors, which we attended in June 2024. The three-day event in Rome included speakers from local and regional government, representatives from NGOs, political leaders, artists, and other celebrities. One, the Spanish model Jon Kortajarena, calls himself an "imperfect environmentalist",[45] but he focuses on giving visibility to projects that promote second-hand fashion items in an attempt to change his industry from the inside. Kortajarena is self-aware; his followers "engage" with him primarily because he is a charismatic celebrity, but acknowledging his environmental imperfections is an endearing trait. Even so, it was Gore everyone had come to see. His training as a public speaker was evident, revealing the rhetorical skills of a seasoned professional. Shaping his speech with texture and dynamics, he used performance to inspire and galvanize his audience. The audience felt a frisson when he was on stage, acknowledging the charisma of his presence, and the cachet of attending his programme. Gore is genuinely passionate about climate change, and his goals to raise awareness, to get people to act, and to hold governments responsible for leading the way, especially by stopping the funding that goes to fossil fuels, are precisely the actions that are needed for us to live in a healthier world and to make it sustainable.

Just as conservation can benefit from the profile of charismatic species, then, it can also benefit from the focus that charismatic leaders bring to the debate.

Unintended Consequences

Beneficial though such focus might be, there are nevertheless critiques of ecological charisma, which can disproportionately benefit some species above others. "Programmes to save the tiger, the giant panda, or the California condor are well publicized examples of [...] single-species conservation priorities", writes Heywood[46]; the same might be said of high-profile campaigns like "Save the Whale". This means that there's an imbalance: some species get attention, donations, and support, while few care for less charismatic species, for those that aren't as visible, or even those that are just plain small. There is an associated imbalance caused by other emphases that are disproportionately advantaged: high-profile events or emergencies (wildfires; floods) can dominate media exposure at the expense of other environmental needs, for example, while Krause and Robinson discuss certain "charismatic landscapes" that are privileged (coral reefs, rainforests), and "charismatic solutions" that receive priority (reforestation) over others.[47] This ultimately reflects what researchers have called "an anthropocentric, two-tiered conservation agenda favoring a very small percentage of biodiversity".[48] More than this, support for such causes, and especially donations, can sometimes be misplaced, as although the charismatic may get lots of money, it's not always clear what they might need by way of support. "Citizens' support for a specific conservation policy tends to be based on superficial characteristics of an animal rather than its ecological value and taxonomic uniqueness", writes Gunnthorsdottir.[49] It may be cute, but is it really threatened? What about far less cuddly species, such as the African Wild Dog (endangered) the Humphead Wrasse (endangered), the Bigeye Tuna (vulnerable), or the Mountain Plover (near- threatened), species whose vulnerability is overshadowed by more charismatic examples who receive the attention?[50] And what about even less familiar species whose plight is not featured even by the WWF (though they are listed on the IUCN Red List of Threatened Species): the Island Oak (endangered), a tree found only on a handful of islands off the coast of Mexico[51]; Spawls' Boolee (critically endangered), a forest-dwelling worm-like amphibian for which there is no systematic monitoring scheme or action recovery plan[52]; Bob's Robber Frog (possibly extinct), a native of Panama, of whom there are thought to be fewer than 50 specimens remaining[53]; or the Lost Spathoglottis (possibly extinct), a pretty yellow flower from India whose "known population is now potentially lost due to habitat destruction by road construction and the expansion of shifting agriculture"?[54] There are many other examples of critically endangered but "lower trophic level"[55] or "uncharismatic"[56] species, though even this chapter has to be selective, elevating some and diminishing others.

Just as problematically, a focus on individual species can distract resources away from "more subtle ecosystem conservation priorities",[57] broader but perhaps less visible concerns that may affect a whole habitat or biome. This alerts us to two different conservation approaches: "one is to focus on species of special interest, and the other is to focus on ecosystems".[58] Single species conservation is an "expedient" conservation strategy, note Runge et al., an "umbrella" approach with the commendable consequence that many species might benefit from its trickle-down effect.[59] On the other hand, research considering such effects has also revealed downsides: one study of a giant panda preservation in China found that the panda's protection came at the detriment of the endangered Asiatic black bear and forest musk deer[60]; another noted a decline in tigers, wolves, leopards, and snow leopards. "A more tailored conservation approach for large carnivores is urgent", it concluded, noting "the inadequacy of the current panda-centric conservation strategy in preserving ecosystem functionality comprehensively".[61] Writing about the issue, Daniel Simberloff agrees: "The only way to deal with this challenge is then seen to be to manage at least entire ecosystems, if not whole landscapes, by unified methods designed to save all their inhabitants at one time".[62] As he says, "for some persons, species richness itself is ecosystem health—the tacit goal of all conservation should be species richness, and, ipso facto, a rich community is a healthy one".[63]

Charismatic Species in the Arts

It's time for us to apply our ecological perspective to the arts.

Consider how we have come to view charismatic species: they are popular and appealing plants and animals that benefit from wide visibility; this brings them support and resources which means their conservation status is secured; they can also serve as attractive figureheads that bring our attention to environmental concerns. On the other hand, they can gain disproportionate focus that distracts us from other less charismatic plants and animals; those species in turn can be disproportionately deprived of support and resources, which can lead to an imbalance that precipitates environmental problems.

Let's take a look at two examples which might be considered charismatic species—or at least, flagship emphases—in the arts: the case of Arts Council funding in the UK, and the case of Shakespeare as a national institution.

The Case of Arts Council Funding

Many countries have policies supporting the arts, and the UK's has been underpinned by the work of the Arts Council, operating since 1946. In principle, its mission is to support artistic practice, and it has frequently turned to an ethos of "great art for everyone" to guide its vision. From the beginning, the Arts Council championed flagship events or organizations as a way of celebrating the cultural legacy of the UK. It was central to supporting the Festival of Britain in 1951, and since the mid-1950s, it has included significant organizations such as the Royal Opera House and the Royal Court Theatre as key beneficiaries of its funding. Its strategic framework for 2010–2020, initially titled *Great Art for Everyone*,[64] identified a suite of National Portfolio Organisations (NPOs).

These are flagship organizations, and according to the Arts Council's definition, "leaders in their areas, with a collective responsibility to protect and develop our national arts and cultural ecology".[65] But this raises questions: are these flagship emphases just charismatic species in another setting? And if they are, what's at stake? Can we rely on the trickle-down of "collective responsibility" to protect and develop our arts ecology? And if not, who misses out? What is the cultural equivalent to the Lost Spathoglottis, whose "known population is now potentially lost due to habitat destruction?".

The Arts Council funding for 2023–2026 disburses an annual £444m among 985 NPOs. A quarter of this (£114m) is allocated to just the top ten beneficiaries (see Fig. 6.1). These, we might suggest, could be the charismatic species of the arts: much-loved edifices of cultural attraction that receive the lion's share of the funding.

Among these organizations, there is a significant emphasis on particular art forms perennially favoured in ACE funding: opera (£37m out of that £114m), theatre (£31.4m), and ballet (£14m). Meanwhile, the proportion of funds disbursed around the country privileges London (34% of the total), then the metropolitan centres around Birmingham

Annual allocation of ACE funding, 2023-2026			
Organization	**Amount**	**Region**	**Proportion**
Royal Opera House	£22,268,584	London	34.12%
Southbank Centre	£16,828,042	West Midlands	12.64%
National Theatre	£16,156,916	Yorks and the Humber	12.03%
Royal Shakespeare Company	£15,259,706	North West	11.20%
Opera North Limited	£10,677,102	South East	7.21%
Manchester International Festival	£9,908,150	South West	6.53%
Birmingham Royal Ballet	£8,036,194	North East	5.49%
English National Ballet	£6,011,921	East Midlands	5.08%
BookTrust	£5,763,099	East of England	4.65%
Welsh National Opera	£4,000,000	Wales	1.06%

Fig. 6.1 Top ten recipients of annual ACE NPO funding, 2023–2026; regional proportion of funds, 2023–2026

(West Midlands, 12.6%), Leeds/Sheffield/Bradford (Yorkshire and the Humber, 12%), and Manchester (North West, 11.2%), with the remainder (just 30%) scattered around other regions.[66] In terms of this particular model of arts funding, then, there are two further types of charismatic profiling: art forms themselves (opera, theatre, and ballet, but not basket-weaving or origami), and regional priorities (London and the metropolitan centres, but not the provinces).[67]

Why does this matter? Aside from noting some evident imbalances in the funding landscape, what can we learn from our ecological perspective about the significance of these charismatic species in the arts? Let's not forget that studies of panda conservation in China revealed that prioritizing a single species led to an adverse impact on several other species populations, from black bears to tigers and leopards. Although we're not suggesting that privileging opera in the UK arts funding model will necessarily lead to the direct collapse of other specific forms of art, we recollect the ecological call for "more comprehensive and effective conservation decisions",[68] and the more holistic ecological warnings about damage done to "ecosystem functionality",[69] "species richness", and "ecosystem health".[70]

The Case of Shakespeare

Let's take another example, this time of the artworks themselves. Notice that among the recipients of funding in the table above is the Royal Shakespeare Company, dedicated to the legacy of "our national playwright".[71] The RSC is the fourth largest recipient of ACE NPO funding, receiving an annual subsidy of £15.3m, roughly equivalent to the entire country's literature allocation (£16m), and more than three times the allocation given to libraries (£4.2m). Simplistically, we could say that the work of this single literary figure is given a status comparable to that of all the other writers the country has ever had put together. Shakespeare, by this reckoning, is undoubtedly a charismatic species; and so too are Shakespeare plays—*Hamlet*, *Othello*, *Macbeth*, *R&J*—much-loved stories perennially performed on the theatre circuit, studied for school exams, and referenced throughout our culture. Certainly an "ecology" of Shakespeare might be said to have developed over the last four hundred years, especially since the nineteenth century: Shakespeare and his plays feed into the curriculum at every level; they dominate the theatrical landscape not only of the RSC but also of other regional and national organizations;

they underpin our idioms and cultural points of reference; and they even offer an alternative access to our early modern history. "Shakespeare is a key pillar of English teaching in the UK (and in other countries around the world)", write Victoria Elliott and Sarah Olive; "he has been the only compulsory author since the inception of the National Curriculum in 1989".[72] Meanwhile, according to Elizabeth Schafer in 2000, he is "the playwright who is still the most produced, high status and high profile in British theatre".[73]

Yet to focus on Shakespeare means also to marginalize other voices. This is not a new revelation, but it is a criticism corroborated by our ecological perspective. Nor is the singular prioritization of Shakespeare, from that point of view, the image of a functional or healthy ecosystem. That is why a slew of revisionist histories has sought to reclaim the contributions of other voices, such as those of women to the early modern stage.[74]

But restoring ecosystem health is not just about seeing beyond these charismatic species; it's also about seeing past the equally charismatic patterns of the form. For just as pathways get ingrained into a landscape and communities sedimented into a place, the rhythms, textures, and drives of artistic practice become standardized. It's partly thanks to the charismatic legacy of Shakespeare, for example, that we have a largely literary tradition on (our: UK) stage, that we privilege speech and drama as performance skills in the theatre, and that we have a particular way of behaving on stage and off; likewise, it's due to the charismatic legacy of Mozart and Beethoven that (our: European) music is diatonic, and because of the Hollywood Studio System that (our: Western) cinema is structured as a narrative form. Equally, it's because "our" (UK, European, Western) culture more widely gets privileged that other nations become marginalized in our perceptions. We saw something very similar in chapter two when we noticed how the language and form of academia has slipped into an "ideology": it's thanks to the imbalanced profiling of Western mindsets that epistemologies from the global South have been marginalized in multiple areas of our cultural discourse.[75]

For Crystal Abidin and her colleagues, celebrity environmentalists have a similar effect: they "foster emotional responses in the audience and wider public ecosystems" which "implicitly and explicitly frame how the public should think about, feel, and relate to the environment". Conservation becomes a matter transacted through impulses of emotional response and charitable donation focused on single species or single issue campaigns, rather than through strategic, holistic planning. Ultimately, suggest Abidin et al., this "contributes to the rearticulation of human environment relationships that encourage individual consumer action over state-supported structural change".[76] Transfer that argument to UK arts and we encounter exactly the sort of neoliberal attitudes towards culture that have led to what numerous commentators have signalled to be a "crisis".[77] If the state absolves itself of the responsibility to protect the cultural ecosystem, as William Rees-Mogg advocated when he was Chairman of the Arts Council in the 1980s, the landscape is at the mercy of market forces.[78] There's a risk of eroding anything that doesn't attract charismatic attention, or that doesn't fit an economic interpretation of success.

We can never know how theatre, music, or film might have developed differently had we been able to see beyond the charismatic patterns of form. There's nothing intrinsically "wrong" with them per se, just as there's nothing intrinsically wrong with our instinct to save the giant panda, or our recourse to Shakespeare or Mozart as gateways to the arts. But moving away from a single species model to an ecosystem approach might be a way of unlocking that which has remained unseen: the potential of the alternative, the experimental, the innovative, and the different; the synergistic, the collaborative, the interdisciplinary, and the dialogic. This is in part what we are trying to unlock through this book, and as we will see, it's ecologies of community—such as the contemporary art ecosystem we encountered in St Ives—that open up such creative possibilities, potentially rebalancing the species richness of our arts. Would Alfred Wallis have been discovered had it not been for the support of the artist community? Would the primitive work of a 67-year-old hobbyist have been championed in a market forces model? Would a characteristic style of abstraction have ever evolved?

We've seen so far in this chapter how the ecological phenomenon of the charismatic or flagship species impacts conservation, positively in that it boosts general conservation awareness, but problematically in that it imbalances the different species gains; we've seen that the tendency of such a scenario is to encourage a single species model of conservation rather than an ecosystem approach, and we've seen that the charismatic leadership of reputational entrepreneurs can weight focus further on to single species awareness; finally, we've noted that these phenomena proliferate a neoliberal approach to state support such that governments push responsibility for conservation on to NGOs, charities, and ultimately the public. We're going to turn now to our central case study of this chapter, the Swedish activist Greta Thunberg. Thunberg's case is significant because of the greater complexity it brings to this debate: on the one hand she can clearly be viewed as a charismatic figure who reflects all the pros and cons of charismatic leadership: she raises the general level of conservation awareness but focuses on a single issue; she attracts disproportionate attention to the extent that others become less visible; and she performs a division with governments and world leaders such that conservation is ingrained as the responsibility of the people. On the other hand, we'll also see that Thunberg's approach introduces other dynamics into the landscape: she exploits contemporary technologies in a way that enhances network gains; she engages, inspires and enlists such that a meaningful and considerable community emerges; she establishes a new generational mindset that challenges existing patterns of form; and in doing these things she effects a conceptual ecosystem approach that demonstrates awareness, capacity, and scale.

Greta Thunberg

In the summer of 2018, a young student frustrated with the lack of progress around climate action decided to go on strike. She sat outside the Swedish parliament in Stockholm every Friday for the next five years, with a placard bearing the slogan Skolstrejk för Klimatet (school strike for the climate). Her intention was to force the Swedish government to commit to the agreement it had signed at the Paris Climate Change Conference in 2015 to reduce carbon emissions in order to keep global warming from rising any more than 2 degrees above the 1990 baseline. Increasingly, it was becoming clear that major governments in the global North were reneging on their promises, and Thunberg was determined to take a stand. As she began to gain prominence, she was invited to appear at climate rallies and events, where she proved herself to be a powerful public speaker.

Thunberg's messaging has been consistent—we're heading for climate disaster, and if we don't act now, it will be too late. She frequently berates politicians for their inaction, making the point that she isn't qualified, old enough, or powerful enough to change things by herself. She has the scientific facts at her fingertips, and regularly cites statistics to carry her message. But she also calls on her emotions: sometimes anger, sometimes fear, and sometimes, a sense of betrayal. "Our lives are in your hands", she cried desperately at the Climate March in Stockholm in 2018[79]; "Our house is on fire", she warned dramatically at the World Economics Forum in Davos in 2019[80]; "Can you hear me?", she asked incredulously at the Houses of Parliament later that year.[81] "We now need a whole new way of thinking", she concluded at the Austrian World Summit.[82] Over five years of campaigning, she rallied a community of support as more students came together under the banner of #FridaysForFuture. By March 2019, 1.4 million across 110 countries were participating in the strike.[83] In September 2019, 6 million people marched onto the streets.[84] By 2023 it had become a movement boasting 14 million participants in 7500 cities across all continents.[85]

Thunberg has also attracted criticism. She suffers frequent online abuse "for being young, mentally sick, ignorant, irrational, or idealistic",[86] and she even received trolling from US President Trump after having been nominated as Person of the Year in 2019. Nevertheless, she has gained rapid recognition for her climate activism, and although she is frequently presented as a single (young, feminized) individual battling against the (mighty, daunting) phenomenon of climate change and the (powerful, hypermasculine) intransigence of state governments, she has inspired many followers and leaders for change. For this, she has been recognized as a role model and icon[87]—indeed, a "bio-icon".[88] She's even been made the subject of several children's books, which position her "as an environmental hero".[89]

One of the reasons for Thunberg's success is that she has harnessed the support of the wider public, and in particular a younger generation. Without appearing to "lead" a movement for climate action, she has attracted a considerable number of followers, forging a community of those young people which is geographically dispersed and global in scope and vision. Abidin et al. attribute a lot of this success to Thunberg's use of social media, and Molder et al. point out that she is "one of the first political icons to have been fully formed within the new social media ecology".[90] "Greta's success in motivating such an unimaginable number", writes Lelia Green, "provides persuasive evidence of horizontal connection and young people's capacity to self-organize and collaborate".[91]

Thunberg's actions, especially between 2018 and 2023, have frequently hit the headlines: for making a lone stand against the Swedish parliament in her school strike; for sailing rather than flying to America to attend a climate summit; and for repeatedly calling out world leaders in talks to the UN, the World Economic Forum, the European Parliament, and the American Congress. "Thunberg rips through the normalized and often ingratiating charm offensive of the typical environmental celebrities comfortable with the trappings and networks of social, cultural, economic, and political power", write Abidin et al.[92] In 2019 she too was nominated for the Nobel Peace Prize,[93] and named Person of the Year by *Time Magazine* "for sounding the alarm about humanity's predatory relationship with the only home we have", and "for showing us all what it might look like when a new generation leads".[94] But her iconicity has brought people together in their millions; and part of this is because she seems so ordinary, so much in concert with the people, and so understanding of their needs. "Having become iconic through the mass distribution of her image and speeches", Quesada and Coughlin assert, "Thunberg became a shorthand for understanding the global interconnections of the climate crisis".[95] Rather than being seen as a self-serving celebrity, Thunberg became "a mobilising political icon"[96]—and the notion of "the Greta Effect"—her ability to galvanize global youth communities across the world into collective action—has been much admired.[97] Thunberg hit the zeitgeist, and the fact that she stood outside of party politics enabled her to appeal to more people.

Greta Thunberg as Charismatic Species

Even so, an obsession with Thunberg has distracted the media from other environmental activists, notably young female campaigners from the global South, especially if they are seen to be following in her footsteps. Ana Paulina Morera Quesada and Jenna M. Coughlin reclaim the work of some of these young women, both in their article "The Other Greta Effect" and in "The Global Activists Map", created to showcase climate activists that have been kept in Thunberg's shadow.[98] Their article considers Autumn Peltier ("Canada's Own Greta Thunberg"), Anuna de Wever (the "Belgian Greta Thunberg"), Vanessa Nakate (a "Greta of the South"), and Xiye Bastida ("the Greta Thunberg of America"),[99] though many others appear on their map. Drawing attention to these women as alternative Greta Thunbergs is an ironic play, though it goes some way to emphasizing their symbolic positioning as figureheads of a cause. As the article points out, "activists who are compared to Thunberg gain beneficial recognition while also experiencing a lack of attention".[100] In this, there is an obvious correlation with the way in which charismatic species gain conservation profile that both benefits and disadvantages other species. Just as the majority of people for whom Thunberg's name is familiar might not have heard about the environmental initiatives of Peltier, de Wever, Nakate, or Bastida, neither might they have heard about the plight of the Lost Spathoglottis or other species whose vulnerability is overshadowed.

Nakate was initially inspired by Thunberg, though she was concerned about taking action because of strict regulations in Uganda. There are rules about what women can do, bureaucratic obstacles to holding public demonstrations, and constant fears about the reaction of the authorities, or even corruption. Still, in January 2019, Nakate rallied her younger brothers and cousins and announced to her mother that they were going to protest for climate action. They painted placards: "Trees are important for us"; "Nature is life"; "When you plant a tree, you plant a forest"; "Climate strike now"; and "Thanks for the global warming". They took up positions at four intersections in Kampala, where to their surprise, public response was positive. Soon Nakate had initiated her own #FridaysForFuture activities outside the parliament, penning a letter to the Ugandan President. In 2021 she published her book, *A Bigger Picture*. She continues to be an influential voice for climate action in Uganda.[101]

One of the benefits of listening to the voices of activists such as Nakate is that she comes with a different perspective from her more high-profile peers. And because she offers this different viewpoint, she comes with different climate solutions, complementing those of Thunberg or other climate activists. For example, she cites "educating girls and family planning" as a key solution.[102] That may not seem immediately obvious as a way of tackling climate change, compared to many of the approaches in the global North, but this is the point: Nakate's is not the "charismatic" solution of the solar panel, the wind farm, or the recycling scheme; it's an altogether different take.

In proposing this solution, Nakate is citing a study by the American NPO Project Drawdown, which since 2017 has been compiling a list of the most effective approaches for minimizing climate change. It publishes two sets of figures: scenario 1 accounting for a 2 degree temperature rise by 2100; scenario 2 for a 1.5 degree rise. Currently, "Family Planning and Education" is number three on the scenario 1 list and number seven for scenario 2 (see Fig. 6.2). In other words, scientists calculate the education of women to be a highly effective means of mitigating climate change.

Scenario 1: to achieve no more than 2° rise		Scenario 2: to achieve no more than 1.5° rise	
Solution	**Gigatons CO^2 equivalent**	**Solution**	**Gigatons CO^2 equivalent**
1. Reduced food waste	88.50	1. Onshore wind turbines	143.56
2. Plant-rich diets	78.33	2. Utility-Scale Solar Photovoltaics	111.59
3. Family planning and education	68.90	3. Plant-rich diets	103.11
4. Refrigerant management	57.15	4. Reduced food waste	102.20
5. Tropical forest restoration	54.45	5. Tropical forest restoration	85.14
6. Onshore wind turbines	46.95	6. Clean cooking	76.34
7. Alternative refrigerants	42.73	7. Family planning and education	68.90
8. Utility-Scale Solar Photovoltaics	40.83	8. Distributed solar photovoltaics	64.86
9. Clean cooking	31.38	9. Refrigerant management	57.15
10. Distributed solar photovoltaics	26.65	10. Alternative refrigerants	48.75

Fig. 6.2 Top ten solutions for minimizing global climate change in December 2024, according to Project Drawdown[103]

The benefits are significant: "Including and funding girls' education and family planning in country-level holistic climate adaptation approaches will boost the ability of vulnerable populations to adapt to climate change", reports a policy brief for Project Drawdown.[104] It continues: "closing the gender gap around access to education and land tenure could increase crop production by 19%, boost gross domestic product, and lift thousands out of poverty, thereby empowering communities and countries to better manage climate-related natural hazards".[105] Such headlines have been picked up by other global South voices: "we're all familiar with the arguments for solar energy and electric vehicles", remarks Shabana Basij-Rasikh, founder of Afghanistan's only all-girl boarding school; "both get plenty of attention", she observes. "But why hasn't more been said about [girls' education]? Why hasn't this been the topic of media coverage worldwide?". She answers herself: "I can only imagine it's because the climate change discussion has been largely driven by Western nations—and in Western nations, girls' education is a given. It's not a topic most Westerners think about".[106] And therein lies the rub: the West has been distracted by its own charismatic solutions.

As we saw at the beginning of this chapter, this sort of educational empowerment was at the heart of Wangari Maathai's tree-planting philosophy, and noted by the Nobel committee when it recognized her contribution to "education, family planning, nutrition and the fight against corruption".[107] As such, Maathai and Nakate both set valuable examples to moving beyond the most obvious charismatic features of ecological awareness. Indeed, as Schell records in writing about GBM, it has in part been Maathai's focus on the non-charismatic that has enabled her to work on behalf of ordinary citizens. Schell calls attention to the Woman of Yaounde, "a sort of African 'everywoman'"[108] who Maathai invokes in her book, *The Challenge for Africa* (2009). The woman was a Congolese hill farmer Maathai encountered who was digging seed furrows on her land. Maathai noticed that the furrows were oriented up and down the hill rather than side to side. When it rained, this meant that rather than the water pooling in the furrows and sinking into the water table, it would gush down the furrows and away from the crops. More problematically, it would take with it the topsoil and nutrients. "Of course, the farmers I observed, and others like them, aren't the primary threats to the Congo Basin forests", writes Maathai; "but as I stood there that morning, the woman on the hillside in Yaounde came to represent to me the collective challenges that face many African countries".[109] Maathai recognized that the conservation policies instituted by the Congolese government needed not only to be targeting challenges like wholesale deforestation, but also the concerns of grassroot communities, which in this case flagged up a lack of basic farming education among subsistence hill farmers. "Sometimes when we do our work in high level meetings, we're not making changes where they really matter", she reflects.[110] Such a perspective, Schell concludes, "allows her to see and understand the woman of Yaounde while other development experts might not even notice her".[111] In the end, solutions like improving girls' education, voices like Maathai's and Nakate's, people like the woman of Yaounde, and global South countries like those referred to in the Project Drawdown policy brief all risk being overshadowed if louder, more dominant, charismatic solutions, voices, challenges, or countries get prioritized.

While this overshadowing might be cause enough for consideration (remember what happened with the Giant Panda?), the shocking treatment of some of these lesser-known activists illustrates how they contend with multiple other challenges. Not only are there in-country barriers to visibility, less developed infrastructures, and state obstruction, but there are also enduring colonial and racist frameworks which persist. In one incident, a photograph of climate activists at the Davos World Economic Forum in 2019 was cropped prior to publication by the Associated Press to remove Nakate from the image. "You didn't just erase a photo. You erased a continent", she later tweeted.[112] "It laid bare what African activists have claimed for a long time", remark Daniella Rafaely and Brendon Barnes in their analysis of the incident: "that African women not only experience the brunt of climate change but that they are discriminated against by the media in their activism".[113] This reminds us that "as a child of the wealthy global North, Thunberg is hardly the 'universal' face of climate injustice"; indeed, Quesada and Coughlin note, "She represents societies that are most responsible for the climate crisis, rather than those most impacted".[114] More than this, it also reminds us that those same (Western) societies have dominated the ownership of environmental narratives, whether these have been about defining relationships with the land, ideologies of conservation and preservation, or choices about what species or issues matter.[115] The same codes of erasure that have operated historically throughout the development of ecological thought are still those unbalancing contemporary conversations and solutions.

Greta Thunberg and Community

The wide support for Thunberg's agenda, though—a recognition of the need for climate action—demonstrates one of the concepts to which our own study is most attuned: the concept of "community". Specifically, it demonstrates a very specific feature we discussed in chapter three, and as we come towards the end of this chapter we'd like to turn back to Durkheim's "collective representation".

Durkheim argues that one way in which community is formed is that individuals collectively perform rituals in relation to shared symbolic referents. He uses religion as the obvious example of this behaviour, with ritual worship, prayer, or communion having a direct relation to "symbols"—referents such as God or the Prophet. Such behaviour becomes "the currency by which members of the group organize their interactions with the world", explains David Bloor, and he notes that "that interaction is not merely the sum of many independent interactions, but a truly collective exercise".[116] To complexify this, Durkheim's "symbol" is what Tim Olaveson has called "bipolar": on the one hand it stimulates highly-charged emotions, while on the other hand it is redolent with ideological affect. "These symbols effectively bind strong emotional content with 'higher', more abstract cognitive content such as behavioural norms, values, and cultural ideals", Olaveson explains; "This is the mechanism through which ritual makes and remakes the moral community".[117]

In relation to #FridaysForFuture, we can see that Thunberg balances a persuasive combination of emotional energy and cognitive ideals, even in her own performance. Her strength as a symbol therefore offers "a heightened sense of shared collective identity with other protest participants", reason Mattias Wahlström and Katrin Uba: "While the physical individual herself can be conceptualised as the background person, the icon is, in a sense, a collectivised individual as a symbol for collective values". This flies in the face of concerns "that Greta's individual iconicity might downplay the need for collective mobilisation", and perhaps even challenges our concerns about her charismatic identity. Far from reaping the privileges of singular attention (as "single species"), Thunberg's iconicity acts as a symbol of communal proliferation (an "ecosystem approach"): Wahlström and Uba found "a positive correlation" between "perceived influence by Greta" and "a general sense of collective identity with others in the movement and a tendency to regard one's participation as an expression of solidarity"[118]; another team concluded that "exposure to inspirational young public figures like Greta" links to "stronger collective efficacy beliefs, and greater intentions to take collective action".[119] Indeed, Durkheim's thinking goes on to conceptualize a state of "collective effervescence" which has been compared to Victor Turner's *communitas*—and this is where the magic happens: "It is an experience of intense creativity, definitively outside of normal social life, from which new myths and ideas spring".[120] And new ideas have sprung: after all, while they may not have the global visibility of Thunberg, other regional activists and even ordinary people are nevertheless gaining awareness, making lifestyle changes, and sometimes having impact themselves. In these instances, the trickle-down effect of Thunberg galvanizes proactive dynamics throughout the community—ecosystem health, if you will.

In the end, while Thunberg is evidently a figurehead, the climate change movement has far more the feel of a collective than an organization, decentring the notion of the leader. Sinéad Moriarty picks up on the concept of "community as hero" from children's literature, seeing it as "a more productive model for environmental narratives [...] as it circumvents the individualism of the conventional hero narrative in favor of a celebration of the power and responsibility of the group".[121] And Quesada and Coughlin suggest something similar: "By utilizing this collective narrative, youth climate activists strengthen the polyphonic nature of their rhetoric and amplify the voice of not a single individual, but a diverse generation that has coalesced around a common goal".[122]

Notes

1. Anon. (2004), "The Woman who Planted Trees", *The Independent*, 14 October, https://www.economist.com/leaders/2004/10/14/the-woman-who-planted-trees. Accessed 28 December 2024. Maathai is also "affectionately known as Mama Miti, 'the mother of trees'". Maria Popova (n.d.), "Planting Trees as Resistance and Empowerment: The Remarkable Illustrated Story of Wangari Maathai, the First African Woman to Win the Nobel Peace Prize", *The Marginalian*, https://www.themarginalian.org/2019/06/04/wangari-maathai-the-woman-who-planted-millions-of-trees/. Accessed 5 January 2025.
2. Wangari Maathai (2004), "Trees of Hope", *ABC News*, 19 December, https://abcnews.go.com/ThisWeek/TheList/story?id=343771. Accessed 28 December 2024.
3. Aisha Karanja (2017), "The Green Belt Movement's Bamboo Biomass Entrepreneurship Project", *The International Journal on Green Growth and Development*, 3:1, pp. 67–69, p. 67.
4. Green Belt Movement (2024), "Who We Are", *Green Belt Movement*, https://www.greenbeltmovement.org/. Accessed 28 December 2024.
5. Maathai, *Green Belt*, pp. 93–101.
6. Stuart Jeffries (2007), "Planting the Future", *The Guardian*, 16 February, https://www.theguardian.com/environment/2007/feb/16/ethicalliving.lifeandhealth. Accessed 5 January 2025.
7. On the other hand, Michaelson also notes that GBM was not a grassroots, community-initiated project, but that it emerged thanks to the "educated female elite" of the NCWK, supported by organizations like the UN Environment programme and the Kenyan Forestry Department, and—ironically—funded in part by the largesse of the petrochemical industry through early sponsorship from Mobil Oil. Oil companies globally have of course been

active in offsetting their carbon contributions, not least through tree-planting initiatives, some of which are discussed in Schell, "Transnational Environmental Justice", pp. 606–607.

8. Michaelson, "Wangari Maathai", pp. 541–557.
9. Karanja, "Green Belt", p. 67.
10. Heather Ball (2004), *Great Women Leaders*, Toronto: Second Story Press, p. 73.
11. Ball, *Great Women*, p. 74.
12. Wangari Maathai (2004), *The Green Belt Movement: Sharing the Approach and the Experience*, New York: Lantern Books, p. 35.
13. Karanja, "Green Belt", p. 67.
14. Eileen E. Schell (2013), "Transnational Environmental Justice Rhetorics and the Green Belt Movement: Wangari Muta Maathai's Ecological Rhetorics and Literacies", *JAC*, 33:3/4, pp. 585–613, p. 586.
15. Maathai, *Green Belt*, p. 21.
16. Maathai, *Green Belt*, pp. 35–41.
17. Maathai, *Green Belt*, pp. 42–47.
18. Maathai, *Green Belt*, pp. 47–49.
19. Maathai, *Green Belt*, pp. 50–51.
20. Kimari Ngugi (2023), "Wangari Mathai's legacy lives on in the tree planting program", *LinkedIn*, 3 June, https://www.linkedin.com/pulse/wangari-mathai-legacy-live-tree-planting-program-kimani-ngugi. Accessed 5 January 2025.
21. Schell, "Transnational", p. 588.
22. Reed F. Noss (1990), "Indicators for monitoring biodiversity: a hierarchical approach", *Conservation Biology* 4:4, pp. 355–364, p. 361.
23. Céline Albert, Gloria M. Luque, and Franck Courchamp (2018), "The twenty most charismatic species", *PloS one* 13:7, e0199149.
24. Clayton Fordahl (2024), "Beyond Animal Charisma: A Sociological Approach to Charismatic Species", *Conservation and Society*, 22:1, pp. 14–24, p. 15.
25. "Pictures of attractive animals help generate support for their cause", writes Gunnthorsdottir, while Jaric et al. remark on "messages emphasized with emotive language". Anna Gunnthorsdottir (2001), "Physical attractiveness of an animal species as a decision factor for its preservation", *Anthrozoös*,14:4, pp. 204–215, p. 206; Ivan Jarić et al. (2020), "The role of species charisma in

biological invasions", *Frontiers in Ecology and the Environment*, 18, pp. 345- 353, p. 348.
26. Fordahl, "Beyond", p. 20.
27. Fordahl, "Beyond", p. 15.
28. Fordahl, "Beyond", p. 20.
29. Albert, Luque, and Courchamp, "The twenty", p. 7. See also Gunnthorsdottir, "Physical attractiveness", p. 205.
30. Maxim Isaac et al. (2024), "How is the concept of charisma used in the academic literature about biodiversity conservation? A systematic map protocol", *Environmental Evidence*, 13:29, pp. 1–6, p. 2.
31. Jamie Lorimer (2007), "Nonhuman charisma", *Environment and Planning D: Society and Space*, 25:5, pp. 911–932, p. 915.
32. Lorimer, "Nonhuman", p. 917.
33. Lorimer, "Nonhuman", p. 919.
34. Lorimer, "Nonhuman", p. 923.
35. Lorimer, "Nonhuman", p. 927.
36. Lorimer, "Nonhuman", p. 926.
37. Jeroen F. Warner and Richard Meissner (2008), "The politics of security in the Okavango River Basin: From civil war to saving wetlands (1975–2002)—a preliminary security impact assessment", in Nevelina I. Pachova, Mikiyasu Nakayama, and Livor Jansky (eds) (2008), *International water security: Domestic threats and opportunities*, Tokyo: United Nations University Press, pp. 252–286, p. 259.
38. World Wildlife Fund (2024), "Adopt a Panda", *World Wildlife Fund*, https://support.wwf.org.uk/adopt-a-panda?utm_source=Google-Brand-Plus&utm_medium=PaidSearch-Brand&pc=AWS014002&ds_rl=1262308&ds_rl=1263317&gad_source=1&ds_rl=1262308&ds_rl=1263317&gclid=EAIaIQobChMIisqPm5ikigMVK5tQBh3B-DmHEAAYASAAEgIHUfD_BwE&gclsrc=aw.ds. Accessed 13 December 2024. The WWF programme foregrounds just a handful of species for adoption, giving prominence alongside the giant panda to the gorilla, tiger, snow leopard, and rhino, and emphasizing their low remaining numbers. According to WWF, as of December 2024 there are only 1860 giant pandas left in the wild.
39. Gary K. Meffe and C. Ronald Carroll (1997), "The Species in Conservation", in Gary K. Meffe and C. Ronald Carroll

(eds) (1997), *Principles of Conservation Biology* (second edition), Sunderland: Sinauer Associates, pp. 57–86, p. 83.
40. Vernon H. Heywood (ed.) (1996), *Global Biodiversity Assessment*, Cambridge: Cambridge University Press, p. 935.
41. Nigel Leader-Williams and Holly T. Dublin (2000), "Charismatic Megafauna as 'flagship species'", in Abigail Entwistle and Nigel Dunstone (eds) (2000), *Priorities for the Conservation of Mammalian Diversity: Has the Panda Had its Day?*, Cambridge: Cambridge University Press, pp. 53–81, p. 56.
42. Gary K. Meffe and C. Ronald Carroll (1997), "Conservation Reserves in Heterogeneous Landscapes", in Meffe and Carroll, *Principles*, pp. 305–344, p. 311.
43. Leader-Williams and Dublin, "Charismatic", p. 55.
44. Simon Clark (2023), "The Many Errors of An Inconvenient Truth", *YouTube*, https://www.youtube.com/watch?v=uxiHzjH-uIo. Accessed 22 December 2024.
45. Jon Kortajarena (2024), untitled post, *Instagram*, https://www.instagram.com/jonkortajarena/reel/DCSDMa8OfXY/. Accessed 5 January 2025.
46. Heywood, *Global*, p. 935.
47. Monika Krause and Katherine Robinson (2017), "Charismatic species and beyond: how cultural schemas and organisational routines shape conservation", *Conservation and Society*, 15:3, pp. 313–321, pp. 318–319.
48. M. Root-Bernstein, L. Douglas, A. Smith, D. Veríssimo (2013), "Anthropomorphized species as tools for conservation: utility beyond prosocial, intelligent and suffering species", *Biodiversity and Conservation*, 22:8, pp. 1577–1589, p. 1578.
49. Gunnthorsdottir, "Physical attractiveness", p. 211.
50. Opportunities to support these species are available through the WWF website; from these examples, at the time of writing, only the African Wild Dog is available for "adoption": https://www.worldwildlife.org/species/african-wild-dog. Accessed 22 December 2024. Dozens of species that are critically endangered or near extinction don't even feature in the relative profile of the WWF.
51. E. Beckman and D. Jerome (2017), "*Quercus tomentella*", *The IUCN Red List of Threatened Species* 2017: e.T30959A2799049.

https://www.iucnredlist.org/species/30959/2799049. Accessed 23 December 2024.
52. IUCN SSC Amphibian Specialist Group (2021), "*Boulengerula spawlsi*", *The IUCN Red List of Threatened Species* 2021: e.T121385062A121385865. https://www.iucnredlist.org/species/121385062/121385865, Accessed 23 December 2024.
53. IUCN SSC Amphibian Specialist Group (2020), "*Craugastor punctariolus*", *The IUCN Red List of Threatened Species* 2020: e.T56888A3052456. https://www.iucnredlist.org/species/56888/3052456, Accessed 23 December 2024.
54. J. Tsering (2022), "*Spathoglottis arunachalensis*", *The IUCN Red List of Threatened Species* 2022: e.T186303279A186805869. https://www.iucnredlist.org/species/186303279/186805869, Accessed 23 December 2024.
55. Frédéric Ducarme, Gloria M. Luque, and Franck Courchamp (2013), "What are 'charismatic species' for conservation biologists?", *BioSciences Master Reviews*, 1, pp. 1–8, p. 6.
56. Robert Home et al. (2009), "Selection criteria for flagship species by conservation organizations", *Environmental Conservation*, 36:2, pp. 139–148; Joaquin Munoz (2007), "Biodiversity conservation including uncharismatic species", *Biodiversity and Conservation*, 16:7, pp. 2233–2235.
57. Heywood, *Global*, p. 935.
58. C. Ronald Carroll and Gary K. Meffe (1997), "Management to Meet Conservation Goals: General Principles", in Meffe and Carroll, *Principles*, pp. 347–384, p. 353.
59. Claire A. Runge et al. (2019), "Single species conservation as an umbrella for management of landscape threats", *PLoS ONE*, 14:1, e0209619, p. 2.
60. Fang Wang et al. (2021), "The hidden risk of using umbrella species as conservation surrogates: A spatio-temporal approach", *Biological Conservation*, 253: 108913.
61. Ying Yuea et al. (2024) "The effectiveness of using giant panda as a surrogate for protecting sympatric species", *Journal of Environmental Management*, 351: 119803, p. 4.
62. Daniel Simberloff (1998), "Flagships, umbrellas, and keystones: is single-species management passé in the landscape era?", *Biological conservation*, 83:3, pp. 247–257, p. 247.
63. Simberloff, "Flagships", p. 248.

64. The ten-year strategic framework for 2010–2020, *Achieving Great Art for Everyone*, was rebranded *Great Art and Culture for Everyone* in 2013. Arts Council England (2010), *Achieving Great Art for Everyone: A Strategic Framework for the Arts*, London: Arts Council England; Arts Council England (2013), *Great Art and Culture for Everyone: 10-Year Strategic Framework* (second edition), Manchester: Arts Council England.
65. Arts Council England (2018), *Relationship Framework: How Arts Council England works with National Portfolio Organisations*, https://www.artscouncil.org.uk/sites/default/files/download-file/NPO_2018-22_Relationship_Framework.pdf. Accessed 18 December 2024.
66. The imbalance between London and the regions was recognized in *The Glory of the Garden* report in 1984. Subsequent studies have led to strategic approaches to balance (or latterly, to "level up") the arts landscape. Even so, it has not really been until recently that substantial movement has addressed the imbalance: ACE figures from 2010 to 2011 show London receiving a whopping 51% of the total allocation, with 13% to Birmingham (West Midlands), 7.8% to Leeds/Sheffield/Bradford (Yorkshire and the Humber), and 7.5% to Manchester (North West), and the remainder allocated elsewhere. In short, where London's funding has dropped by about 26% between 2010–2011 and 2023–2026 (£206m to £151m), and the West Midlands has been maintained (£53m to £56m), Yorkshire's has risen by around 70% (£7.8m to £12m) and the North West's by 66% (£7.5m to £11.2m); all other regions have had modest rises from 4 to 7%. Even so, the national picture, and particularly the urban-rural discrepancy, is hardly equitable. Source data from Arts Council England (2011), *Regularly funded organisations: Key data from the 2009/10 annual submission*, London: Arts Council England; Arts Council England (2023), "ACE NPO Investment Programme 2023–26 Data—Offered organisations NPO IPSO Transfer (V3)", https://www.artscouncil.org.uk/how-we-invest-public-money/2023-26-Investment-Programme/2023-26-investment-programme-data. Accessed 18 December 2024.
67. Controversially, English National Opera has been left out of the NPO allocations, with the Arts Council effectively forcing it to relocate outside of London.

68. Joanna M. Burgar, A. Cole Burton, and Jason T. Fisher (2019), "The importance of considering multiple interacting species for conservation of species at risk", *Conservation Biology,* 33:3, pp. 709–715, p. 712.
69. Yuea et al., "The effectiveness", p. 4.
70. Simberloff, "Flagships", p. 248.
71. BBC (2024), "William Shakespeare: The life and legacy of England's bard", *BBC,* https://www.bbc.co.uk/teach/articles/zfw6d6f. Accessed 18 December 2024.
72. Victoria Elliott and Sarah Olive (2021), "Secondary Shakespeare in the UK: what gets taught and why?", *English in Education,* 55:2, pp. 102–115, p. 102.
73. Elizabeth Schafer (2000), *Ms-Directing Shakespeare: Women Direct Shakespeare,* New York: St Martin's Press, p. 1.
74. See for example Pamela Allen Brown and Peter Parolin (2005), *Women Players in England 1550–1660: Beyond the All-Male Stage,* Burlington: Ashgate; Sophie Tomlinson (2005), *Women on Stage in Stuart Drama,* Cambridge: Cambridge University Press.
75. Jennifer Berneda Grenz (2020), "Healing the Land by Reclaiming an Indigenous Ecology: A journey exploring the application of the Indigenous worldview to invasion biology and ecology", Ph.D. thesis, Vancouver: University of British Columbia, p. 16.
76. Crystal Abidin et al. (2020), "The tropes of celebrity environmentalism", *Annual Review of Environment and Resources,* 45:1, pp. 387–410, pp. 403.
77. John Last (2017), *A crisis in the creative arts in the UK?,* London: Higher Education Policy Institute; Jonathan Savage and David Barnard (2019), *The State of Play—a review of music education in England,* London: Musicians' Union, p. 3; All Party Parliamentary Group for Music Education (2019), *Music Education: State of the Nation,* London: All-Party Parliamentary Group for Music Education, the Incorporated Society of Musicians and the University of Sussex, p. 29; Kim Solga (2019), "Theatre and performance, crisis and survival", *Research in Drama Education: The Journal of Applied Theatre and Performance,* 24:3, pp. 251–256.
78. Rees-Mogg famously compared arts financing with that of the NHS when he queried the subsidy of experimental art: "Ought

the health service to exist for the medical and administrative staff, or for the patients? Obviously for the patients", he reasoned; "Ought arts funding to be for the artists or for the audiences?" he then asked, following his question up with some equivocation; but his case had already been stated: "I was always convinced that arts grants should be primarily a consumer and not a producer subsidy", he asserted, reinforcing his perennial mantra, "subsidy weakens the sinews of self-help". Reflecting on the state of the arts a generation later, theatre critic Lyn Gardner's observation is to the point: "Well, we can all look around and see where self-help has got some of our biggest arts institutions. On their knees". See William Rees-Mogg (1993), "Affording Art for All", *The Times*, 15 July, p. 16; Lyn Gardner (2020), "Panto is off, but we can stop political grinches stealing theatre", *The Stage*, 20 August, p. 7.

79. Greta Thunberg (2019), *No One is Too Small to Make a Difference*, London: Allen Lane, p. 9.
80. Thunberg, *No One*, p. 37.
81. Thunberg, *No One*, p. 81.
82. Thunberg, *No One*, p. 99.
83. Beckie Leach (2019), "School Strike for Climate #FridaysForFuture", *Climate Museum UK*, https://climatemuseumuk.org/2019/07/10/school-strike-for-climate-fridaysforfuture/. Accessed 20 December 2024.
84. Matthew Taylor, Jonathan Watts, and John Bartlett (2019), "Climate Crisis: 6 Million People Join Latest Wave of Global Protests", *The Guardian*, 27 Sept, https://www.theguardian.com/environment/2019/sep/27/climate-crisis-6-million-people-join-latest-wave-of-worldwide-protests. Accessed 20 December 2024.
85. Muhammad Hassan Dajana and Syeda Hamna Shujat (2023), "How today's youth are taking the lead in global climate action", *World Economic Forum*, https://www.weforum.org/stories/2023/07/empowering-tomorrows-climate-leaders-how-youth-influence-climate-action/. Accessed 22 December 2024.
86. Mucha Mkono, Karen Hughes, and Stella Echentille (2020), "Hero or Villain? Responses to Greta Thunberg's activism and the implications for travel and tourism", *Journal of Sustainable Tourism*, 28:12, pp. 2081–2098, p. 2089.

87. Mattias Wahlström and Katrin Uba (2024), "Political Icon and Role Model: Dimensions of the Perceived 'Greta Effect' among Climate Activists as Aspects of Contemporary Social Movement Leadership", *Acta Sociologica*, 67:3, pp. 301–316.
88. Ana Paulina Morera Quesada and Jenna M. Coughlin (2024), "The Other Greta Effect (OGE): Recognizing Youth Climate Activists Beyond Thunberg", *Kvinder, Køn & Forskning*, 37:1, pp. 49–63, p. 50.
89. Sinéad Moriarty (2021), "Modeling Environmental Heroes in Literature for Children: Stories of Youth Climate Activist Greta Thunberg", *The Lion and the Unicorn*, 45:2, pp. 192–210, p. 192.
90. Amanda L. Molder et al. (2022), "Framing the Global Youth Climate Movement: A Qualitative Content Analysis of Greta Thunberg's Moral, Hopeful, and Motivational Framing on Instagram", *The International Journal of Press/Politics*, 27:3, pp. 668–695, 672.
91. Lelia Green (2020), "Confident, Capable and World Changing: Teenagers and Digital Citizenship", *Communication Research and Practice*, 6:1, pp. 6–19, p. 11.
92. Abidin et al., "The tropes", p. 400.
93. Damien Carrington (2019), "Greta Thunberg Nominated for Nobel Peace Prize", *The Guardian*, 14 March, https://www.theguardian.com/world/2019/mar/14/greta-thunberg-nominated-nobel-peace-prize. Accessed 20 December 2024.
94. Charlotte Alter, Suyin Haynes and Justin Worland (2019), "Time 2019 Person of the Year—Greta Thunberg", *Time Magazine*, https://time.com/person-of-the-year-2019-greta-thunberg. Accessed 22 December 2024.
95. Quesada and Coughlin, "The Other", p. 52.
96. Wahlström and Uba, "Political Icon", p. 304.
97. Anandita Sabherwal et al. (2021), "The Greta Thunberg Effect: Familiarity with Greta Thunberg Predicts Intentions to Engage in Climate Activism in the United States", *Journal of Applied Social Psychology*, 51:4, pp. 321–33; Wahlström and Uba, "Political Icon"; Quesada and Coughlin, "The Other".
98. Anon. (2024), *The Global Activists Map*, (website) stolaf.maps.arcgis.com/apps/instant/basic/index.html?appid-51fc810775a64ee583198089c36c60ec. Accessed 22 December 2024.

99. Quesada and Coughlin, "The Other", p. 55.
100. Quesada and Coughlin, "The Other", p. 51.
101. Vanessa Nakate (2021), *A Bigger Picture: My Fight to Bring a New African Voice to the Climate Crisis*, London: Pan Macmillan.
102. Vanessa Nakate (2021), "Walking the Talk: Vanessa Nakate", *International Monetary Fund*, pp. 41–42, p. 42.
103. Anon., "Table".
104. Kristen P. Patterson et al. (2021), "Policy Brief: Girls' education and family planning: essential components of climate adaptation and resilience", *Project Drawdown*, https://drawdown.org/sites/default/files/Drawdown_Lift_Policy_Brief_Girls_Education_122121.pdf. Accessed 23 December 2024, p. 1.
105. Patterson et al., "Policy Brief", p. 2.
106. Shabana Basij-Rasikh (2018), "Want to fight climate change? Educate a girl", *Ideas.TED.com*, https://ideas.ted.com/want-to-fight-climate-change-educate-a-girl/. Accessed 23 December 2024.
107. Maathai, *Green Belt*, p. x.
108. Schell, "Transnational", p. 597.
109. Wangari Maathai (2009), *The Challenge for Africa*, New York: Pantheon Books, pp. 14–15.
110. Matthai, *Challenge*, p. 15.
111. Schell, "Transnational", pp. 595–597.
112. Quoted in Quesada and Coughlin, "The Other", p. 57.
113. Daniella Rafaely and Brendon Barnes (2020), "African climate activism, media and the denial of racism: The tacit silencing of Vanessa Nakate", *Community Psychology in Global Perspective*, 6:2/2, pp. 71–86, p. 72.
114. Quesada and Coughlin, "The Other", p. 52.
115. Prominent among such narratives have been those in the Western canon, including several of Shakespeare's plays (*As You Like It* with its pastoral husbandry; *A Midsummernight's Dream* with its wildwood magic; *The Tempest* with its man vs. nature debate). Yet perhaps the charisma of Shakespeare as a figure, and of these texts as cultural classics, has distracted us from our actual relationship with the natural world. In January 2025, Just Stop Oil protesters interrupted a staging of *The Tempest* in London's West End with a sign announcing that "over 1.5 degrees is a global shipwreck", referencing both the play and the

pledges of the Paris Climate agreement from 2015. This particular demonstration, which disrupted a very familiar instance of cultural reverence towards Shakespeare, revealed multiple levels of irony, as the audience focused on a stage-world narrative about colonial and environmental threats while concerns outside the theatre about real-world colonial and environmental threats were arguably going unacknowledged.

Hannah Roberts (2025), "Two arrested after Just Stop Oil activists disrupt The Tempest starring Sigourney Weaver with confetti cannon", *The Independent*, 28 January. https://www.independent.co.uk/arts-entertainment/theatre-dance/news/just-stop-oil-sigourney-weaver-the-tempest-west-end-b2687453.html. Accessed 10 March 2025.

116. David Bloor (2000), "Collective Representations as Social Institutions", in William S.F. Pickering (2000), *Durkheim and Representations*, London: Routledge, pp. 157–166, p. 162.
117. Tim Olaveson (2001), "Collective Effervescence and Communitas: Processual Models of Ritual and Society in Emile Durkheim and Victor Turner", *Dialectical Anthropology*, 26, pp. 89–124, p. 98.
118. Wahlström and Uba, "Political Icon", pp. 303–304.
119. Sabherwal et al., "The Greta", p. 329.
120. Olaveson, "Collective Effervescence", p. 110.
121. Moriarty, "Modeling", p. 193.
122. Quesada and Coughlin, "The Other", p. 57.

Bibliography

Abidin, Crystal, Dan Brockington, Michael K. Goodman, Mary Mostafanezhad, and Lisa Ann Richey (2020), "The tropes of celebrity environmentalism", *Annual Review of Environment and Resources*, 45:1, pp. 387–410.

Albert, Céline, Gloria M. Luque, and Franck Courchamp (2018), "The twenty most charismatic species", *PloS one* 13:7, e0199149.

All Party Parliamentary Group for Music Education (2019), *Music Education: State of the Nation*, London: All-Party Parliamentary Group for Music Education, the Incorporated Society of Musicians and the University of Sussex.

Allen Brown, Pamela, and Peter Parolin (2005), *Women Players in England 1550–1660: Beyond the All-Male Stage*, Burlington: Ashgate.

Alter, Charlotte, Suyin Haynes and Justin Worland (2019), "Time 2019 Person of the Year—Greta Thunberg", *Time Magazine*, https://time.com/person-of-the-year-2019-greta-thunberg. Accessed 22 December 2024.

Anon. (2004), "The Woman who Planted Trees", *The Independent*, 14 October, https://www.economist.com/leaders/2004/10/14/the-woman-who-planted-trees. Accessed 28 December 2024.

Anon. (2024), "Table of Solutions", *Project Drawdown*, https://drawdown.org/solutions/table-of-solutions. Accessed 23 December 2024.

Anon. (2024), *The Global Activists Map*, (website) stolaf.maps.arcgis.com/apps/instant/basic/index.html?appid=51fc810775a64ee583198089c36c60ec. Accessed 22 December 2024.

Arts Council England (2010), *Achieving Great Art for Everyone: A Strategic Framework for the Arts*, London: Arts Council England.

Arts Council England (2011), *Regularly funded organisations: Key data from the 2009/10 annual submission*, London: Arts Council England.

Arts Council England (2013), *Great Art and Culture for Everyone: 10-Year Strategic Framework* (second edition), Manchester: Arts Council England.

Arts Council England (2018), *Relationship Framework: How Arts Council England works with National Portfolio Organisations*, https://www.artscouncil.org.uk/sites/default/files/download-file/NPO_2018-22_Relationship_Framework.pdf. Accessed 18 December 2024.

Arts Council England (2023), "ACE NPO Investment Programme 2023–26 Data—Offered organisations NPO IPSO Transfer (V3)", https://www.artscouncil.org.uk/how-we-invest-public-money/2023-26-Investment-Programme/2023-26-investment-programme-data. Accessed 18 December 2024.

Ball, Heather (2004), *Great Women Leaders*, Toronto: Second Story Press.

Basij-Rasikh, Shabana (2018), "Want to fight climate change? Educate a girl", *Ideas.TED.com*, https://ideas.ted.com/want-to-fight-climate-change-educate-a-girl/. Accessed 23 December 2024.

BBC (2024), "William Shakespeare: The life and legacy of England's bard", *BBC*, https://www.bbc.co.uk/teach/articles/zfw6d6f. Accessed 18 December 2024.

Beckman, E. and D. Jerome (2017), "*Quercus tomentella*", *The IUCN Red List of Threatened Species* 2017: e.T30959A2799049. https://www.iucnredlist.org/species/30959/2799049. Accessed 23 December 2024.

Burgar, Joanna M., A. Cole Burton, and Jason T. Fisher (2019), The importance of considering multiple interacting species for conservation of species at risk", *Conservation Biology*, 33:3, pp. 709–715.

Carrington, Damien (2019), "Greta Thunberg Nominated for Nobel Peace Prize", *The Guardian*, 14 March, https://www.theguardian.com/world/2019/mar/14/greta-thunberg-nominated-nobel-peace-prize. Accessed 20 December 2024.

Clark, Simon (2023), "The Many Errors of An Inconvenient Truth", *YouTube*, https://www.youtube.com/watch?v=uxiHzjH-uIo. Accessed 22 December 2024.

Dajana, Muhammad Hassan, and Syeda Hamna Shujat (2023), "How today's youth are taking the lead in global climate action", *World Economic Forum*, https://www.weforum.org/stories/2023/07/empowering-tomorrows-climate-leaders-how-youth-influence-climate-action/. Accessed 22 December 2024.

Ducarme, Frédéric, Gloria M. Luque, and Franck Courchamp (2013), "What are 'charismatic species' for conservation biologists?", *BioSciences Master Reviews*, 1.

Elliott, Victoria and Sarah Olive (2021), "Secondary Shakespeare in the UK: what gets taught and why?", *English in Education*, 55:2, pp. 102–115.

Entwistle, Abigail and Nigel Dunstone (eds) (2000), *Priorities for the Conservation of Mammalian Diversity: Has the Panda Had its Day?*, Cambridge: Cambridge University Press.

Fordahl, Clayton (2024), "Beyond Animal Charisma: A Sociological Approach to Charismatic Species", *Conservation and Society*, 22:1, pp. 14–24.

Gardner, Lyn (2020), "Panto is off, but we can stop political grinches stealing theatre", *The Stage*, 20 August, p. 7.

Green Belt Movement (2024), "Who We Are", *Green Belt Movement*, https://www.greenbeltmovement.org/. Accessed 28 December 2024.

Green, Lelia (2020), "Confident, Capable and World Changing: Teenagers and Digital Citizenship", *Communication Research and Practice*, 6:1, pp. 6–19.

Grenz, Jennifer Berneda (2020), "Healing the Land by Reclaiming an Indigenous Ecology: A journey exploring the application of the Indigenous worldview to invasion biology and ecology", Ph.D. thesis, Vancouver: University of British Columbia.

Gunnthorsdottir, Anna (2001), "Physical attractiveness of an animal species as a decision factor for its preservation", *Anthrozoös*,14:4, pp. 204–215.

Heywood, Vernon H. (ed.) (1996), *Global Biodiversity Assessment*, Cambridge: Cambridge University Press.

Home, Robert, Claudia Keller, Peter Nagel, Nicole Bauer, and Marcel Hunziker (2009), "Selection criteria for flagship species by conservation organizations", *Environmental Conservation*, 36:2, pp. 139–148.

Isaac, Maxim, Caroline S. Fukushima, Biljana Macura, Enrico Di Minin, Ricardo A Correia (2024), "How is the concept of charisma used in the academic literature about biodiversity conservation? A systematic map protocol", *Environmental Evidence*, 13:29.

IUCN SSC Amphibian Specialist Group (2020), "*Craugastor punctariolus*", *The IUCN Red List of Threatened Species* 2020: e.T56888A3052456. https://www.iucnredlist.org/species/56888/3052456, Accessed 23 December 2024.

IUCN SSC Amphibian Specialist Group (2021), "*Boulengerula spawlsi*", *The IUCN Red List of Threatened Species* 2021: e.T121385062A121385865. https://www.iucnredlist.org/species/121385062/121385865, Accessed 23 December 2024.

Jarić, Ivan, Franck Courchamp, Ricardo A. Correia, Sarah L. Crowley, Franz Essl, Anke Fischer, Pablo González-Moreno, Gregor Kalinkat, Xavier Lambin, Bernd Lenzner, Yves Meinard, Aileen Mill, Camille Musseau, Ana Novoa, Jan Pergl, Petr Pyšek, Klára Pyšková, Peter Robertson, Menja von Schmalensee, Ross T. Shackleton, Robert A. Stefansson, Kateřina Štajerová, Diogo Veríssimo, and Jonathan M. Jeschke (2020), "The role of species charisma in biological invasions", *Frontiers in Ecology and the Environment*, 18, pp. 345–353.

Jeffries, Stuart (2007), "Planting the Future", *The Guardian*, 16 February, https://www.theguardian.com/environment/2007/feb/16/ethicalliving.lifeandhealth. Accessed 5 January 2025.

Karanja, Aisha (2017), "The Green Belt Movement's Bamboo Biomass Entrepreneurship Project", *The International Journal on Green Growth and Development*, 3:1, pp. 67–69.

Kortajarena, Jon (2024), untitled post, *Instagram*, https://www.instagram.com/jonkortajarena/reel/DCSDMa8OfXY/. Accessed 5 January 2025.

Krause, Monika and Katherine Robinson (2017), "Charismatic species and beyond: how cultural schemas and organisational routines shape conservation", *Conservation and Society*, 15:3, pp. 313–321.

Last, John (2017), *A crisis in the creative arts in the UK?*, London: Higher Education Policy Institute.

Leach, Beckie (2019), "School Strike for Climate #FridaysForFuture", *Climate Museum UK*, https://climatemuseumuk.org/2019/07/10/school-strike-for-climate-fridaysforfuture/. Accessed 20 December 2024.

Lorimer, Jamie (2007), "Nonhuman charisma", *Environment and Planning D: Society and Space*, 25:5, pp. 911–932.

Maathai, Wangari (2004), "Trees of Hope", *ABC News*, 19 December, https://abcnews.go.com/ThisWeek/TheList/story?id=343771. Accessed 28 December 2024.

Maathai, Wangari (2004), *The Green Belt Movement: Sharing the Approach and the Experience*, New York: Lantern Books.

Maathai, Wangari (2009), *The Challenge for Africa*, New York: Pantheon Books.

Marc Michaelson (1994), "Wangari Maathai and Kenya's Green Belt Movement: Exploring the Evolution and Potentialities of Consensus Movement Mobilization", *Social Problems*, 41:4, pp. 540–561, pp. 541, 550.

Meffe, Gary K. and C. Ronald Carroll (eds) (1997), *Principles of Conservation Biology* (second edition), Sunderland: Sinauer Associates.

Michaelson, Marc (1994), "Wangari Maathai and Kenya's Green Belt Movement: Exploring the Evolution and Potentialities of Consensus Movement Mobilization", *Social Problems*, 41:4, pp. 540–561.

Mkono, Mucha, Karen Hughes and Stella Echentille (2020), "Hero or Villain? Responses to Greta Thunberg's activism and the implications for travel and tourism", *Journal of Sustainable Tourism*, 28:12, pp. 2081–2098.

Molder, Amanda L., Alexandra Lakind, Zoe E. Clemmons, and Kaiping Chen (2022), "Framing the Global Youth Climate Movement: A Qualitative Content Analysis of Greta Thunberg's Moral, Hopeful, and Motivational Framing on Instagram", *The International Journal of Press/Politics*, 27:3, pp. 668–695.

Moriarty, Sinéad (2021), "Modeling Environmental Heroes in Literature for Children: Stories of Youth Climate Activist Greta Thunberg", *The Lion and the Unicorn*, 45:2, pp. 192–210.

Munoz, Joaquin (2007), "Biodiversity conservation including uncharismatic species", *Biodiversity and conservation*, 16:7, pp. 2233–2235.

Nakate, Vanessa (2021), "Walking the Talk: Vanessa Nakate", *International Monetary Fund*, pp. 41–42.

Nakate, Vanessa (2021), *A Bigger Picture: My Fight to Bring a New African Voice to the Climate Crisis*, London: Pan Macmillan.

Ngugi, Kimari (2023), "Wangari Mathai's legacy lives on in the tree planting program", *LinkedIn*, 3 June, https://www.linkedin.com/pulse/wangari-mathai-legacy-live-tree-planting-program-kimani-ngugi. Accessed 5 January 2025.

Noss, Reed F. (1990), "Indicators for monitoring biodiversity: a hierarchical approach", *Conservation Biology* 4:4, pp. 355–364.

Olaveson, Tim (2001), "Collective Effervescence and Communitas: Processual Models of Ritual and Society in Emile Durkheim and Victor Turner", *Dialectical Anthropology*, 26, pp. 89–124.

Pachova, Nevelina I., Mikiyasu Nakayama, and Livor Jansky (eds) (2008), *International water security: Domestic threats and opportunities*, Tokyo: United Nations University Press.

Patterson, Kristen P., Yusuf Jameel, Mamta Mehra, and Carissa Patrone (2021), "Policy Brief: Girls' education and family planning: essential components of climate adaptation and resilience", *Project Drawdown*, https://drawdown.org/sites/default/files/Drawdown_Lift_Policy_Brief_Girls_Education_122121.pdf. Accessed 23 December 2024.

Pickering, William S.F. (2000), *Durkheim and Representations*, London: Routledge.

Quesada, Ana Paulina Morera, and Jenna M. Coughlin (2024), "The Other Greta Effect (OGE): Recognizing Youth Climate Activists Beyond Thunberg", *Kvinder, Køn & Forskning*, 37:1, pp. 49–63.

Popova, Maria (n.d.), "Planting Trees as Resistance and Empowerment: The Remarkable Illustrated Story of Wangari Maathai, the First African Woman to Win the Nobel Peace Prize", *The Marginalian*, https://www.themarginalian.org/2019/06/04/wangari-maathai-the-woman-who-planted-millions-of-trees/. Accessed 5 January 2025.

Rafaely, Daniella, and Brendon Barnes (2020), "African climate activism, media and the denial of racism: The tacit silencing of Vanessa Nakate", *Community Psychology in Global Perspective*, 6:2/2, pp. 71–86.

Rees-Mogg, William (1993), "Affording Art for All", *The Times*, 15 July, p. 16.

Roberts, Hannah (2025), "Two arrested after Just Stop Oil activists disrupt The Tempest starring Sigourney Weaver with confetti cannon", *The Independent*, 28 January. https://www.independent.co.uk/arts-entertainment/theatre-dance/news/just-stop-oil-sigourney-weaver-the-tempest-west-end-b2687453.html. Accessed 10 March 2025.

Root-Bernstein, M., L. Douglas, A. Smith, D. Veríssimo (2013), "Anthropomorphized species as tools for conservation: utility beyond prosocial, intelligent and suffering species", *Biodiversity and Conservation*, 22:8, pp. 1577–1589.

Runge, Claire A., John C. Withey, David E. Naugle, Joseph E. Fargione, Kate J. Helmstedt, Ashley E. Larsen, Sebastian Martinuzzi, Jason D. Tack (2019), "Single species conservation as an umbrella for management of landscape threats", *PLoS ONE*, 14 :1, e0209619.

Sabherwal, Anandita, Abel Gustafson, Matthew T. Ballew, Matthew H. Goldberg, John E. Kotcher, Janet K. Swim, Sander van der Linden, Edward W. Maibach, Seth A. Rosenthal and Anthony Leiserowitz (2021), "The Greta Thunberg Effect: Familiarity with Greta Thunberg Predicts Intentions to Engage in Climate Activism in the United States", *Journal of Applied Social Psychology*, 51:4, pp. 321–33.

Savage, Jonathan, and David Barnard (2019), *The State of Play—a review of music education in England*, London: Musicians' Union.

Schafer, Elizabeth (2000), *Ms-Directing Shakespeare: Women Direct Shakespeare*, New York: St Martin's Press.

Schell, Eileen E. (2013), "Transnational Environmental Justice Rhetorics and the Green Belt Movement: Wangari Muta Maathai's Ecological Rhetorics and Literacies", *JAC*, 33:3/4, pp. 585–613.

Simberloff, Daniel (1998), "Flagships, umbrellas, and keystones: is single-species management passé in the landscape era?", *Biological conservation*, 83:3, pp. 247–257.

Solga, Kim (2019), "Theatre & performance, crisis & survival", *Research in Drama Education: The Journal of Applied Theatre and Performance*, 24:3, pp. 251–256.

Taylor, Matthew, Jonathan Watts, and John Bartlett (2019), "Climate Crisis: 6 Million People Join Latest Wave of Global Protests", *The Guardian*, 27 Sept, https://www.theguardian.com/environment/2019/sep/27/climate-crisis-6-million-people-join-latest-wave-of-worldwide-protests. Accessed 20 December 2024.

Thunberg, Greta (2019), *No One is Too Small to Make a Difference*, London: Allen Lane.

Tomlinson, Sophie (2005), *Women on Stage in Stuart Drama*, Cambridge: Cambridge University Press.

Tsering, J. (2022), "*Spathoglottis arunachalensis*", *The IUCN Red List of Threatened Species* 2022: e.T186303279A186805869. https://www.iucnredlist.org/species/186303279/186805869, Accessed 23 December 2024.

Wahlström, Mattias, and Katrin Uba (2024), "Political Icon and Role Model: Dimensions of the Perceived 'Greta Effect' among Climate Activists as Aspects of Contemporary Social Movement Leadership", *Acta Sociologica*, 67:3, pp. 301–316.

Wang, Fang, Julie Winkler, Andres Vina, William Joseph McShea, Sheng Li, Thomas Connor, Zhiqiang Zhao, Dajun Wang, Hongbo Yang, Sheng Li, Ying Tang, Jindong Zhang, and Jianguo Liu (2021), "The hidden risk of using umbrella species as conservation surrogates: A spatio-temporal approach", *Biological Conservation*, 253: 108913.

Yuea, Ying, Zhisong Yang, Wei Weid, Biao Yang, Dunwu Qie, Xiaodong Guf, Xuyu Yang, Song Luh, Wen Zhang, Qiang Daia, and Zejun Zhang (2024) "The effectiveness of using giant panda as a surrogate for protecting sympatric species", *Journal of Environmental Management*, 351: 119803.

CHAPTER 7

Seedbank

David Milarch's is a replanting scheme with a difference. He started it in 1992 not only to promote reforestation but also to preserve superlative tree specimens—the best of the best. Milarch's story is extraordinary in every way, and everything in it seems larger than life. Having spent years in his home state of Michigan drinking big, smoking big, and arm-wrestling anyone who would compete, Milarch was spooked by a near-death experience in which he was visited by angels who tasked him with saving the world. He transformed his life, devoted his energies to conservation, and started to clone America's trees. Milarch's vision was to create "charismatic mega-forests"[1] from the oldest, tallest, and most unique specimens he could find. The thinking was simple: "If you are going to plant a tree", he argued, "why not plant trees genetically you know will survive?"[2] By selecting trees that had proven their genetic worth, Milarch hoped to provide future forests that were resilient, long-living, healthy, and disease-free. Over the last thirty years, his organization Archangel Ancient Tree Archive (AATA) has gathered offcuts from some of the most extraordinary trees on the planet: they have set about cloning America's national and state "champion" trees—the prime examples of the country's 826 tree species in terms of their height, crown size, and diameter; they have also cloned copies of Hyperion, the tallest Coastal Redwood; Methusaleh, a Bristlecone Pine and the oldest-living tree in the

K. Savage and D. Symonds, *Ecologies of Community in Performance*,
https://doi.org/10.1007/978-3-031-94343-0_7

world; the Waterfall Sequoia, the largest-diameter single-stem tree; and the Fieldbrook Stump, the remains of the largest ever Giant Redwood.

"Trees are most often cloned by taking a six-inch or so cutting of the newest growth", explains Jim Robbins, whose book *The Man Who Plants Trees* tells Milarch's story. Cuttings are carefully packaged and sent to a specialist nursery, where they are propagated in controlled conditions to create clones of the originals. "A small opening is scraped into the cambium layer at the base of the cutting. It's then immersed in a mix of soil and hormones, and if the procedure is successful, new roots emerge from the small scrape".[3] Once the clones sprout roots, they are grown into tiny saplings, and from there, they are sent to be planted in designated sites around the world. Clones have been planted in 9/11 memorial gardens in New York and at the Pentagon; in Salt Lake City as part of the 2002 Winter Olympic celebrations; in the grounds of the US Capitol, Arlington National Cemetery, and Mount Vernon; in locations across Canada, Ireland, Sweden, Germany, Australia, and New Zealand; and in countless community spaces throughout America. Meanwhile, AATA has partnered with other organizations including the Interlochen Center for the Arts in Michigan; the community wellbeing organization Teravana in California; and the much-admired Eden Project in Cornwall, UK.

The focus of this chapter is the *archive*, and through the prism of the ecological principles of *conservation* and *preservation*, we're going to consider the cultural practices we associate with archives, libraries, and special collections. By way of example, we'll take a look at Milarch's Archangel Ancient Tree Archive and artist Katie Paterson's *Future Library* as two projects in which the ecological and the archival overlap. We'll consider some of the ways in which the natural world preserves its own heritage through processes of archiving, and we'll recognize a different metaphor by exploring three different types of a particular archive known as a *seedbank*. We'll pick up on some of the issues that inform archival practice such as selection and curation, and towards the end of the chapter we'll frame the shared stewardship of community archives as "archives of the people, by the people, for the people".

Conservation and Preservation

We noted in chapter two that environmental practices have been guided by slightly different visions. We looked at the approaches of two environmentalists who sought to protect the great forests of America: John Muir campaigned to leave swathes of forest untouched, while Gifford Pinchot believed in protecting the land but using it as a resource. At heart, these mindsets reflect the different approaches of *preservation* (the "advocacy of nature unaltered by man") and *conservation* ("the wise use or planned development of resources").[4]

Perhaps surprisingly, there is a corollary between these environmental differences and the competing concepts of archival practice, suggests Erik A. Moore. According to him, notions of "holistic preservation" on the one hand, and "utilitarianism" on the other, also present a "theoretical clash" between the differing views of archival theory laid out by two prominent luminaries in the field.[5] Hilary Jenkinson's *A Manual of Archive Administration* (1922) stated that an archive should consist of documentation pertaining to a particular matter and kept by the original custodian exclusively for their own use[6]; while Theodore R. Schellenberg's *Modern Archives: Principles and Techniques* (1956) stated that "to be archives, materials must be preserved for reasons other than those for which they were created or accumulated"[7]—such as research. Archival theory, of course, goes into detail about far more than just this bone of contention, considering matters of selection and destruction, organization and arrangement, description and classification. But in all of these matters there is a persistent pull towards principles drawn from the natural sciences, whether this is through the implicit reminder of botanical classification schemes along the lines of which archives tend to be organized, or through explicit references to archives as if they were actually living things: as Jenkinson put it and Schellenberg agreed, archives are "almost, you might say, as much an organism as a tree or an animal".[8]

If the notion of the archive being a tree is an evocative idea, it's matched in reverse by the idea of Milarch's newly-cloned trees becoming an archive, as the name of his organization proposes (Archangel Ancient Tree Archive). Still, as a term, it works well; where the more common sivicultural term "nursery" suggests the farming of new, young tree stock, the idea of an "archive" hints at holdings that are historical, befitting the ancient old-growth specimens Milarch seeks to save. The Lady Liberty Florida Cypress was dated at 2400 years of age when AATA cloned it, while Methusaleh was estimated to be 4800. The term also speaks to protection and preservation, which is equally appropriate: like the fragile material in more conventional archives, these trees are vulnerable, and cuttings have sometimes been taken only weeks or months before the original trees have died.[9] In preserving such specimens and saving them from irrevocable loss, AATA draws on the "core archival principles" of "provenance, appraisal, arrangement, and description",[10] but extends beyond the basic premises of preservation to work as a dynamic resource. "The [AATA] warehouse hosts thousands of cuttings, tissue samples, and cloned saplings of sequoias and redwoods",[11] reports Marguerite Bouvard, "where they can safely grow into the future and be available for further study and preservation".[12] As such, the concept of an archive captures well the aims of Milarch's organization: to "propagate the world's most important old growth trees before they are gone"; to "archive the genetics of ancient trees in living libraries around the world for the future"; and to "reforest the Earth with the offspring of these trees to provide the myriad of beneficial ecosystem services essential for all life forms to thrive".[13]

The other phrase Milarch uses to describe the ancient tree archive is a "living library". Here, he toys with the sense of the materiality of the collection while at the same time suggesting the educational value of its initiatives. The Archangel Education Initiative "is now partnering with Michigan schools and universities to offer hands-on learning programs in nature awareness, agro-forestry, and permaculture design",[14] Bouvard reports, while "the greenhouse is frequently open to young people […] so that a new generation can learn how to create new forests".[15] Right across the world in reforestation or conservation projects such as this, education is a key initiative in ensuring that new generations are introduced to environmental awareness, and that communities feel able to take ownership of their relationships with the planet. We'll see in due course how this dynamic sense of conservation has increasingly become a feature of contemporary practices. Other projects with similar aims to Milarch's, such as the University of Tennessee's "Champion Tree Program", also have pillars of conservation, information, and education at their core. "By communicating the value of these trees, we will change hearts and minds, shift priorities, and expand nationally the focus of tree conservation work beyond the simple planting of seedlings and young trees", states its director Jaq Payne.[16]

Future Library

Across the world, another archive is in the making. "A forest in Norway is growing", teases the website; "in 100 years it will become an anthology of books".[17] But wait: a forest? An anthology? Is this about books or is it about trees? This is *Future Library*, the brainchild of Scottish artist Katie Paterson. It's another project that makes a conceptual link between the archive and ecology.

Paterson's ruse is to build a repository of manuscripts over the next hundred years. Every year she invites a celebrated author to deposit a document into the archive, which will remain unpublished until 2114, by which time 100 books will have been collected. Meanwhile, 1000 spruce saplings have been planted to provide the paper on which to print the books. In this way, remarks Paterson, "*Future Library* has nature, the environment at its core". For her, its engagement with ecology is about "the interconnectedness of things, those living now and still to come".[18]

The first contributor was Margaret Atwood, author of *The Handmaid's Tale* (1985), whose manuscript *Scribbler Moon* was consigned to *Future Library* in 2015. "I imagine her words growing through the trees, an unseen energy, activated and materialised, the tree rings becoming chapters in a book", Paterson commented.[19] Since then a string of international writers has been approached, including Turkish novelist Elif Shafak, whose *The Last Taboo* was submitted in 2018; South Korean Han Kang, who wrote *Dear Son, My Beloved* for the archive in 2019; and Zimbabwean author Tsitsi Dangarembga, who contributed *Narini and Her Donkey* in 2022. The manuscripts are consigned to a specially built room called the "Silent Room" in Oslo's Deichman Library. "It feels like being inside a tree", Paterson observes; "It's quite magical, because it's very small and intimate: surrounded by tree rings, with light shining through the manuscript drawers".[20] Here they wait, unread, and warded by the Future Library Trust under the eye of project collaborator Anne Beate Hovind. "Until that day, only small groups of visitors may sit shoeless and in silent contemplation of the growing family tree of authors surrounding them", Rosie Goldsmith notes.[21]

Future Library has received much attention, partly because of its romantic conceit, and partly because of the way it seems to weave the ideas of the archive and ecology together. "This is a living, breathing, organic artwork that spans time, literature, ecology and human mortality", remarks the influencer Professor Wu.[22] Each year a ritual "handover ceremony" takes place in the Nordmarka forest, in which the author and supporters parade to the site. "I feel like now I belong to a circle, a family of very special souls, beautiful souls, who come from all walks of life", Shafak reflects in a short film celebrating the handover of her manuscript; "and what unites us is our faith, is our hope about the future, and also our commitment to books, to culture, to the future, to the art of story-telling".[23] A flute plays wistfully; the small community sits among the trees; she announces the title of her book—the only clue that will be given about its contents for the next hundred years.

Like David Milarch with his extraordinary backstory, Paterson has become a charismatic figure, and the project has been magnified thanks to our contemporary passion for environmental awareness. She cites on the one hand a conceptual rather than ecological impulse for the project: "making a connection with tree rings to chapters—the material nature of paper, pulp and books, and imagining the writer's thoughts infusing themselves, 'becoming' the trees".[24] On the other hand, she also sees inspiration in the spiritual practices of Shintoism in Japan, where "shrines have been built and rebuilt in exactly the same form every 20 years, for a few thousand years. And the community have been growing the wood for the shrines in a special forest, over this vast expanse of time".[25]

For *Future Library*, an area of the local Nordmarka forest has been cleared; 1000 trees have been planted, and in 100 years, these will be felled to produce the books. It's what commentator Paul Benzon calls "a kind of Möbius strip between the space of the forest and the space of the library over the next century".[26] He notes that "the project replenishes and accumulates wood in advance, in a manner that simultaneously repays the environmental debt incurred by the library and saves towards the future moment of publication".[27] This, he argues, is "a conservational project in both the archival and environmental senses of the word",[28] and other observers poetically agree. "By watering the saplings now—writing the manuscripts—we are tending to the future", suggests Arifa Akbar[29]; "The environmental message goes far beyond a library and a forest: this duality of preservation and creation asks us to take action, to conserve and protect, for the sake of tomorrow".[30]

Such a sentiment is surely honourable, though quite how actively participants take on environmental responsibilities as a result of engaging with *Future Library* might be called into question. There can often be something rather "eco-lite"[31] in such assumptions. The same might be said of comments evoking "the timeless nature of writing", or noting that "writing and literature can stretch across centuries; just as trees and forests do", as other observers have put it.[32] Some of the contributors voice emotive or spiritual reflections about the project: the South Korean author Han Kang enigmatically calls it "a century-long prayer"[33]; Atwood calls it a "meditation on time"[34]; while Jay Griffiths poetically acknowledges that "a book is not a dead tree but living wood turned into speaking leaves".[35] Meanwhile, Paterson is rather coy in addressing the theme of time, which has been a fascination for her throughout her oeuvre. "Why I'm drawn to time is hard to describe", she admits; "it's to do with being outside myself, and being inside a more universal network where distance and time might not necessarily even exist".[36]

If our critique of such remarks queries the ecological significance of this project, how do we otherwise gauge its value? Some of the more reasoned reflections on *Future Library* refer us to the way it causes us to reconsider the legacy we are handing down to—or storing up for—the next generation, and they see this in environmental terms. "The project asks that we delay our need for instant gratification", suggests Joanna Pascoe[37]; "not only does it encourage a more 'long-minded' perspective", writes Richard Fisher, "it also invites all those that learn about it to reflect on their duty to posterity and what they are leaving behind for future generations". He goes on:

> In the 21st Century, we are passing on so many malignant heirlooms to unborn people: a heated atmosphere, plastic fibres in the oceans, nuclear waste beneath the ground. It's rare that our generation is genuinely giving something to unborn descendants.[38]

Others frame the environmental resonances of the project more conceptually: "its time or rhythm is seed time, one of suspended becoming", writes Lars Bang Larsen. He playfully imagines *Future Library* not as a body of texts being sent into the future, but as an organism growing rhizomatically, such that "all the literary seeds it will contain will sprout at varying rates and in numerous directions".[39] Meanwhile, Joanne Pascoe picks up on the posthuman scholar Rosi Braidotti in seeing *Future Library* as "a transversal alliance between the bio of a forest, the geo of the land, the techno of a library and the writers of stories".[40]

As such, the project invites us to think in more critical terms about the nature of conservation, or at least of what we are conserving and how we are conserving it for the future[41]:

> Will any human beings be waiting there to receive it? Will there be a 'Norway?' Will there be a 'forest'? Will there be a 'library'? It's hopeful to believe that all of these elements—despite climate change, rising sea levels,

> forest insect infestations, global pandemics, and all of the other threats, real or not, that trouble our minds today—will still exist.[42]

Atwood's hope here is poignant, and laced with a contemporary anxiety about our environmental future. Such hope is perhaps what *Future Library* most evokes, which Joanna Pascoe refers to as "hope in practice, a patient transformation", or rather more esoterically "hopepunk praxis".[43] For 2015 author David Mitchell, "the Future Library project is a vote of confidence in the future. Its fruition is predicated upon the ongoing existence of Northern Europe, of libraries, of Norwegian spruces, of books and of readers".[44] And these kinds of sentiment are also articulated in the reflections of other commentators: "It is a project of magical, magnificent ambition, trust and hope", Goldsmith chimes; "hope that there will be books, readers and forests in 2114".[45] In hope, there is of course a sense of optimism about the future, though in terms of conservation the dynamics of *Future Library* might seem rather passive: a community gathering; a wistful flute; an enigmatic manuscript sealed into a time capsule; a waiting game. What happens if the world as we know it does not survive into the future? What happens if we can't ever read these books? Do we still take solace in the project's "meditation on time"?

Archives in Context

If the use of terms such as "archive" and "library" in these projects is esoteric, it's worth considering some of the ways in which archives have been theorized to note the recurring attractions of ecology and conservation within this area of study. In fact, Moore's article "Birds of a Feather", which applies an understanding of ecology to the practices of archivism, is not only a good example of such conceptual thinking, but also an example of the sort of ecological approach we have been taking throughout this book. Principally, the argument Moore makes is that the study of an archive ought not to be detached from an understanding of its context. It's a point that Jenkinson and Schellenberg also both made in their earlier treatises, so to contextualize the archive in itself is nothing new. But Moore considers this topic using the metaphors of natural science. If taken in isolation, he warns, "each collection becomes like an animal at the zoo"; "Each is safe and well cared for by the custodial archivist, each can be studied and researched, and yet each is in an artificial environment, forever separated from its habitat". By contrast, "If we understand archives to be small ecosystems containing dynamic chains of energy and representations of its non-privileged members, a new perspective comes into play". Just like ecosystems, he suggests, "archival records and their metadata are continually shifting, transforming, and gaining new meanings, rather than remaining fixed, static objects". More than this, as we are now able to acknowledge, even human intervention in a system is a part of its dynamic, as we affect the ecosystems with which we engage. Thus he is able to characterize archivists as "active shapers of the archival heritage" rather than merely "passive keepers of the entire documentary residue".[46] Maybe *Future Library* is not such a passive waiting game, then—though perhaps in assuming that we are waiting for the manuscripts to be unlocked, we are missing the point. It is through witnessing the active shaping of figures such as Paterson and Milarch that we can contextualize their archives and acknowledge their performativity.

To be sure, Milarch's initiatives have been critiqued in these terms. "The logic of the group's well-meaning program derives more from American culture than botanical science", suggests Jared Farmer, noting that "There is something distinctively American—and masculinist—about the practice: bigger means better, biggest wins first prize, and the winner gets a photo shoot".[47] This reminds us not only of our discussion of charismatic species preservation in the previous chapter, but also of those abiding mythologies of the frontier that have rhapsodized America's dominion over the land ("from sea to shining sea"). In this light, Milarch implicitly moves on from arm-wrestling other humans to hunting big game, conquering giant trees, and owning America's majestic wilderness. Farmer is among several commentators who have queried the scientific credentials of the project: "Champion trees are certainly amazing", he writes, "but they are not, in fact, genetically superior, simply lucky: the beneficiaries of optimal conditions".[48] Meanwhile, as Jens Staal submits, the Champion Tree Project is "arguably inspired by the aesthetic value of the immense redwood trees" rather than scientific evidence per se. Moreover, "introducing those trees to areas where they do not grow naturally", he argues, is "introducing an anthropogenic component".[49] Perhaps such reservations have good grounds: strip away the fabulous origin story of Milarch's project, his high-profile photo-ops, and even his green credentials, and you have a latter-day Prospero: the conquest of nature; the science of cloning; the power of man to manipulate the natural world. And this raises a number of ethical questions about our human engagement with nature even when that is driven by the ostensible virtues of conservation: how far are efforts merely perpetuating the interference of human beings with the natural world? To what extent should we intervene with natural processes, even if our approaches are scientifically informed? How authentic are our green actions, and how much are they simply performances of care, custodianship, or hope?

If Milarch performs in his activities the frontier narratives of America's fascination with the land, Paterson and her collaborators perform other assumptions, not least about the value of books. Not content with preserving texts that already occupy the canon, as a conventional time capsule might pass on the work of poets whose work we see as "timeless", *Future Library* archives texts that have not yet even been proven, attributing to them automatically and without question a mystical power. Framed by the authority of lavish artistic rituals (with ceremonies, wistful flutes, silent rooms), invocations of environmental good ("A forest in Norway is growing"), and heavily weighted anticipation (earnest commentary about the passing of time and the gifting to future generations), *Future Library* elevates its status as a repository of timeless literature, whether or not its content has any literary merit. This equally freighted mythology contextualizes *Future Library* and highlights qualities we value today: the magic of the written word; the spiritual power of writing; the sanctity of art.

Yet there is also a further performance in the engagement of the project's followers and the expanding literature of *Future Library*'s secondary archive. We can't of course consider the actual texts of this library: the majority have not been written, and those that have are locked away like the stuff of fairytale, or perhaps more appropriately, hibernation. All we have are the enigmatic titles, a silent room of contemplation, a ritual handover, the comments of participants and observers, and press attention generating articles and articles and articles. Yet in that significant digital archive of metatexts (press files, videos, podcasts, interviews, and essays), in the engagement of followers at the handover events, in the involvement of contemplators in the silent room, and even in the apparently passive statements of hope for the future, we can sense another performance—one of community and solidarity. These are all individuals, aware that their own actions may be insignificant against the hyperobjects of climate change and global warming.[50] In coming together in conceptual accord, in quiet contemplation, and even in digital affinity, these individuals envisage the greater potential of speaking with like-minded communities around the world. In this sense both *Future Library* and the Champion Tree Project, in their widest sense, are hubs for our communal ecological focus, beacons drawing our collective will, opportunities inviting us to perform our connection to the planet, and even portals to our future world.

We'll return to a discussion of that future world towards the end of this chapter, but first it is worth considering some of the ways in which nature has its own means of archiving for the future, and some of the ways in which it pulls us collectively in to that endeavour through the establishment of seedbanks.

Seed Dispersal and Seedbanks

"Evolution has the effect of making all living beings forward-thinking", writes Sarah Spencer, noting that "all trees will attempt to provide their offspring with the most ideal conditions". After all, she suggests, "their aim is for their 'children' to survive, regardless of whether they are around to see them grow".[51] Plants will try to engineer the best chances for their offspring through various means of dispersing their seeds as widely as possible. Through *anemochory*, or dispersal by the wind, a tree will allow its seeds to be carried beyond the span of its branches; using *hydrochory*, or dispersal in water, seeds will float downstream to lodge in fertile soil further away; thanks to *zoochory*, or dispersal by animals, seeds will be carried greater distances as they become trapped in fur or hooves; and in some cases, known as *anthropochory*, they will be dispersed by humans as they cling to our boots or our vehicles, or find themselves among goods we pack and ship abroad.[52] The different means of dispersal thereby scatter seeds at varying distances from the mother plant: even in the nineteenth century, Darwin reported seeds found in mud clinging to the legs of migrating birds, deposited in the dung of locusts, and transported in organic matter found in ice floes[53]; Clement Reid reported animal carcasses washed up by the sea with seeds in their stomachs, pigeon remains with undigested seeds in their craw, and uneaten acorns dropped by greedy rooks[54]; and Charles Lyell even reported seeds having been carried across oceans in the cavities of pumice.[55] Some plants have adapted their seeds to maximize this potential for dispersal: the seeds, for example, of birch trees have developed "wings"—an advantage known as *pterometeorochory*, which allows them to be caught by the wind and deposited as far from the mother tree as possible; maple and sycamore seeds (commonly known as "helicopter seeds") have evolved even more complex wing structures enabling their seeds to "autorotate and generate lift as they fall".[56] Through such mechanisms their seeds can fly up to 150 metres from their origin.[57]

Still, the success of even these methods in producing a new plant relies to an enormous extent on chance: the seed must land in suitable soil; it must avoid being eaten by birds or animals; it will need sufficient light and water to develop; and even if it takes root it will remain vulnerable during its early stages. Adult plants try their hardest to maximize success against such odds, and one way of doing that is through the sheer numbers of seeds they produce: "Every five years, a beech tree produces at least thirty thousand beechnuts", writes Peter Wohlleben, calculating a total of 1.8 million seeds being dispersed over the lifespan of a tree. Yet the proportion that successfully grow into adult trees is extremely low: on average, he figures, out of the 1.8 million seeds "exactly one will develop into a full-grown tree".[58]

It therefore seems a miracle that trees produce any offspring at all, but another of the ploys that a mother plant will use to ramp up its chances is to create a store of seeds which lie dormant until their growth mechanisms are activated further down the line. Underneath the soil around an origin plant, whether that is a tree, a bush, or even a weed, can be found a rich "bank" of seeds that are waiting and biding their time. Some of these seedbanks are deemed to be "transient"— seeds build up annually during fertile periods but germinate relatively quickly and then deplete the seedbank; while others are "persistent", storing seeds for long periods—"buried seeds of some arable weeds can retain viability in undisturbed soils for periods in excess of 50 years".[59] Research in a 1930s winter wheatfield revealed as many as 39,000 seeds per square metre, though quantities can vary considerably; in woodland the number might only reach a few hundred or maybe a thousand in the same space. Meanwhile, the seeds themselves are diverse: the wheatfield research revealed 47 species of seed, which were two-thirds dominated by poppy seeds, while in woodland samples, a surprizing proportion come from species not even present in the vegetation above ground. Among a wealth of bramble seeds, potentially dropped by birds, and other species that are likely "relics of previous vegetation", or "carried in as part of the seed rain from other areas", the seeds of tree species themselves, with the common exception of silver birch, are relatively rare. "Part of the reason may be that the larger tree seeds are so heavily predated and decompose rapidly", Harper speculates, "but another factor is the delayed reproduction of climax tree species which may be present in the vegetation for 20 or more years before they produce seed".[60]

More recent research has revealed some of the adaptations and ecosystem processes that are at work in the operations of soil seedbanks. Seeds again develop adaptations to ensure being securely deposited in the bank: "active" or "passive" awns enable them respectively to push into surface cracks, or to fall in such a way that they will penetrate the surface of the soil; meanwhile, many of the seeds are extremely small, meaning that they are able to be ingested by earthworms close to the surface, transported down to safe levels, and then returned to the surface in earthworm casts, where they may eventually grow roots and sprout.[61]

Through various approaches to reproduction, dispersal, and storage, plants thus ensure species survival into the future. Even so, numerous threats to their long-term survival, from the impacts of global warming to the activities of deforestation, and from the contamination of the land through pollution to the destruction of habitat through development, can all pose existential risks to whole species of flora. Indeed, scientists have used AI to calculate that up to 45% of the world's 328,565 known flowering plant species are susceptible to extinction.[62] Small wonder that conservationists like David Milarch offer a human hand. Other initiatives the world over have created artificial seedbanks for a similar purpose.

The Global Seed Vault

The icy archipelago of Svalbard, between Norway and the North Pole, is home to the world's most comprehensive archive of such seeds. The Global Seed Vault, housing 1.3 m seed samples from over 6,000 species of plants, is a storage facility for seedbanks from across the world to keep copies of their seeds. Its purpose is to "offer safe, free and long-term storage of seed duplicates from all genebanks and nations participating in the global community's joint effort to ensure the world's future food supply".[63] Working just like a financial bank, customers—who in this instance are generally nation states or conservation organizations—can deposit their savings. Over 100 different seedbanks have contributed to the seedbank,[64] which is located in the perfect place, in climate-safe conditions 150 metres inside the rock of this arctic archipelago: "even though Svalbard is affected by global warming", the website admits, "it is still expected to remain one of the world's coldest places". This will ensure the continuing safe storage of seeds as the world warms. However, the real value of the seedbank, argues the facility's operator NordGen, is "in providing back-up to individual collections in the event that the original samples, and their duplicates in conventional genebanks, are lost due to natural disasters, human conflict, changing policies, mismanagement, or any other circumstances".[65] Opened in 2008, the withdrawal services of the Seed Vault have so far only been demanded twice, when it was called upon by the International Institute for Agricultural Research in Dry Areas (ICARDA) to repopulate seed collections that had been destroyed by conflict in the Middle East.

"The seeds lying in the deep freeze of the vault include wild and old varieties, many of which are not in general use anymore", reports Jennifer Duggan; "But the genetic diversity contained in the vault could provide the DNA traits needed to develop new strains for whatever challenges the world or a particular region will face in the future".[66] Wheat, rice, cereal, vegetable, and forage crops are heavily represented, with over 200,000 species of rice alone in the repository. In this sense the Svalbard Seed Vault is a compelling example of global civil spirit, with the Norwegian government footing the bill for the rest of the world's nations to protect their food security. To honour those commitments, Svalbard maintains tight restrictions over its contents. "The vault does not include research facilities", Sheryl D. Breen reports; "its governing bodies have no subsidiary contracts with research institutions, and its High Arctic location does not allow for propagation".[67] Even so, the geopolitical complexities are measurable, relating to issues of food sovereignty and scarcity, ownership and property rights, ethics, and access. Ethical concerns are common enough to any archive, regardless of its contents, and reinforce the need for coherent policy decisions and robust management practices to be well-implemented from the start. The ex situ nature of this particular archive though, which translocates its stock from all over the world, and its sub-zero storage conditions, which artificially force dormancy onto the seeds, are both concerns for some communities, especially first nations communities whose relationships with heritage and the land are strongly embedded in their beliefs. "Because seeds are living beings that exist within a web of relationships", explains Breen, "they are connected to the human who plants the seed, the microbes that live in the soil alongside the seed, the soil itself, the harvester, and those who use and/or consume the plant".[68] Any artificial means of preserving materials, no matter how well-intentioned, poses particular ethical dilemmas for some communities. In terms of respecting the living spirit of our ecological biome, "genebanks negate the essence of 'seedness'", they argue.[69]

Invoking such a broad web of relationships returns us to indigenous worldviews, and the precepts of deep and dark ecology, all of which understand humans to be part of the ecosystems they inhabit. From such a perspective, the very idea of an archive existing in a conceptual bubble like a vault or a silent room is fundamentally flawed. Just as "the human who plants the seed, [...] the harvester, and those who use and/or consume the plant" are intrinsic to their systems, so too are Milarch and his team, Paterson and her contributors, and Svalbard's scientists to theirs. It could be argued that such conservationists provide crucial ecosystem services; that Svalbard enhances the organic processes of a "natural" soil seedbank, maximizing storage conditions, improving yield ratios, and ultimately sustaining the new evolutionary ecosystem of the Anthropocene. It might even be argued that the seeds are complicit in this process, adapting to use human prostheses in their quest for survival. Perhaps it is not after all David Milarch who holds dominion over his trees, but those venerable champions who exploit the services of a useful worker within their ecosystem. That might sound far-fetched, though it's worth drawing attention to another more organic archiving example common to many regions in the global South, in which the natural and human ecosystem processes of conserving and storing seeds are far more integrated.

Community Seedbanks

"Community seed banks tend to be small-scale local institutions that store seed on a short term basis, serving individual communities or several communities in surrounding villages", writes Bhuwon Sthapit.[70] Initiatives to develop such community resources began to emerge in the 1980s, enabling smallholder farms, particularly in the global South, to "address local conservation needs", "back up conventional gene banks", "ensure local seed supply", and maintain "food sovereignty"—the right of communities to take ownership over food cultivation and provision.[71] As Sthapit adds, these initiatives also enable "community [and especially] women empowerment", "Farmers' Rights (recognition, access, participation in decision making and benefit sharing)", "community resilience", and "the sharing of information and knowledge".[72] Although some seedbanks exist on individual farms,[73] many localities use cooperative practices to pool seeds in the interest of community food security. "Key functions of these networks are saving and exchanging local seeds and keeping them under the control of the farming community, allowing easy access and use at an affordable price and in the right time", Sthapit explains.[74]

Lewis and Mulvany identify several different types of community-based seedbank, three of which are particularly relevant. First, *de facto seedbanks*, which are the most ad hoc, are operated by individuals keeping reserves of seeds on their premises and informally exchanging or trading them with neighbours; locally-held seed is often trusted more by farmers in the global South, thanks to the well-documented reliance on family or neighbourhood networks in many communities. On the other hand, individual farmers sometimes lack the skills to keep the seed adequately or to combat contamination, meaning that locally-held seed can be prone to disease or rot. In some neighbourhoods, Lewis and Mulvany's second type of seedbank, a *community seed exchange* will therefore be set up to facilitate local trade and sometimes competition.[75] This is "a community-managed approach that expands local practices from the household seed store to the community level through collective actions",[76] offering farmers the chance to trade seeds grown in excess of their subsistence requirements, often in home gardens, where they may experiment with small quantities of different plant or vegetable types.[77] "Seed exchanges usually do not involve monetary transactions", note Vernooy et al.—"seeds borrowed are paid back with seeds; thus, they allow greater participation, even of cash-poor households".[78] In this way, agency is given to communities to participate equitably with others in their network. The third type of seedbank is the *organised seedbank*, a more formally established collection for distribution and exchange. This may be established with the support of an outside organization, perhaps arising from a research or breeding programme, or being driven by a charity or NGO. Once established, the initiative will often be handed over to communities such that seedbanks become "organised as community undertakings", and "owned and managed by community assemblies".[79] Ongoing interventions by charities or NGOs may continue to contribute to the welfare of the seedbank by "training farmers in plant breeding techniques, pre-harvest selection and post-harvest storage",[80] and it may be through more formal arrangements that community farms are helped to prosper.

By way of examples, rural development researcher Ronnie Vernooy and his colleagues have outlined the community seedbank approaches of a number of areas. For instance, they have documented a project set up by the non-governmental organization LI-BIRD (Local Initiatives for

Biodiversity, Regeneration, and Development) in Nepal to conserve local crop varieties. This began in 2012 and has since been handed over to communities. "At the beginning of each growing season, the leaders of the community seed bank organize a meeting with villagers to determine which seeds they would like to borrow", they explain; "anyone who borrows from the fund must grow at least one traditional crop variety", and "each village takes care of the regeneration and seed multiplication of all varieties of one crop". Meanwhile, in Guatemala, a recognition that traditional crop-farming knowledge was beginning to disappear as farmers bought in commercially available seeds which did not perform as well in local soils, led the agricultural cooperative Asocuch to work alongside the government to create a participatory breeding programme. "Farmers were trained in selection techniques", write the researchers, "whose application gradually improved the performance of local varieties according to farmers' preferences". With the conservation of local seeds and an increase in crop yields, this has gone some way to "achieving resilience of local communities and agro-ecosystems" for at least 5000 people in the region. Finally, in Bangladesh, the Mamudpur Nayakrishi Seed Hut was set up in 2001 by UBINIG, a non-governmental organization specializing in policy research for alternative development, and handed over again to communities. "A group of farming households jointly takes responsibility for looking after their seeds, which are collected and regenerated on behalf of a wider community of farmers", the report observes. In particular, the shared knowledge of this arrangement has enabled bespoke selection and breeding of seeds to withstand the variable climatic conditions of the region: "Farmers are now selecting rice varieties that can survive dry conditions and adapt if flooding occurs".[81] It's clear that these practices are workable solutions for food security, operating at grassroots levels but as dynamic interactions between different parts of the ecosystem. While such a system might not be what is typically explained by ecologists as a "food web", it nevertheless bears many similarities in terms of how processes flow, how services are exchanged, and how needs are resolved, especially from the perspective of "process-functional" thinking.[82]

There are many synergies between these examples of "archival" seedbank practice and the archiving of cultural traditions through the preservation of oral traditions among indigenous communities (or in the more formal holdings of museums and collections). In many ways, such cultural archives fulfil similar ecosystem functions to seedbanks, and can also be seen in terms of sustainability, conservation, access, and sovereignty.[83] Traditions, ways of life, interpersonal relationships, and communities themselves are maintained (conservation, sustainability) as long-established expressions of self-identity owned by the communities within which they take place (access, sovereignty). Perhaps these synergies between community seedbank practices and cultural preservation approaches become especially poignant at this particular moment of space–time: a heightened time of ecological anxiety, and a magnified moment of decolonial awareness. Perhaps they also stand out to us in the identities we have grown into and which we now perform: liberal intellectuals within the knowledge tradition; inheritors of the economic and social privileges of the global North. This—the web that informs our worldview—is just another example of an ecosystem that has evolved to be the way it is: things will change, things will be differently perceived elsewhere, and things will operate in ecologically varied ways in other circumstances, habitats, and environments. Such is ecological flux, recognized elsewhere as attitudes shift or paradigms change.

As it happens, the archival scholar Terry Cook gives us a useful stop-motion overview of this sort of flux: "I want to suggest that since the later nineteenth century, archival identity has shifted, or has been in the process of shifting, through four [...] paradigms or frameworks or mindsets", he writes.[84] First it was the quest for *evidence* that was a guiding principle, overwhelming any other concerns and dominating our attitudes towards the archives; next (from about 1930) he suggests that *memory* emerged as the arbiter of archival selection; third (from the 1970s) came a focus on *identity*, offering a reference point for a rapidly changing world, and acknowledging the archivist as "a conscious mediator"[85]; and finally (Cook was writing in 2013), "a fourth archival mindset is on the horizon", enabled by twenty-first-century attitudes, awareness, and capabilities (media saturation, Web 2.0, co-production). Nowadays, "*Community* is the key concept", Cook writes, "a democratizing of archives suitable for the social ethos, communication patterns, and community requirements of the digital age".[86]

Cook's focus on archives in this way recognizes the "process-functional" dynamics that inform our relationships within the ecosystems we inhabit, which is why he has elsewhere referred to archives as "dynamic virtual concepts", "active agents themselves in the formation of human and organizational memory".[87] We work with them, we respond to them, they call upon us, we are synergy. Gone are any previous assumptions that archivists are "passive guardians of an inherited legacy"; in the new paradigm, they can be "celebrating their role in actively shaping collective (or social) memory".[88] This is "a shift in core principles, from exclusive custodianship and ownership of archives to shared stewardship and collaboration".[89] In the words of another archival scholar invoking the nation-building ideals of an earlier generation, these are "archives of the people, by the people, for the people".[90]

Selectivity and Curation

As activities, traditions, or collections become increasingly formalized, and especially as they grow, decisions define parameters, mark out boundaries, and identify material for inclusion. We have seen already that wheat, rice, cereal, vegetable, and forage crops are all heavily represented in the Svalbard Seed Vault, a decision determined by the food security need for these crops in the event of planetary emergency; the collection also houses old and wild varieties of seed, reflecting a general consensus mitigating against species loss. In many ways, these pragmatic curatorial decisions reflect no more than a tangible example of Raymond Williams' "selective tradition", an abstract concept that he used to explain how cultures develop characteristic identities: "from the whole body of activities, certain things are selected for value and emphasis", he wrote.[91] Such emphasis becomes sustained, and it is this that influences how we represent those cultures to ourselves and to the world in the stories we tell, the attitudes we perform, and the archives we preserve. Other theorists have taken up the same idea: "social memory is [...] highly selective", writes Hall about the way we understand "our" heritage; "it highlights and foregrounds, imposes beginning, middles and ends on the random and contingent. Equally, it foreshortens, silences, disavows, forgets and elides many episodes which—from another perspective—could be the start of a different narrative", he writes.[92] Elsewhere, Cook has made a similar point about archives: "With memory comes the inevitable privileging of certain records and records creators, certain functions, activities, and groups in society, and the marginalizing or silencing of others".[93] This is by no means a process of *natural selection*, and nor is it quite a *survival of the fittest*, though such Darwinian terms in ecology signal that parameters of selection operate in the natural world as much as the cultural. In cultural terms these concepts may hew close to the adage that history is written by the victors, or at least by the dominance of Western worldviews, and for reasons determined by that significance: somebody makes selective decisions about what gets saved into archives, how communities are represented in culture, and how dynamics of ownership, agency, and access are controlled.[94] Such dynamics are defined by the contexts in which they operate: the exigencies of the now; the paradigm of the present; the performativity of the players.

In his essay "Constituting an Archive", cultural theorist Stuart Hall reflects on the words of Jamaican anthropologist David Scott. He recognizes "a relation among past, community and identity" that enacts, in Scott's evocative phrase, "the meaning in the present of our past".[95] And this is a fascinating way to conceptualize the various archives we have introduced in this chapter, since it brings to light the sort of focus, fascination, even obsession, that guides us in our acts, whether those be of celebration or conservation. We've noted already that one of the impulses guiding David Milarch to clone the most spectacular "champion" specimens of tree is perhaps rooted in the cultural fascination of America with strength, power, and its legacy of dominion over the land. We can also see in the *Future Library* project a fascination with writing as precious, sanctified, even spiritual. In David Scott's terms, these archives are articulating resonances of the past that hold formative meaning for us in the present. As we've also seen, these collections are also performances of our present which we are archiving for the future world.

Notes

1. Archangel Ancient Tree Archive (2015), *Archangel Ancient Tree Archive Press Kit*, https://www.ancienttreearchive.org/wp-content/uploads/2015/12/ATA1955_DigitalPressKit_Update_Dec9.pdf. Accessed 27 December 2024, p. 10.
2. Jim Robbins (2012), *The Man Who Plants Trees*, London: Profile Books, p. 202.
3. Robbins, *The Man Who*, pp. 18–19.
4. Roderick Nash (1982), *Wilderness and the American Mind*, New Haven: Yale University Press, p. 129.
5. Erik A. Moore (2007), "Birds of a Feather: Some Fundamentals on the Archives–Ecology Paradigm", *Archivaria*, 63, pp. 103–119, p. 107.
6. Hilary Jenkinson (1937), *A Manual of Archive Administration* (new and revised edition), London: Percy Lund, Humphries & Co., p. 11.

7. Theodore R. Schellenberg (1956), *Modern Archives: Principles and Techniques*, Chicago: University of Chicago Press, p. 13.
8. Schellenberg, *Modern Archives*, p. 19, quoting from a 1947 lecture given by Jenkinson and entitled "The English Archivist: A New Profession".
9. Robbins, *The Man Who*, pp. 19–20, 43.
10. Laura Helton (2021), "Archive", in Michelle Kennerly, Samuel Frederick, and Jonathan E. Abel (eds) (2021), *Information: Keywords*, New York: Columbia University Press, pp. 45–56.
11. Marguerite Guzman Bouvard (2016), *Social Justice and the Power of Compassion: Meaningful Involvement of Organizations Improving the Environment and Community*, Lanham: Rowman & Littlefield, p. 16.
12. Archangel Ancient Tree Archive (AATA) (2024), "AATA Living Archives Save Genetics for Future Generations", *Archangel Ancient Tree Archive*, https://www.ancienttreearchive.org/our-work/archive/. Accessed 15 October 2024.
13. Archangel Ancient Tree Archive (AATA) (2024), "About us", *Archangel Ancient Tree Archive*, https://www.ancienttreearchive.org/about-us/. Accessed 15 October 2024.
14. Bouvard, *Social Justice*, p. 21.
15. Bouvard, *Social Justice*, p. 13.
16. Jaq Payne (2024), "The National Champion Tree Program: An Honor for Tennessee, a Boon for the Nation", *Tennessee Turfgrass*, February/March, pp. 22–27, pp. 26–27. https://theturfzone.com/tta/?ascat=7&sub=issue&issue_id=8627&rti=true&rel=2. Accessed 22 October 2024.
17. Anon. (2014), "Intro", *Future Library*, https://www.futurelibrary.no/. Accessed 5 November 2024.
18. Katie Paterson (2014), "The Artwork", *Future Library*, https://www.futurelibrary.no/#/the-artwork. Accessed 5 November 2024.
19. Alison Flood (2014), "Margaret Atwood's new work will remain unseen for a century", *The Guardian*, 5 September, https://www.theguardian.com/books/2014/sep/05/margaret-atwood-new-work-unseen-century-future-library. Accessed 28 August 2024.
20. Paterson in Richard Fisher (2022), "The Norwegian library with unreadable books", *BBC*, https://www.bbc.com/future/article/20220630-the-norwegian-library-with-unreadable-books. Accessed 5 November 2024.

21. Rosie Goldsmith (2022), "Future Library opens secret archive of unseen texts in Oslo", *The Guardian*, 14 June, https://www.theguardian.com/books/2022/jun/14/future-library-opens-secret-archive-of-unseen-texts-in-oslo. Accessed 28 August 2024.
22. Professor Wu (2018), "Creatives in profile—interview with Katie Paterson", *Nothing in the Rule Book*, 23 May, https://nothingintherulebook.com/2018/05/23/creatives-in-profile-interview-with-katie-paterson/. Accessed 28 August 2024.
23. Katie Paterson (2018), "Future Library Handover Day with Elif Shafak, 2018", https://vimeo.com/282327747. Accessed 7 November 2024.
24. Claire Fallon (2015), "'Future Library' Reminds Us How Connected Books And The Environment Really Are", *Huffington Post*, 10 June, https://www.huffpost.com/entry/future-library_n_7532012. Accessed 28 August 2024.
25. Paterson in Professor Wu, "Creatives in profile", n.p.
26. Paul Benzon (2016), "On Unpublishing: Fugitive Materiality and the Future of the Anthropocene Book", in Annette Gilbert and Priscilla Posada (2016), *Publishing as Artistic Practice*, Berlin: Sternberg Press, pp. 282–294, p. 291.
27. Benzon, "On Unpublishing", p. 291.
28. Benzon, "On Unpublishing", p. 291. Benzon self-consciously uses economic language ("transaction, debt, repayment") in discussion of this archive, not only to remark on how "climate change is inextricably bound up with the longue durée of globalization and market capitalism", but also to note Norway's role in REDD, the Reduced Emissions from Deforestation and Forest Degradation scheme, "in which developed nations pay aid to developing nations in return for an agreement to slow or cease deforestation".
29. Arifa Akbar (2018), "The Future is not a Foreign Country", p. 3, https://assets.ctfassets.net/9sa97ciu3rb2/7yFifl9N9SAmia2QMcIaSi/dbc30d23424f72a31f3e2f3d6176078f/The_Future_is_not_a_Foreign_Country-Arifa_Akbar.pdf. Accessed 7 November 2024.
30. Akbar, "The Future".
31. The term "eco-lite" is typically used in the context of eco-tourism to critique the pull of light-touch eco-awareness as a way of commercializing travel. See among others Martha Honey (2003),

"Protecting Eden: Setting green standards for the tourism industry", *Environment: Science and Policy for Sustainable Development*, 45:6, pp. 8–20.
32. Professor Wu, "Creatives in profile", n.p.
33. Han Kang (2019), Text for *Future Library bBooklet*, https://assets.ctfassets.net/9sa97ciu3rb2/4UxWvLzP7sBusxQykz6dCp/b435e9c8fd5e65735126575ecd37ffd1/Han_Kang_English.pdf. Accessed 11 November 2024.
34. Margaret Atwood (2014), "Future Library", n.p., https://assets.ctfassets.net/9sa97ciu3rb2/2hdAyLQYmESc0eYemIEcm2/09772ac1c62defc7ccf50fe6ea207a83/Margaret_Atwood.pdf. Accessed 6 November 2024.
35. Jay Griffiths (2020), "Future Library" (ed. Faye Fournasier), in Katie Paterson (ed.) (2020), *Future Library Booklet 2021: Ocean Vuong*, n.p., https://assets.ctfassets.net/9sa97ciu3rb2/6mhd4RgwmA0Cj5b12JmcPw/19ad8d8bf67ce06d0db9aa7381f146b7/2616_KatiePaterson_FutureLibrary_Vuong_Booklet_VISUAL_2.pdf. Accessed 7 November 2024.
36. Paterson in Alison Hugill (2015), "Science: An Interview with Katie Paterson", *Berlin Artlink*, 23 December, https://www.berlinartlink.com/2015/12/23/science-an-interview-with-katie-paterson/. Accessed 9 November 2024.
37. Joanna Pascoe (2024), "Future Library: Hopepunk Heroism", *Heroism Science* 9.1, pp. 1–10, p. 3.
38. Fisher, "The Norwegian Library", n.p.
39. Lars Bang Larsen (2017), "The Manuscripts Stored in Oslo", n.p., https://assets.ctfassets.net/9sa97ciu3rb2/taYrxz0CaGWkMuws2UEGs/11e3ffbb41faac94e6b4eeeda80a2967/Katie_Paterson_Lars_Bang_Larsen_essay2017.pdf. Accessed 7 November 2024.
40. Pascoe, "Future Library", p. 4.
41. One way or another, the notion of time recurs in contributions to *Future Library*: the word "Narini", used in Dangarembga's title, derives from the Zimbabwean word for "infinity" (Fisher, "The Norwegian Library", n.p.); meanwhile, the second contributor David Mitchell's entry is called *From Me Flows What You Call Time* (2015), referencing Japanese composer Tōru Takemitsu's 1990 concerto of the same name, itself taken from the poem "Clear Blue Water" by Makoto Ōoka. In Ōoka's poem, the phrase is called out by a Swiss glacier as it melts into the Rhone.

42. Atwood, "Future Library".
43. Pascoe, "Future Library", p. 2.
44. David Mitchell (2015), "The Ayes Have It", n.p., https://assets.ctfassets.net/9sa97ciu3rb2/1RU9pw3kiUmSmS8i60sciW/e7e326d546dbe4d88a018ef8b6009f75/David_Mitchell.pdf. Accessed 6 November 2024.
45. Goldsmith, "Future Library", n.p.
46. Moore, "Birds of a Feather", pp. 115–118.
47. Jared Farmer (2019), "American Champion Trees and the Wye Oak of Maryland", *Arcadia*, 3:1, Rachel Carson Center for Environment and Society.
48. Farmer, "American Champion Trees".
49. Jens Staal (2019), "Artistic research in breeding: The Bifrost Eucalyptus project", *Research Catalogue*, 9. https://www.researchcatalogue.net/view/392580/392581. Accessed 15 October 2024.
50. Timothy Morton (2013), *Hyperobjects: Philosophy and Ecology After the End of the World*, Minneapolis, University of Minnesota Press.
51. Sarah Spencer (2019), *Think Like a Tree: the Natural Principles Guide to Life*, Swarkestone: Swarkestone Press, p. 183.
52. Pascal Vittoz and Robin Engler (2007), "Seed dispersal distances: a typology based on dispersal modes and plant traits", *Botanica Helvetica*, 117, pp. 109–124.
53. Charles Darwin (1900), *The Origin of Species*, New York: H.M. Caldwell & Co., pp. 358–367.
54. Clement Reid (1899), *The Origin of the British Fauna*, London: Dulao & Co., pp. 20–32.
55. Charles Lyell (1875), *Principles of Geology* (twelfth edition), London: John Murray, p. 379.
56. Madeleine Seale and Naomi Nakayama (2020), "From passive to informed: mechanical mechanisms of seed dispersal", *New Phytologist*, 225:2, pp. 653–658, p. 654.
57. Vittoz and Engler, "Seed Dispersal", p. 112.
58. Peter Wohlleben, *The Hidden Life of Trees*, London: William Collins, p. 29.
59. Ken Thompson and John Philip Grime (1979), "Seasonal variation in the seed banks of herbaceous species in ten contrasting habitats", *The Journal of Ecology*, 67, pp. 893–921, p. 894. Thompson and Grime's typology of four different types of soil seedbank has become the standard point of reference. Elsewhere, Thomspon

has observed the seeds of some variations of St John's Wort being stored for up to 200 years. Ken Thompson (1987), "Seeds and Seed Banks", *New Phytologist*, 106, pp. 23–34, p. 26.

60. John L. Harper (1977), *Population Biology of Plants*, London; Academic Press, pp. 85–95.
61. Thompson, "Seeds and Seed Banks", pp. 26–27.
62. Steven P. Bachman et al. (2024), "Extinction risk predictions for the world's flowering plants to support their conservation", *New Phytologist*, 242, pp. 797–808.
63. Svalbard Global Seed Vault (2024), "Purpose, Operations, and Organisation", *Svalbard Global Seed Vault*, https://www.seedvault.no/about/purpose-operations-and-organisation/. Accessed 8 October 2024.
64. NordGen (2024), "The Seed Portal", *NordGen*, https://www.nordgen.org/our-work/svalbard-global-seed-vault/the-seed-portal/. Accessed 8 October 2024.
65. Svalbard Global Seed Vault, "Purpose".
66. Jennifer Duggan (2024), "Inside the 'Doomsday' Vault", *Time Magazine*, https://time.com/doomsday-vault/. Accessed 4 November 2024.
67. Sheryl D. Breen (2015), "Saving seeds: The Svalbard Global Seed Vault, Native American seed savers, and problems of property", *Journal of Agriculture, Food Systems, and Community Development*, 5:2, pp. 39–52, p. 42.
68. Breen, "Saving Seeds", p. 46.
69. Breen, "Saving Seeds", p. 47. Ironically, although Milarch's processes come across as rather casual compared to the formal restrictions of Svalbard's more clinical approach, the responses of First Nation communities to his exploits seem more supportive than those directed towards resources like Svalbard.
70. Bhuwon Sthapit (2013), "Emerging theory and practice: Community seed banks, seed system resilience and food security", in Pitambar Shrestha, Ronnie Vernooy and Pashupati Chaudhary (eds) (2013), *Community Seed Banks in Nepal: Past, Present, Future*, Proceedings of a National Workshop, LI-BIRD/USC Canada Asia/Oxfam/The Development Fund/IFAD/Bioversity International, 14–15 June 2012, Pokhara, Nepal, pp. 16–40, p. 24.
71. Ronnie Vernooy et al. (2014), "The multiple functions and services of community seedbanks", *Resources*, 3:4, pp. 636–656, p. 639.

These drivers in the global South differ from those of community seedbanks in the North, which may be more intrigued by preserving heritage or "heirloom" seed varieties.

72. Sthapit, "Emerging theory", pp. 20, 18.
73. "Overall, some 80–90% of seed used by smallholder farmers in developing countries is sourced from informal farmers' seed systems on farm". Sthapit, "Emerging theory", p. 17.
74. Sthapit, "Emerging theory", p. 20.
75. High status accords to those who win awards in competition, and Lewis and Mulvany report that in some communities women especially gain prestige by displaying more types and varieties of crops than men. V. Lewis and P.M. Mulvany (1997), *A Typology of Community Seed Banks*; Chatham: Natural Resources Institute, p. 16.
76. Sthapit, "Emerging theory", p. 28.
77. Lewis and Mulvany, *A Typology*, pp. 15–17.
78. Vernooy et al., "The multiple", p. 644.
79. Lewis and Mulvany, *A Typology*, pp. 18–19.
80. Lewis and Mulvany, *A Typology*, p. 32.
81. Vernooy et al., "The multiple", pp. 643–648.
82. Robert V. O'Neill et al. (1986), *A hierarchical concept of ecosystems*, Princeton: Princeton University Press, p. 1.
83. Vernooy et al., "The multiple", p. 641. On the other hand, it may be simplistic to assume that any of these vectors are as neutral as we imply, since such traditions—like the seedbanks—will be guided by the complex dynamics that underpin the formation of cultures and communities in the first place: belief systems, social hierarchies, kinship patterns, and decision-making. As Lewis and Mulvany acknowledge, "there is a hierarchy of access" in terms of gender, wealth, social status and ethnicity which can make de facto seedbanks "less than egalitarian, with restricted access for the poor, especially women and marginalised ethnic groups". Lewis and Mulvany, *A Typology*, p. 15. Similarly, traditions that are reiterated in cultural performances of community can reinscribe pre-existing disparities into the structures of the activities themselves.
84. Terry Cook (2013), "Evidence, memory, identity, and community: four shifting archival paradigms", *Archival science*, 13, pp. 95–120, p. 105.
85. Cook, "Evidence", p. 113.

86. Cook, "Evidence", p. 116.
87. Terry Cook (2001), "Archival Science and Postmodernism: new formulations for old concepts", *Archival Science*, 1, pp. 3–24, p. 4.
88. Cook, "Archival Science", p. 4.
89. Cook, "Evidence", p. 115.
90. Eric Ketellar (1992), "Archives of the People, By the People, For the People", *Archives Journal*, 34, pp. 5–16. For an early proposal of how to use the commons-based model of Web 2.0 to peer produce archives, see also Max Evans (2007), "Archives of the People, by the People, for the People", *The American Archivist*, 70:2, pp. 387–400.
91. Raymond Williams (2001), *The Long Revolution*, Peterborough: Broadview Press, p. 67.
92. Stuart Hall (1999), "Un-settling 'the heritage', re-imagining the post nation: Whose heritage?", *Third Text*, 13:49, pp. 3–13, p. 5.
93. Cook, "Evidence", p. 101.
94. The issue of control in respect of archives has been much discussed, not least in Michel Foucault (2012), *The Archaeology of Knowledge* (trans. A.M. Sheridan Smith), London: Tavistock Publications, and Jacques Derrida (1996), *Archive Fever: A Freudian Impression* (trans. Eric Prenowitz), Chicago: University of Chicago Press.
95. Stuart Hall (2001), "Constituting an archive", *Third Text*, 15:54, pp. 89–92, p. 89. Originally addressed to "The Living Archive", a 1997 conference at London's Tate Gallery to mark the opening of the African and Asian Visual Artists' Archive (AAVAA). Hall is citing David Scott (1999), *Refashioning Futures: Criticism after Postcoloniality*, Princeton: Princeton University Press, pp. 125, 123.

Bibliography

Akbar, Arifa (2018), "The Future is not a Foreign Country", p. 3, https://assets.ctfassets.net/9sa97ciu3rb2/7yFifl9N9SAmia2QMcIaSi/dbc30d23424f72a31f3e2f3d6176078f/The_Future_is_not_a_Foreign_Country-Arifa_Akbar.pdf. Accessed 7 November 2024.

Anon. (2014), "Intro", *Future Library*, https://www.futurelibrary.no/. Accessed 5 November 2024.

Archangel Ancient Tree Archive (2015), *Archangel Ancient Tree Archive Press Kit*, https://www.ancienttreearchive.org/wp-content/uploads/2015/12/ATA1955_DigitalPressKit_Update_Dec9.pdf. Accessed 27 December 2024, p. 10.

Archangel Ancient Tree Archive (AATA) (2024), "AATA Living Archives Save Genetics for Future Generations", *Archangel Ancient Tree Archive*, https://www.ancienttreearchive.org/our-work/archive/. Accessed 15 October 2024.

Archangel Ancient Tree Archive (AATA) (2024), "About us", *Archangel Ancient Tree Archive*, https://www.ancienttreearchive.org/about-us/. Accessed 15 October 2024.

Atwood, Margaret (2014), "Future Library", n.p., https://assets.ctfassets.net/9sa97ciu3rb2/2hdAyLQYmESc0eYemIEcm2/09772ac1c62defc7ccf50fe6ea207a83/Margaret_Atwood.pdf. Accessed 6 November 2024.

Bachman, Steven P., Matilda J. M. Brown, Tarciso C. C. Leão, Eimear Nic Lughadh, and Barnaby E. Walker (2024), "Extinction risk predictions for the world's flowering plants to support their conservation", *New Phytologist*, 242, pp. 797–808.

Bouvard, Marguerite Guzman (2016), *Social Justice and the Power of Compassion: Meaningful Involvement of Organizations Improving the Environment and Community*, Lanham: Rowman & Littlefield.

Breen, Sheryl D. (2015), "Saving seeds: The Svalbard Global Seed Vault, Native American seed savers, and problems of property", *Journal of Agriculture, Food Systems, and Community Development*, 5:2, pp. 39–52.

Cook, Terry (2001), "Archival Science and Postmodernism: new formulations for old concepts", *Archival Science*, 1, pp. 3–24.

Cook, Terry (2013), "Evidence, memory, identity, and community: four shifting archival paradigms", *Archival Science*, *13*, pp. 95–120.

Darwin, Charles (1900), *The Origin of Species*, New York: H.M. Caldwell & Co.

Derrida, Jacques (1996), *Archive Fever: A Freudian Impression* (trans. Eric Prenowitz), Chicago: University of Chicago Press.

Duggan, Jennifer (2024), "Inside the 'Doomsday' Vault", *Time Magazine*, https://time.com/doomsday-vault/. Accessed 4 November 2024.

Evans, Max (2007), "Archives of the People, by the People, for the People", *The American Archivist*, 70:2, pp. 387–400.

Fallon, Claire (2015), "'Future Library' Reminds Us How Connected Books And The Environment Really Are", *Huffington Post*, 10 June, https://www.huffpost.com/entry/future-library_n_7532012. Accessed 28 August 2024.

Farmer, Jared (2019), "American Champion Trees and the Wye Oak of Maryland", *Arcadia*, 3:1, Rachel Carson Center for Environment and Society.

Fisher, Richard (2022), "The Norwegian library with unreadable books", *BBC*, https://www.bbc.com/future/article/20220630-the-norwegian-library-with-unreadable-books. Accessed 5 November 2024.

Flood, Alison (2014), "Margaret Atwood's new work will remain unseen for a century", *The Guardian*, 5 September, https://www.theguardian.com/books/2014/sep/05/margaret-atwood-new-work-unseen-century-future-library. Accessed 28 August 2024.

Foucault, Michel (2012), *The Archaeology of Knowledge* (trans. A.M. Sheridan Smith), London: Tavistock Publications.

Gilbert, Annette and Priscilla Posada (2016), *Publishing as Artistic Practice*, Berlin: Sternberg Press.

Goldsmith, Rosie (2022), "Future Library opens secret archive of unseen texts in Oslo", *The Guardian*, 14 June, https://www.theguardian.com/books/2022/jun/14/future-library-opens-secret-archive-of-unseen-texts-in-oslo. Accessed 28 August 2024.

Hall, Stuart (1999), "Un-settling 'the heritage', re-imagining the post nation: Whose heritage?", *Third Text*, 13:49, pp. 3–13.

Hall, Stuart (2001), "Constituting an archive", *Third Text*, 15:54, pp. 89–92.

Harper, John L. (1977), *Population Biology of Plants*, London; Academic Press.

Honey, Martha (2003), "Protecting Eden: Setting green standards for the tourism industry", *Environment: Science and Policy for Sustainable Development*, 45:6, pp. 8–20.

Hugill, Alison (2015), "Science: An Interview with Katie Paterson", *Berlin Artlink*, 23 December, https://www.berlinartlink.com/2015/12/23/science-an-interview-with-katie-paterson/. Accessed 9 November 2024.

Jenkinson, Hilary (1937), *A Manual of Archive Administration* (new and revised edition), London: Percy Lund, Humphries & Co.

Kang, Han (2019), Text for *Future Library Booklet*, https://assets.ctfassets.net/9sa97ciu3rb2/4UxWvLzP7sBusxQykz6dCp/b435e9c8fd5e65735126575ecd37ffd1/Han_Kang_English.pdf. Accessed 11 November 2024.

Kennerly, Michelle, Samuel Frederick, and Jonathan E. Abel (eds) (2021), *Information: Keywords*, New York: Columbia University Press.

Ketellar, Eric (1992), "Archives of the People, By the People, For the People", *Archives Journal*, 34, pp. 5–16.

Larsen, Lars Bang (2017), "The Manuscripts Stored in Oslo", n.p., https://assets.ctfassets.net/9sa97ciu3rb2/taYrxz0CaGWkMuws2UEGs/11e3ffbb41faac94e6b4eeeda80a2967/Katie_Paterson_Lars_Bang_Larsen_essay2017.pdf. Accessed 7 November 2024.

Lewis, V. and P.M. Mulvany (1997), *A Typology of Community Seed Banks*; Chatham: Natural Resources Institute.

Lyell, Charles (1875), *Principles of Geology* (twelfth edition), London: John Murray.

Mitchell, David (2015), "The Ayes Have It", n.p., https://assets.ctfassets.net/9sa97ciu3rb2/1RU9pw3kiUmSmS8i60sciW/e7e326d546dbe4d88a018ef8b6009f75/David_Mitchell.pdf. Accessed 6 November 2024.

Moore, Erik A. (2007), "Birds of a Feather: Some Fundamentals on the Archives–Ecology Paradigm", *Archivaria*, 63, pp. 103–119.

Morton, Timothy (2013), *Hyperobjects: Philosophy and Ecology After the End of the World*, Minneapolis, University of Minnesota Press.

Nash, Roderick (1982), *Wilderness and the American Mind*, New Haven: Yale University Press.

NordGen (2024), "The Seed Portal", *NordGen*, https://www.nordgen.org/our-work/svalbard-global-seed-vault/the-seed-portal/. Accessed 8 October 2024.

O'Neill, Robert V., D.L. DeAngelis, J.B. Waide, and T.F.H. Allen (1986), *A hierarchical concept of ecosystems*, Princeton: Princeton University Press.

Pascoe, Joanna (2024), "Future Library: Hopepunk Heroism", *Heroism Science* 9.1, pp. 1–10.

Paterson, Katie (2014), "The Artwork", *Future Library*, https://www.futurelibrary.no/#/the-artwork. Accessed 5 November 2024.

Paterson, Katie (2018), "Future Library Handover Day with Elif Shafak, 2018", https://vimeo.com/282327747. Accessed 7 November 2024.

Paterson, Katie (ed.) (2020), *Future Library Booklet 2021: Ocean Vuong*, n.p., https://assets.ctfassets.net/9sa97ciu3rb2/6mhd4RgwmA0Cj5b12JmcPw/19ad8d8bf67ce06d0db9aa7381f146b7/2616_KatiePaterson_FutureLibrary_Vuong_Booklet_VISUAL_2.pdf. Accessed 7 November 2024.

Payne, Jaq (2024), "The National Champion Tree Program: An Honor for Tennessee, a Boon for the Nation", *Tennessee Turfgrass*, February/March, pp. 22–27. https://theturfzone.com/tta/?ascat=7&sub=issue&issue_id=8627&rti=true&rel=2. Accessed 22 October 2024.

Reid, Clement (1899), *The Origin of the British Fauna*, London: Dulao & Co.

Robbins, Jim (2012), *The Man Who Plants Trees*, London: Profile Books.

Schellenberg, Theodore R. (1956), *Modern Archives: Principles and Techniques*, Chicago: University of Chicago Press.

Scott, David (1999), *Refashioning Futures: Criticism after Postcoloniality*, Princeton: Princeton University Press.

Seale, Madeleine and Naomi Nakayama (2020), "From passive to informed: mechanical mechanisms of seed dispersal", *New Phytologist*, 225:2, pp. 653–658.

Shrestha, Pitambar, Ronnie Vernooy and Pashupati Chaudhary (eds) (2013), *Community Seed Banks in Nepal: Past, Present, Future*, Proceedings of a National Workshop, LI-BIRD/USC Canada Asia/Oxfam/The Development Fund/IFAD/Bioversity International, 14–15 June 2012, Pokhara, Nepal.

Spencer, Sarah (2019), *Think Like a Tree: the Natural Principles Guide to Life*, Swarkestone: Swarkestone Press.

Staal, Jens (2019), "Artistic research in breeding: The Bifrost Eucalyptus project", *Research Catalogue*, 9, https://www.researchcatalogue.net/view/392580/392581. Accessed 15 October 2024.

Svalbard Global Seed Vault (2024), "Purpose, Operations, and Organisation", *Svalbard Global Seed Vault*, https://www.seedvault.no/about/purpose-operations-and-organisation/. Accessed 8 October 2024.

Thompson, Ken and John Philip Grime (1979), "Seasonal variation in the seed banks of herbaceous species in ten contrasting habitats", *The Journal of Ecology*, 67, pp. 893–921.

Thompson, Ken (1987), "Seeds and Seed Banks", *New Phytologist*, 106, pp. 23–34.

Vernooy, Ronnie, Bhuwon Sthapit, Gea Galluzzi, and Pitambar Shrestha (2014), "The multiple functions and services of community seedbanks", *Resources*, 3:4, pp. 636–656.

Vittoz, Pascal and Robin Engler (2007), "Seed dispersal distances: a typology based on dispersal modes and plant traits", *Botanica Helvetica*, 117, pp. 109–124.

Williams, Raymond (2001), *The Long Revolution*, Peterborough: Broadview Press.

Wohlleben, Peter, *The Hidden Life of Trees*, London: William Collins.

Wu, Professor (2018), "Creatives in profile—interview with Katie Paterson", *Nothing in the Rule Book*, 23 May, https://nothingintherulebook.com/2018/05/23/creatives-in-profile-interview-with-katie-paterson/. Accessed 28 August 2024.

CHAPTER 8

Mycorrhizae

With its towering trunk and iconic fronds, the coconut tree is a ubiquitous symbol of the tropics. Standing up to 30 m in height, it can live for 100 years, producing around 50–100 coconuts in a season. It is a hardy evergreen, requiring humid temperatures and ample rainfall, but around the equator, it is abundant. The coconut tree grows across India, Indonesia, the Philippines, Vietnam, Brazil, and the South Pacific,[1] where it is a crop of immense economic as well as ecological significance. There are literally dozens of ways in which the tree and its fruit can be put to use. Indeed, the coconut is reportedly "among the ten most useful trees in the world",[2] leading to an Indonesian belief that "there are as many uses for the coconut as there are days in the year",[3] and prompting a recognition of the tree as "the great provider".[4] Not only does the coconut tree offer a source of timber, food, oil, cosmetics, and medicines, but its cultivation also provides considerable employment throughout the global South. For reasons such as this, it has been called "the most valuable tree in the world", and one report from 1928 suggested that "without it, the future of tropical lands would be dark indeed".[5] Not to be confused with the oil palm, whose industrial cultivation has led to mass deforestation throughout similar regions, the coconut palm is largely sustainable, providing "important ecosystem services such as coastline protection, carbon sequestration, and biodiversity conservation", researchers report.[6]

K. Savage and D. Symonds, *Ecologies of Community in Performance*,
https://doi.org/10.1007/978-3-031-94343-0_8

No wonder the tree is widely referred to by communities as the "tree of life".[7]

Nowhere is this more the case than in the tiny archipelago of the Admiralty Islands, a part of Papua New Guinea situated around 300 miles north of New Guinea itself. Centred around the largest of its own islands, Manus, which measures around 60 miles by 20 miles, the Admiralty Islands rely on their coconut trees, which have offered a staple trading commodity since the Germans colonized the region at the end of the nineteenth century. Coconut trees were extensively planted, especially on the smaller islands, and copra—the dried, white flesh of the coconut—was harvested for soap and other products, at one point representing 90–95% of the off-island trade.[8] "Copra has been the backbone of Manus' economy for the past 100 years", suggests one report, "providing the main means for the island's population to pay for children's education, health, transport and essential needs".[9] Nowadays, as the traditional industry declines, entrepreneurs are developing a "sustainable local and provincial market of coconut bio-fuel",[10] and striving to do this without falling into the same traps as the palm oil industry twenty or thirty years previously[11]: in order to avoid the blights of palm oil monoculture, "coconut farms, grown primarily by smallholders, are mixed with crops like banana, cacao, and coffee, and farmers then often integrate them with the surrounding tropical landscapes"[12]; meanwhile, rather than being burned or simply disposed of, coconut by-products such as the shell are converted into organic fertilizer or alternative fuels to maximize a circular economy.[13] Naturally sourced bio-fuel, after all, will help island communities combat the sort of "diseconomies of scale" that come with setting up expensive infrastructure and importing fossil fuels to tiny islands.[14] Here, as with other concerns, coconuts can offer a solution. The coconut trees do indeed loom large on Manus Island.

Coconut trees also loom evocatively in the work of Kurdish-Iranian poet Behrouz Boochani: "they are all striving to reach higher, higher up into the sky, all competing for oxygen, all trying to ascend, all trying to breathe", he writes in his poignant memoir *No Friend but the Mountains* (2018).[15] Boochani's award-winning account documents his incarceration in the Manus Regional Processing Centre, or as he refers to it, "Manus Prison".[16] This was set up by the Australian government in 2001 to combat immigration. It was where Boochani found himself detained between 2013 and 2017, a victim of the West's hostile sanctions against refugees. Through his writing, Boochani spoke out beyond the confines of his prison camp and over the fences of immigration bureaucracy. "I learn over time to stand alone", he writes; "like the coconut trees inside the prison—how to exist in seclusion like them".[17] The image of the coconut tree recurs throughout his writing, as if the tree observes his situation; at once sentinel and protector: "The tall coconut trees that line the outskirts of the camp have grown naturally in rows / But unlike us they are free / Their grand height allows them to peep into the camp at all times / To know what is going on in the camp / To see what is happening in the camp / To witness the anguish suffered in the camp".[18] In the coconut tree Boochani appears to find support: "the beauty and abundance of the tree have an effect far beyond being soothing and comforting", suggests Kathrin Bartha-Mitchell; "they serve as a reminder of the innate 'goodness' of the cosmos, therefore, presenting a 'power' that puts the pettiness and scarcity logic of the [detention centre] into perspective".[19] In one touching lament, the decay of a tree leads to the regrowth of fresh young flowers. "I feel that there is a special connection between the flowers and spirit of this coconut tree which has died young", Boochani writes; "It will last out there for years until it rots away, until it deteriorates bit by bit and is absorbed back into the dirt from which it was born".[20]

We've considered in earlier chapters some of the ecological processes that Boochani invokes here: the sense of a self-contained ecosystem; the energy cycle of growth and decay; the aboveground-belowground interactions. In this chapter, we will explore another of these processes, found in the operations of *mycorrhizae*, tiny fronds of fungal growth that intertwine with the roots of trees and extend underneath the soil. As scientists have discovered, it is by working with the mycorrhizae that trees send sustenance, nutrients, and even communications to one another. And the role mycorrhizae play in relation to coconut palms is fundamental to their wellbeing; healthy network relations with arbuscular mychorrhizal fungi boost survival rates and growth patterns, while protecting them from pests.[21] Recent cultivation techniques have taken to using mycorrhizae as conduits to treat disease[22] and to stimulate growth.[23] "The most obvious mechanism for protection against diseases is the barrier that Ectomycorrhizae create when they coat the external surface of the root", explain researchers; "Mycorrhizae compete with pathogenic microorganisms and antibiotics to protect the plants[,] and also root feeding nematodes are less likely to attack".[24] When we consider that the coconut industry in Papua New Guinea directly employs around 20,000 people and indirectly "sustains the livelihood of an estimated 2.6 million people"[25] (25% of the population), we can see that the underground fungal networks of the mycorrhizae and their relationships with the host trees are integral to the lives and livelihoods of PNG communities, including those on Manus. Extrapolate that to rural communities throughout the global South who are reliant on the produce and incomes that their local flora can provide, and we begin to understand the ecological significance of ecosystemic relationships with the land. All of this speaks directly to our project of synthesizing ecological thinking, the arts, and communities.

We're drawing attention to the mycorrhizal networks in such an ecosystem because they seem to compare to the hidden networks through which Boochani was able to communicate beyond the confines of Manus Prison. We will explore these in relation to the dynamics of incarcerated communities as witnessed by Boochani in his writing, film, and poetry. We'll be turning to an ecological understanding of *mycorrhizal networks* to consider how such communities might speak beyond their confinement, and we'll see how Boochani's invocations of some of Manus Island's flora and fauna—such as the coconut tree, and the small Manusian bird the Chauka—serve to magnify the significance of community, solidarity, and resilience. But we first need to turn towards the topic of migration, to consider how individuals become displaced, how they end up incarcerated, and how Boochani came to find himself a detainee of Manus Prison.

Migration

"Since the earliest times, humanity has been on the move", reports the UN, noting varying reasons behind people's decisions to uproot their lives: "Some people move in search of work or economic opportunities, to join family, or to study. Others move to escape conflict, persecution, terrorism, or human rights violations. Still others move in response to the adverse effects of climate change, natural disasters, or other environmental factors".[26] As Cathrine Talleraas notes: "there is no universally accepted definition of human migration. Consequently, multiple interpretations of the term 'migrant' coexist".[27]

In his general theory of human migration from 1966, the American sociologist Everett S. Lee observed four factors determining the flow of migration: the push away from a place of origin; the pull of a destination; intervening obstacles *en route*; and personal circumstances.[28] Each of these factors can be seen in the impulses of twenty-first-century migrants: the push of famine or genocide; the pull of better prospects or human rights; the obstacles of state borders or political policies; and the personal circumstances—race, sexuality, or religion—that define bodies as different. Inevitably, migration theories have become more sophisticated since the 1960s, though as more recent scholars note, "there is no single, coherent theory of international migration, only a fragmented set of theories that have developed largely in isolation from one another, sometimes but not always segmented by disciplinary boundaries".[29]

Nevertheless, migration cannot be conceptualized as a stand-alone journey from origin to destination; nor can it be understood as something undertaken by individuals alone. It happens within what scholars view as systems and networks. In this sense, the dynamics of migration share much in common with the ecological approach we have been exploring. "Migration decisions are not made by isolated individual actors", write Massey et al., "but by larger units of related people—typically families or households—in which people act collectively".[30] Such networks, sometimes extending over considerable geographies and through secondary, tertiary, or even more distant nodes of a chain, support both the mobility of individual migrants and the overall dynamics of migration's organism within and across its geopolitical terrain. As Ivan Light and his colleagues explain, "network-supported migrants have important help in arranging transportation, finding housing and jobs in their place of destination, and in effecting a satisfactory personal and emotional adjustment to what is often a difficult situation of cultural marginality".[31] Network relations thereby provide social capital,[32] and the subsequent settlement experiences of network-supported migrants build on that capital to form bridging and bonding ties that allow for varying levels of integration in the new location, while maintaining strong connections with existing networks.[33] Conceptualizing such dynamics as a transmission of social capital also helps to account for further, more complex modes of migration: *chain migration*,[34] when subsequent members of a community follow an individual's relocation; *step migration*,[35] in which a migrant may move multiple times in stages of an overall journey; and *cyclical migration*,[36] which involves a to-ing and fro-ing between origin and destination. Many such patterns of migration are enabled by the wider network of family and associates that are already part of the diaspora, and speak to an individual's connection to multiple concentric geographic circles of community. Eventually, theorists argue, the pathways and nodes of migratory routes become formalized, perpetuating particular routes, expectations, and trends into a system linking "flow and counter-flows".[37]

Our overview of migration thus far presents its dynamics as rather benign, though we should acknowledge that many drivers of migration reflect hostile conditions or states, many migratory decisions are not made through choice, and many network actors in the migratory ecosystems are predatory. In this sense, migration reflects issues of persecution, violence, exploitation, human rights abuse, criminal gangs, people-trafficking, modern slavery, and more, and it's no exaggeration to say that the landscape of migration has been fully infiltrated by the darkest of network tangles that have led to it becoming one of the most urgent of contemporary global challenges. In addition to these already complex dynamics, though, further networks of government and non-government organizations and charities have developed, with interweaving initiatives and schemes, laws and regulations, processes and procedures that variously protect, restrict, or intercept flows through the system.

One aid, created in 2017 by the UN Migration Agency, was "MigrantApp", a "one-stop shop"[38] providing information on assistance and protection, a way of sharing routes, and a means of sending money.[39] Later rebranding as "MigApp", available in eight languages, the platform recognizes that "migrants from all five continents need a trusted source of information"[40] through which to access the "system". This App offers agency to migrants in terms of communicating with their networks, providing safe and legal passage, and protecting against the risks of exploitation; it also recognizes the mobility that digital literacy and access to mobile technology offers to those who have been uprooted or who are in transit. Nevertheless, it is also a systemic example of a tool serving other purposes and motivations: "MigApp enables the capture, and therefore analysis, of migration trends and patterns at country, region, or global levels",[41] as project manager Alex Dougan admits, recognizing an element of surveillance. However altruistic the development of such tools might be, their systemic coordination also risks providing guiderails to malicious users who could potentially hijack the system to perpetuate the very violence that is trying to be overcome.

Behrouz Boochani

After several years working as a freelance journalist in Iran, Behrouz Boochani became editor of *Werya*, a social and political magazine which sought to promote Kurdish language, culture, and politics. On 17 February 2013 its offices were raided by the authorities, and its journalists detained. Boochani escaped arrest, but published the news internationally in an exposé that caused widespread outrage about the regime. Fearing repercussions, he fled Iran to make his way to Australia. "I thought if I can publish my ideas in a free country then I can help my people", Boochani writes; "I thought that Australia was free. I thought that in Australia I could write freely".[42] It was several years before that could happen. In 2018 he published *No Friend but the Mountains*, a book that documents his journey, his failure to gain entry to Australia, and his subsequent five-year detention on Manus Island. When the book came out, he was one of countless migrants still trapped in the system. Freedom would have to wait.

Boochani's writing style has been described as "anti-genre": it's "a mix of historical document, political critique, on-the-ground reportage, philosophical reflection, psychoanalytic examination, poetic expression and epic".[43] Interspersed with his prose are poetic fragments, lending the text a dream-like quality. He moves from an account of his migrant journey to a series of experiences in detention, and finally, to a heightened, transcendent sequence in which his narrator figure gazes out on the world, and to a future, beyond the prison's confines.

Boochani's account of his migrant journey is evocative and often harrowing. He takes a well-established route from Indonesia to Australia, and captures the images we're accustomed to seeing in the media: overcrowded vessels; boats that have capsized; migrants who have drowned; bodies washed up on the shore. Eventually, he and his companions are picked up by the Australian navy and taken to Christmas Island, from where they are transported by plane to Manus Regional Processing Centre. But this is an ordeal to which familiar media snapshots don't do justice. The act of migration is painfully long, and at several points delayed while the travellers hang in limbo: "three months of wandering displaced and hungry in Jakarta and Kendari"; "forty days of near starvation in the basement of a tiny hotel"[44]; even the processing wait on Christmas Island is a month long.[45]

Still, it is an ordeal shared with others—something that we only glimpse obliquely in media images. On this journey, we begin to get a feel for the small community in which Boochani travels, and a sense of their relationships. His companions are characterized by physical or personality traits. At times these are sympathetic (the Blue Eyed Boy), at other times ridiculous (the Penguin; the Toothless Fool), and occasionally downright unpleasant (the Kurdish Guy). None of the caricatures are explicit portraits, as he is keen to note[46]; instead, they are ciphers for the faceless thousands that find themselves in flight. Men tend to be cast as types (the Young Man with the Ponytail; the Cadaver), while women warrant names (Azadeh; Golshifteh). Boochani holds himself at a distance, an observer in their midst. Only at times does he allow himself to be thrown into the drama. On one occasion he spends the night bailing out water from the small boat; on another, as the boat sinks, he finds himself clinging to the debris and waiting to be rescued. These intimate moments magnify the existential risks of the journey, and the reality of those dangers is brought home when the Blue Eyed Boy with whom he most identifies dies. Only in the image of a young family, the Sri Lankan Couple, does he find a vestige of hope. Still, there are acts of community and companionship: a chain of people bailing water out of the engine house; a community leader, Golshifteh, who allocates the meagre rations between the refugees.

Manus Prison

Despite the fact that Australia has a long history of welcoming immigrants,[47] its recent immigration policies, overseen by successive Labour and Conservative governments have instigated a rigid entry regime, first in a system called the "Pacific Solution" (2001–2008), and then in "an even harsher offshore detention regime",[48] a "military-led border security operation"[49] known as "Operation Sovereign Borders" (2013-present). Just four days before Boochani's arrival it implemented what he calls the "exile policy": "any person seeking asylum in Australia by boat without a valid visa after [19 July 2013] will be banned from entering Australia for life".[50] In unequivocal terms, the Australian Government warns any would-be immigrants that they have "zero chance" of being admitted to the country: "Border protection policies remain", announces the website; "You will fail".[51]

In enacting this legislation, Australia pulls a sleight-of-hand around the terminology, for the problematic ways in which migrants are labelled exposes a complex rhetoric underpinning issues of migration. Alongside the relatively accepted terms "migrant" and "refugee", a cacophany of vocabulary gets pulled into the debate, including "unlawful non-citizens", "illegals", "irregular migrants", "irregular residents", "visa overstayers", "detainees", "irregular maritime arrivals", "unauthorized maritime arrivals", "boat people", and "queue jumpers".[52] When a migrant arrives in Australian territory without documentation, they enter the system with still a different label, the catch-all "asylum seeker", who (in the Australian Parliament's words) "is seeking international protection but whose claim for refugee status has not yet been determined".[53] The legislation to relocate such individuals offshore removes them from visibility, though criminalizes them in the process. Thus "asylum seekers are categorized as an 'out group'", explains Moones Mansoubi; "*others*, a threat to Australia's sovereignty. 'They' are criminals, queue jumpers, thieves of taxpayers' money, or so say politicians".[54] "The asylum seekers are also specimen[s] of the unintegrable other", argues Renata Summo-O'Connell, "to be kept confiscated from history, as a temporary property seized from the inconvenient disasters they flee from". Thus, through politics, policy, and rhetoric, the migrant is dehumanized. "Detention camps are the place where their unintegrability is sanctioned and affirmed as policy for all to see and learn".[55]

Boochani writes passionately about his own incarceration and about the Australian immigration system more generally.[56] He borrows the concept of "Kyriarchy" from the feminist scholar Elisabeth Schüssler Fiorenza to refer to the Manus regime,[57] which "transforms prisoners into machines, erasing their human identities, and stripping individuals of autonomy and selfhood".[58] He describes the system's "relentless and pervasive practices of micro-control and macro-control"[59] and its "interlocking and mutually reinforcing structures of violence obsessed with oppression, domination and submission".[60] Through the middle section of *No Friend but the Mountains*, we witness the Kyriarchal system in action.

The description of Manus Prison is vividly drawn: in one part "nearly four hundred people are kept in an area smaller than a football field",[61] he writes; "the prison is besieged by fences encircling its outer rim"; "from one end of the camp to the other drain pipes protrude from the kitchen and the bathroom"; "I return to my room. It is so tiny. I feel as if I am suffocating inside it".[62] A picture emerges of the hundreds of detainees, bored and restless; the idle guards, bored and restless; and the interminable, hot days in sordid conditions. Boochani illustrates some of the "micro-level and macro-level disciplinary measures designed to create animosity between the prisoners"[63]: the confiscation of personal effects; lengthy queues for meals in the blazing sun; severe restrictions on using the public telephones; and regular beatings. Above all, as one of his most haunting sections of poetry recounts,[64] the detainees have to wrestle with their own internal thoughts, which he refers to as a second prison of the mind: "In addition to the torment produced by the oppressive enclosure of the prison fences", we're told, "every prisoner creates a smaller emotional jail within themselves".[65] In one of his poems, "Love Song of a Caged Bird", he imagines himself trapped inside a cage and unable to escape: "Forgive me my love as I am solely able to love you from a remote island, inside a cage, from the corner of this small room. Forgive me please as the only portion of the world that belongs to me is this".[66] "Only the genre of poetry could express the intensity and volume of the message", he reasons; "Poetry: a form of expression that dismantles all these structures".[67]

Boochani is not alone in his ordeal. Packed into the centre are hundreds of other men, all diverted from their journeys; all from different corners of the world. But in their plight, a fleeting comfort is the bond of brotherhood, the solidarity of a group. "The simplest way to gain status is to identify with a group", he concedes; "That is, to affiliate yourself with other individuals who you think share your identity: people who are going through the same set of circumstances". It's a camaraderie that begins on the boats, he recalls; but over time, other identifiers bring communities closer together: language, nation. "Depending on a group or a collective identity masks loneliness", Boochani writes; "It's a kind of escape route, a shortcut".[68] "However, in all the years in the prison, people who experienced the boat journey together will insist on their bond. […] The collective trauma from the journey is in our veins—each of those boat odysseys founded a new imagined nation".[69]

This is a forced community, to be sure; but even so, solidarity emerges through shared experience, physical togetherness, and rites of community. Such rituals perhaps take us by surprise in the conditions of the prison, but at one point Boochani details the nightly celebrations staged in Corridor L. A plastic table is set up and the crowd gathers, drawn by the rhythmic clapping of a group of men. As they whip the crowd into a frenzy, inhibitions disappear, and a mood of celebration overtakes the throng. "In the minds of the prisoners it is unnecessary to explain why they are happy and why they want to celebrate", maintains Boochani; "The prisoners dance because they have to dance, to spite those people who exiled them to the prison". But "everything is interconnected", he reflects; "joy, fear, hate, envy, revenge, spite, and even kindness". Eventually, the anticipation is rewarded as the star attraction of this informal performance takes the stage: Maysam the Whore—a name attributed not because of any prostitution but because of his uninhibited abandon. He is "the spirit of the community", "a heartwarming and appealing enigma", who for a few brief minutes every night appears to the crowd and dances frenetically, eliciting a mood of collective effervescence or *communitas*. "His presence, his dancing, his singing, helps us forget the violence of the prison for a moment. He is the superstar of the prison".[70]

In the final section of *No Friend but the Mountains*, Boochani's narrative style lifts into a dream-like quality. He makes his way onto the roof of one of the buildings from which he can survey the prison. Like a sapling searching for light, he outgrows everything around him, heading for the canopies; here he can have a clear view and a clarity of vision. He sees the Green Zone of solitary confinement; he sees an escape attempt apprehended by security officers; he sees a medical examination of a man claiming to be a Prophet—it seems like other detainees are also finding ways to transcend their reality. Interspersed with these stories, Boochani reminisces about his previous life: his mother; his childhood memories; his first love. And he notices the sounds, smells, and activities of the natural surroundings: a cat on the roof; the mango trees; the crickets. In this sequence, both his prose and his lines of poetry are filled with evocative images of sight and clarity:

> (Prose): "Upon the zenith one can see the river with ease; it is possible to see the slithering, looping snake that emerges from deep inside the faraway mountain ranges. Those faraway mountain ranges are decorated with a milky colour. Over and beyond those mountain ranges appear other mountain ranges. And over and beyond those mountain ranges, there are other mountain ranges. And it continues, reflecting a chain of mountain ranges, mountains the colour of milk, mountains becoming milkier in colour, mountains becoming more translucent".[71]

> (And poetry): “Down below is scattered with chestnuts and oak forests / Down below is flourishing with forests of wild figs / Down below is the village I became accustomed to in every way / Until I left it ... only to return again”.[72]

In both his spatial and temporal imaginaries, Boochani’s thoughts stretch beyond his current incarcerated condition, something which given his circumstances may not be surprizing, but which is brought vividly into relief through his evocative writing. “His imagination is always preoccupied with the world beyond the fences and in his mind he forms a picture of a world where people are free”,[73] acknowledges Boochani of the general detainee’s desires. On one occasion, he himself stages a minor revolt by climbing a tree in the compound. “Climbing the tree is an escape of sorts”, writes Anne McNevin; “Atop the tree, he literally sees beyond the prison walls and is free from its enclosures for a day”.[74] Boochani also writes in one account about prisoners throwing chocolates to the Manusian kids beyond the fence.[75] There is a life beyond the fence—at times tantalizingly close—yet too often the cries for help from this community meet the silence of the compound walls, confirming the impenetrability of the barriers behind which they are confined.

Eventually, Boochani takes himself to the “abandoned place”, behind the buildings near the sewerage drain: “the quietest place in Fox prison—there by the Flowers resembling Chamomile—where I can be alone for hours”.[76] The only other person he sees here is the Smiling Youth, whom he sits with in companionable silence. In another wave of transcendence, Boochani imagines himself outside the prison walls:

> (Prose): “And I am now part of the jungle. I am the jungle, like the snakes, like the frogs, like the insects, like the birds. I am the jungle itself”.[77]

> (And poetry): “The jungle seems broodingly darker than ever before / But the crickets and frogs reign supreme / This night the chanting of the crickets and frogs entrances the prison/ The prison has a spell on it / The prison is possessed by the jungle / The prison is absorbed into the jungle / The prison is engulfed in a heavy mood / The prison is terrified / The elderly crab is gradually digging up the moist earth with its claws / It is calmly excavating the soft dirt that lies beneath the fences / The old crab digs until it penetrates the prison”.[78]

As he views the island beyond the fence, from here he can imagine himself outside of the cage, and taking flight.

It is through such evocative images—the flowers reaching up beyond the carcass of the coconut tree; the chestnut oak forests rising into the night sky; the distant vista of the mountains; himself being embodied as the jungle—that he imagines the agency of nature in transcending the prison confines. Thus "the critical location of Boochani's work in new conceptual islands off the mainland of thought", writes Rita Sakr, "enables a visible, embodied voicing that goes beyond haunting the oppressor in the struggle for more-than-human rights by proposing sanctuary in terms of relational, indigenously formed imaginaries of resistance disrupting the thanato-political, speciesist border-industrial complex".[79] Arianna Grasso puts a similar thought more simply: "If journalistic language appeared to Boochani as an ephemeral instrument for conveying meaning, he defined the language of art as an everlasting and timeless language, able to surpass temporal, spatial (one could also add digital) limits".[80]

The Ecology of Mycorrhizae

We're taken by the notion that the images in Boochani's writing reach beyond the fences of his detainment, and that he calls into play so many references to nature in enacting his metaphorical escape: glimpses through the fences of the jungle beyond; depictions of the coconut trees gazing over the walls; glances from the roof at the beautiful terrain of the island. In terms of our own ecological perspective, we're reminded of the ways trees use mycorrhizal networks to reach beyond their roots. Peter Wohlleben's popular book *The Hidden Life of Trees* explains these actions in very accessible terms: "These fungi operate like fiber-optic Internet cables", he writes; "Their thin filaments penetrate the ground, weaving through it in almost unbelievable density. [...] The fungal connections transmit signals from one tree to the next, helping the trees exchange news about insects, drought, and other dangers".[81] More nuanced descriptions can be found elsewhere: "Communication between plants, microbes, and soil animals below ground is driven by the exchange of substances released by one organism and detected by another", explains one report.[82] Scientists refer to these substances as "infochemicals", which include phosphorus, nitrogen, carbon, and water, and which are transmitted through hyphal networks made up of mycorrhizal fungi. They identify five ways in which infochemicals are transported: (a) cytoplasmically through fungal cell membranes inside the hyphae; (b) apoplastically by diffusing through the cell walls lining the hyphae; by dissolving in water and being carried along either (c) the surface of the hyphae, or (d) through water collected in channels between closely twisted hyphae; and (e) by modifying the soil environment in order to enable conduction outside the hyphae themselves. They reason that this mycorrhizal transportation serves three purposes: first, as "a pathway for systemic defense signaling within plant populations" (in other words, as a defence mechanism for one plant to warn others of threats)[83]; second, as a way of influencing the establishment, physiology, growth, survival, reproduction, and competitive ability of neighbouring plants, either beneficially or detrimentally[84]; and third as a "kin recognition" system.[85]

Further research has shown that communication between plants also takes place through electrical impulses.[86] Suzanne Simard equates this particular type of communication to the ways in which neural synapses, neurotransmitters, and the central nervous system operate in the human body. "Mycorrhizal network topology [...] is strikingly similar to the topology of neural networks in our brains", she argues, suggesting that the "general intelligence" of plants involves "both 'crystallized intelligence' (similar to memory) [...] and 'fluid intelligence' (similar to learning)".[87] The strong links that form between easy-to-reach individuals in the network establish robust community relations, while weaker links with individuals that are further afield offer a chance for a type of learning to develop. "Memory-based learning", she writes, "drives behaviors such as choice, decision-making, defense, and neighbor recognition", enabling plants to "change their behavior for optimally interacting with these neighbors, learn to respond to concurrent changes in the behavior of the neighbors, and in so doing increase fitness".[88]

Thus despite the fact that trees have neither a means of movement nor a way of escaping their immediate surroundings, they seem able to communicate in what scientists agree to be "a complex adaptive social network".[89] Indeed, various scientists have explored the communicative activities of plants in terms of a network, and using network theory.[90] More than this, the communicative network dynamics are not simply valuable to the individual plants themselves, but also to their wider ecosystems: "Since plants form the basis of terrestrial ecosystems", argue Gorzelak et al., "their behavioural interactions, feedbacks and influences are important in generating the emergent properties of ecosystems".[91] The "collective behavior, learning, and memory of a community", suggests Simard, may allow it "to solve cognitive problems that go beyond the capacity of a single organism, facilitating altruistic behaviors like kin recognition and more generally promoting cooperation for better ecosystem health".[92] Thus from initial connections with direct members of the network, the communications spread out through "ripple effects on plant community and ecosystems ecology".[93] So significant are these dynamics that they have enthused scientists to refer to such communication as "underground 'tree talk'",[94] borrowing phrases like "communication highway"[95] and "superhighway"[96] from the digital sphere, and even speaking of "hyperlinks"[97] and a "wood-wide web".[98]

Reaching Out

For Boochani, to communicate with the outside world and tell the story of his experiences was a fundamental need. At the most basic level, this was true of all the detainees, and that need was assuaged by reaching out to family and friends in the limited phone conversations they were allowed. One section of *No Friend but the Mountains* offers a grim description of the telephone provision: "The telephone room is a large room with seven or eight phones, but several are always broken", explains Boochani; "at the moment, the prisoners from Fox can [...] access the phones every day from morning till noon [...but] the rules and regulations change weekly".[99] Like other features of prison life, Boochani sees this confusion and inadequacy as a part of the Kyriarchal system: "The system fragments and disorients the prisoner to such an extent that he is alienated from his sense of self", he complains.[100] Meanwhile, in Boochani's film *Chauka, Please Tell us the Time*, we see how phone conversations to home are strained as the families and loved ones don't understand the situation. One detainee, Kaven, seems to have difficult conversations every time he reaches out: he can't get his family to understand his predicament—they think his absence from home is a choice, and don't even seem aware that he is locked up. Kaven struggles patiently to explain, but the emotional strain and frustration are palpable.

In another passage, Boochani describes some of the other (futile) ways in which detainees attempt to connect with the outside world. "Fences cordon off the bathrooms", he writes, "and through their wires are tied numerous small pieces of cloth [...] remnants from the people who were held here before us [...]; each one represents a memory, the ribbons are a series of recollections, all of them hark back to another lamentable time".[101] Back in his room, Boochani comes across graffiti scribbled on a wall—pictures and poetry from former detainees, one of whom Boochani later recounts meeting.[102] "The writing is Persian poetry which everyone uses to interpret their destiny, to interpret their future, to interpret their lives", he says. The messages clearly communicate something to him: "I don't know why I feel the presence of that family when I read the poetry / I don't know why I feel the presence of the wife and her daughters when I read the poetry / I feel their presence and their beauty / They were lively and active / They were living life".[103] Even so, these messages, like the ribbons, are communicating their cries for help no further than the walls of the compound. To reach beyond the isolation and confinement of the prison would take another means of crying out.

Boochani compares his entrapment in the prison to being stuck in a "dark cave": "We are bats in a dark cave that react to the slightest vibrations", he writes[104]; "Australia's presence on Manus and Nauru seems like a thread that leads you further into a dark cave with no end".[105] Helena Zeweri interprets this analogy in two ways: the darkness recalls "the depths of violence that the state will go to to prevent undesirable people reaching its shores", she suggests; and "the darkness of Australia's settler colonial history".[106] As such, it is not just personal connection with family and friends, or even a cry for help that animates Boochani in reaching beyond the prison: the role he sees for himself is in "bearing witness" to the dehumanization and suffering of the refugees' detainment.[107] "We are being made examples to strike fear into others, to scare people so they won't come to Australia",[108] he writes; "the deal is that we have to be a warning, a lesson for people who want to seek protection in Australia".[109] But in grasping the agency to communicate his perspective, Boochani can turn the tables on this Kyriarchal strategy. He can become the "Voice of Manus".[110] "I have reached a good understanding of this situation", Boochani realizes at one point:

> the only people who can overcome and survive all the suffering inflicted by the prison are those who exercise creativity. That is, those who can trace the outlines of hope using the melodic humming and visions from beyond the prison fences and the beehives we live in.[111]

And it is through his creativity, his tracing of the outlines of hope, and his expansion of community that Boochani manages to become the "Voice of Manus" and reveal the truth.

He does this through an innovative way of overcoming the incarceration and its isolation. "Initially, he made use of the slow and unreliable computers in the prison, which he was permitted to use for a short time once a week. When he managed to smuggle in a mobile phone, he was able to work more effectively", explains Omid Tofighian; "Moones Mansoubi and I translated the work, receiving each article via WhatsApp. [...] Boochani would send the pieces, whether a single sentence, full paragraph or whole page, to Mansoubi to store in chapter files for me to translate". The process was not easy, partly due to the restrictions and security checks for unpermitted possessions: "He had to operate clandestinely with vacillating internet reception, and was always at risk of having his phone confiscated",[112] Tofighian adds.

Painstakingly, line by line, Boochani communicated his thoughts beyond the fences of the prison camp, using wireless internet connections and radio waves like mycorrhizal hyphae to send his encrypted messages. Connecting with Tofighian, Mansoubi, and others who translated his work and relayed it to newspaper offices, publishers, and the media, he tapped into a wider network. Not only did this allow him to communicate his thoughts beyond the prison, but it also ensured that the plight of the detainees and the shortcomings of Australia's immigration system were broadcast to a wider readership.

Increasingly called upon to process and disseminate the messages, Tofighian and Mansoubi became accomplices in promoting Boochani's work. Tofighian describes the collaborative operation as a "shared philosophical activity".[113] He acknowledges the help of Boochani's cousin Farhad, and his university friends Najem Waysi and Toomas Askari, with whom Boochani also communicated. "Their exchanges and attempts to understand the phenomenon of Manus Prison open up a new discourse regarding collaboration in the context of creative and intellectual work", he recognizes.[114] Wider still, Boochani was able to communicate with sympathetic journalists such as Ben Doherty (*The Guardian*), activists such as Janet Galbraith (*Writing Through Fences*) and Arnold Zable (*PEN International*), filmmakers such as Arash Kamali Sarvestani (*Chauka, Please Tell us the Time*, 2016) and Hoda Afshar (*Remain*), and academics such as Kirrily Jordan (Australian National University) and Anne Surma (Murdoch University). The network spread, with Boochani frequently appearing via videolink at academic conferences, even while still in detention.[115] When his "A Letter from Manus Island" was published by *The Saturday Paper* on 9 December 2017, it was simultaneously performed by the poet Maxine Beneba Clarke at the Malthouse Theatre, Melbourne, where as Erik Jensen recalls, "a few hundred people gathered to hear these words that had escaped from prison".[116] When *Chauka, Please Tell us the Time* was released in 2016 and Boochani posted a message on Facebook, the response from social media users broadened the network further: "Some of those who commented and shared the post also offered to distribute the film or organise screenings when it was ready to be shown". The film has now been seen at numerous festivals and screenings worldwide.[117]

"This demonstrates how social media networks may help to disseminate Boochani's 'witnessing' to wider audiences",[118] write Maria Rae and colleagues, noting that "Social media platforms are conceived as enabling disintermediation in which traditional power structures and players are bypassed, remade or 're-distributed'".[119] In this way, the mycorrhizal network communications enabled by the detainees' access to technology enabled a "window onto their world", as Arianna Grasso puts it; "these digital devices were ultimately crucial to make the invisible visible, the unseen seen, allowing refugees to gain a space of visibility, self-legitimation, and unmediated expression".[120] "'Technology is so important to us', Boochani told journalist Claire Reilly in 2016 via Whatsapp, 'It gives us the power to send out our voice'".[121]

Chauka

The notion of the voice reaching beyond the prison is given added resonance in Boochani's work through the figure of the Manus Firebird, known colloquially as "Chauka". "This special bird sits everyday on a coconut tree and sings, and the local people compose their daily life with its singing", Boochani explains.[122] Throughout his work, the Chauka takes on a totemic role, especially in *Chauka, Please Tell us the Time* but also in *No Friend but the Mountains*.

The Chauka is endemic to Manus because of the territory's supply of fresh water, which isn't available on other islands. It is a significant bird to the local community "as a guide, a timekeeper and a voice of caution and forewarning",[123] and its image is represented on the Manusian flag. The bird makes its characteristic call at dawn and dusk, by which locals are able to keep track of the time, but it is also believed to have special powers of communication: "the Chauka in Manusian culture is known as a bird that can talk with people", writes Boochani; "The bird has the ability to make predictions, and the elders know what Chauka is saying about the future".[124] The Chauka's voice is notoriously loud; according to Rooney, "loud, interjecting and present"[125]; and Tofigihian recalls, "I often hear Chauka chanting in the background when communicating with [Boochani] through WhatsApp".[126]

But equally as important, Chauka is significant to Boochani and the detainees because of the use of its name within the prison. Each of the four camps within Manus has its own solitary confinement wing, which are called "Bravo", "The Green Zone", the "Managed Accommodation Area" ("MAA"), and "Chauka". Boochani himself was imprisoned on two occasions in Chauka, and "its name still causes fearful shaking in refugees".[127] It "consists of three shipping containers arranged in a triangle. Each shipping container is bare save for a single bed", report Ben Doherty and Nick Evershed; "There are no windows and only one way out, past a sentry post manned 24 hours a day".[128] Its conditions are a powerful deterrent to detainees; as Ben Pynt from Humanitarian Research Partners observes, "Even the suggestion of being sent to Chauka is enough to make people comply". Such human rights activists compare the effects of solitary confinement to those of torture: "post-traumatic stress disorder, heightened anxiety and heightened sensitivity to stimuli or to people".[129] In Boochani's film, the symbolism of the bird ties together its multiple strands. "There are two narratives running parallel in the film", recounts Boochani; "One is an ugly chauka which exists in the minds of the asylum seekers, the other a beautiful chauka present for the Manusians. The film ends when they come together".[130]

It was also through WhatsApp that Boochani made his film: "He did so in captivity, escaping guards' supervision, struggling with a poor Internet connection, thus with considerable difficulty, gradually sending images from the Manus Island detention camp, via his cell phone, to Iranian Dutch director Arash Kamali Sarvestani living in the Netherlands".

Technically, the film has all the qualities of a clandestine documentary, comprising shots and interviews hidden behind veils in small rooms, phone calls in noisy rooms, and long fences keeping the migrants in. Many of the detainees talk about the lack of space, personal, and physical. They are watched and harassed, making it impossible to find any peace. The beauty of the island is literally in sight but out of reach, and the boundaries between the outside environment and the fenced-in space in which the detainees live are a recurring feature. The hard fencing is juxtaposed in images with sand, the shore edge, and tropical trees. Children are on screen laughing in what seems to be a jovial shared space until the camera lens adjusts and they are revealed to be outside, with the viewer trapped behind the fence. Images repeatedly show a pair of feet in flip flops up against the fence, as if yearning for a walk on the sand and a paddle in the water that lies a few feet away. Yet the feet also seem resigned; the pace is slow, and the feet unmoving. What's the point? They can't go beyond the fence. There are several long shots of disinfecting with a vapour machine that pumps chemicals into plumes around the camp. The image becomes unclear through the fog of the treatment.

There is a break from the format when we are in a car with a woman—Janet Galbraith from the activist network *Writing Through Fences*—and a young man. They talk about Chauka, and then continue to discuss the bird with two Manusian men, Clement Soloman and Poruan "Sam" Malai, on the shore. Galbraith asks the men about the solitary confinement wing within the detention centre, and the horrific things that have happened. She asks Soloman and Malai how they feel about the Australian people calling that prison "Chauka". The men feel that it is an abuse of the name because of the torturous things that have happened to the detainees. The Manusian men are surprised to hear that torture is happening in the camp, when they believed that the centre was set up to help people until they could find a place to settle. The spiritual connection between the Manusians and this special bird, and the abuse that they feel with the prison being called Chauka interrogates the rationale for the centre and the way that the detainees are being kept and treated. As they observe, Chauka is a small bird that speaks loudly—therefore, they suggest, people should speak loudly about the mistreatments happening in the centre.

There have been many analyses of the film, not least from some of the individuals most closely connected with Boochani's work. Tofighian calls it "a meditation on time and a critical analysis of how it can be used as an instrument of torture. […] A significant amount of time is spent sitting, smoking, feet up on the fences, staring beyond the cage, contemplating, waiting".[131] In recalling the regime, Tofighian closely aligns the film with the development of "Manus Prison Theory" and Boochani's political activism. Taking a different perspective, but one that also reveals the film's political power, Galbraith herself discusses its aesthetic rather than its content. She picks up on the film's "discord between what we see and hear". For example, at one point the sound of the disinfecting machine—like a "massive mechanical mosquito"—is juxtaposed with the beautiful sound of "harmonious singing". "These images and sounds do not seem to fit together", she remarks. Similarly, the visual continuity of the film is disrupted by images of the outside world zooming out to reveal the frame of prison bars. "Such discontinuities emerge often throughout the film, displacing the viewer, preventing us from being able to watch and listen passively". She explains how Boochani "did not wish to use music or sound effects to manipulate viewers into particular emotions", and advocates a type of "haptic listening" in engaging with the film not as a "documentary-as-unmediated-unconstructed-reality", but instead as a work of art. "Divisive and dehumanising practices and language produced by the detention regime assign particular functions and positions to refugees", she concludes; "*Chauka, Please Tell Us the Time* enacts a refusal of (some) of these practices, not only through its function as a piece of cinematic art, but also through its process".[132]

But the symbolism of Chauka extends much further than this, both in Boochani's film and in the bird's significance to local culture, as Michelle Nayahamui Rooney's poem "Chauka Yu We? Chauka Where Are You" explores. Rita Sakr discusses the rich imagery of the bird as it features in both the film and poem. She writes of "the importance of the Chauka bird in informing an indigenously, ecologically inspired vision of sanctuary", and argues that "Chauka's call resonates in the film both as a form of haunting since it never physically appears and as a leitmotif that evokes feelings of solidarity and indignant resistance". In Sakr's interpretation, "Chauka's call emerges as a dialogical leitmotif of the potential of a more-than-human politics of liberation in Boochani's work and in 'Manus'".[133] And as Rooney acknowledges, in Boochani's film, "Chauka is a symbol of freedom, of how the detainee's voice spread word of the Manus detention centre".[134]

Community

Remarkably, then, the purgatory of Boochani's experience gains a wider audience, and from that connection beyond the prison walls emerges the solidarity of other communities. At first: community with fellow passengers on the boats, within the confines of the prison, within the brotherhood of the ordeal. But then: connection with the Manusians who share this same small place in the world, but who now also share a sense of wrong. And even more distant: community with those reached by the infochemicals of the hyphae, those who read Boochani's words, hear the plight of the detainees, and seek to further their call. And all of these communities—systems in themselves—are also a part of the ecosystem of the "border-industrial complex".[135] That the Chauka is able to fly and call beyond, through, and between the boundaries of this ecosystem is at once a reminder of its entropy and an inspiring motivation that promises successional change.

Or to turn to another analogy, is Boochani's journey reminiscent of the passage of which Timothy Morton speaks? A descent through environmental guilt, shame, melancholy, and horror, and into the realm of toys; a swim down through the hollow of the "uncanny valley" to the sadness, longing, and joy of "the interior of life"; a "fight [...] back from the brink (species-cidal and suicidal) toward the comedy"?[136] Remember that from Morton's perspective "ecosystems are made not only of trees, rock formations, and pigs (seemingly 'external' to the human), but also of thoughts, wishes, fantasies (seemingly 'inside' our heads)"[137]; remember too that "since lifeforms are made of nonlife, and since what counts as a lifeform is very much a performative act down to the DNA level, I see no big reason not to extend the concept of the strange stranger to cover all entities".[138] From an ecological perspective informed by dark ecology, the environment in which Boochani finds himself is the ecosystemic entirety of the fences, the roof, the coconut trees, both the Chauka (the bird) and the Chauka (the prison), and indeed the uncomfortable splicing of Australia's regime with the landscape of the beautiful Manusian island. Boochani tunes into the demonic energy of the migration infrastructure, the thanato-politics, the detention centre, and his island home; he articulates its hypocrisy (weak and ironic) through his poetry and prose; and collaborates with the world around him to distil a genuinely asymmetric art. Of course, Boochani is just one man, a solitary individual; yet as the "voice of Manus" he also embodies the plight and the opportunity of the whole community: he is the despair and the hope; the fight and the surrender; the incarceration and the liberation.[139]

Notes

1. Mike Foale and Hugh Harries (2011), "Farm and Forestry Production Marketing Profile for Coconut (*Cocos nucifera*)", in C.R. Elevitch (ed.) (2011), *Specialty Crops for Pacific Island* Agroforestry, Holualoa: Permanent Agriculture Resources, p. 5.
2. Karandeep Kaur et al. (2019), "Coconut Meal: Nutraceutical importance and food industry application", *Foods and Raw Materials*, 7:2, pp. 419–427, p. 419.
3. Subhash Chander Ahuja, Uma Ahuja, and Siddharth Ahuja (2014), "Coconut-History, Uses, and Folklore", *Asian Agri-History*, 18:3, pp. 221–248, p. 229.
4. Anon. (2024), "Papua New Guinea: Coconut (*cocos nucifera*)", *World Sensorium / Conservancy: Plantings*, https://worldsensorium.com/papua-new-guinea/. Accessed 23 September 2024.
5. P.J. Searles (1928), "The Most Valuable Tree in the World", *The Scientific Monthly*, 27:3, pp. 271–280, p. 271.
6. Anjana J. Atapattu, and Shashi S. Udumann (2024), "Leveraging Agroforestry Principles for Nature-Based Climate-Smart Solutions for Coconut Cultivation", in Walter Leal Filho, Gustavo J. Nagy, and Desalegn Yayeh Ayal (eds) (2024), *Handbook of Nature-Based Solutions to Mitigation and Adaptation to Climate Change*, Cham: Springer International, pp. 1–28, p. 2.
7. Mike Foale (2003), *Coconut Odyssey: The Bounteous Possibilities of the Tree of Life*, Canberra: Australian Centre for International Agricultural Research.
8. Ellen Wrchota Demerath (1997), "The impact of economic modernization on adult nutritional status in Manus, Papua New Guinea", Ph.D. thesis, University of Pennsylvania, p. 76.
9. Anon. (2012), "Manus project fills farming extension service gap", *Our Pacific Ways*, 5 October, https://ourpacificways.wordpress.com/2012/10/05/manus-project-fills-farming-extension-service-gap/. Accessed 20 September 2024.
10. John Brooksbank, Simon Towle, and Paul Lynch (2017), "Sustainable energy for Manus Province", paper delivered at *PNG Impact Conference*, University of Papua New Guinea, 12–13 December 2017. https://researchonline.jcu.edu.au/53140/1/Sustainable_energy_for_Manus_Province.pdf. Accessed 20 September 2024.

11. Nithin Coca (2020), "Coconut Oil Could Be the Next Disastrous Biofuel", *Rainforest Journalism Fund*, 23 February, https://rainforestjournalismfund.org/stories/coconut-oil-could-be-next-disastrous-biofuel. Accessed 23 September 2024.
12. Nithin Coca (2020), "Palm Vs. Coconut Oil: What's the More Sustainable Choice?", *Rainforest Journalism Fund*, 8 April, https://rainforestjournalismfund.org/stories/palm-vs-coconut-oil-whats-more-sustainable-choice. Accessed 23 September 2024.
13. Fabrícia Vieira et al. (2024), "Coconut Waste: Discovering Sustainable Approaches to Advance a Circular Economy", *Sustainability*, 16:7, 3066.
14. Carola Betzold (2015), "Coconut Power: Biofuel from Coconuts in the Solomon Islands", in Godfrey Baldacchino (ed.) (2015) *Entrepreneurship in Small Island States and Territories*, New York: Routledge, pp. 164–179, p. 164.
15. Behrouz Boochani (2018), *No Friend but the Mountains: The True Story of an Illegally Imprisoned Refugee* (trans. Omid Tofighian), London: Picador, p. 101.
16. Omid Tofighian (2018), "No Friend But the Mountains: Translator's Reflections", in Boochani, *No Friend*, p. 369.
17. Boochani, *No Friend*, p. 129.
18. Boochani, *No Friend*, p. 112.
19. Kathrin Bartha-Mitchell (2024), *Cosmological Readings of Contemporary Australian Literature: Unsettling the Anthropocene*, Abingdon: Routledge, p. 161.
20. Boochani, *No Friend*, p. 294.
21. N.E. Gómez-Falcón et al. (2023), "Arbuscular mycorrhizal fungi increase the survival and growth of micropropagated coconut (*Cocos nucifera* L.) plantlets", In Vitro *Cellular & Developmental Biology—Plant*, 59, pp. 401–412.
22. S.E. Nurzannah et al. (2022), "Potential of *Trichoderma* and mycorrhizae as biological agents for controlling *Ganoderma boninense* in oil palm", *IOP Conference Series: Earth and Environmental Science*, 974:012097.
23. Gertrude Scynthya Nirukshan, Sanathanie Ranasinghe, and Steven Sleutel (2022), "The effect of biochar on mycorrhizal fungi mediated nutrient uptake by coconut (*Cocos nucifera* L.) seedlings grown on a sandy Regosol", *Biochar*, 4:68.

24. S.H.S. Senarathne and I.M.P.S. Ilangamudali (2018), "Effect of arbuscular mycorrhizal fungi based biofertilizer on coconut seedlings growth in nursery", *Cord*, 34:2, pp. 30–41, p. 31.
25. KIK (2019), "Overview of the Coconut Industry", *Kokonas Indastri Koporesen*, 22 February, https://www.kik.com.pg/kokonas-tok-column-12/#:~:text=From%20the%20total%20households%20engaging,full%20time%20or%20part%20time. Accessed 23 September 2024.
26. United Nations (n.d.), "Migration", *United Nations: Global Issues*, https://www.un.org/en/global-issues/migration#:~:text=Some%20people%20move%20in%20search,disasters%2C%20or%20other%20environmental%20factors. Accessed 28 July 2023.
27. Cathrine Talleraas (2020), "Migration", in Antonio De Lauri (ed.) (2020), *Humanitarianism: Keywords*, Leiden and Boston: Brill, pp. 133–135, p. 133.
28. Everett S. Lee (1966), "A Theory of Migration", *Demography*, 3:1, pp. 47–57, p. 50.
29. Douglas S. Massey, et al. (1993), "Theories of International Migration: A Review and Appraisal", *Population and Development Review*, 19:3, pp. 431–66, p. 432.
30. Massey, et al., "Theories", p. 436.
31. Ivan Light, Parminder Bhachu, and Stavros Karageorgis (1989), "Migration Networks and Immigrant Entrepreneurship", *UCLA: Institute for Social Science Research*, 5: California Immigrants in World Perspective.
32. See for example Alberto Palloni et al. (2001), "Social capital and international migration: A test using information on family networks", *American Journal of Sociology*, 106:5, pp. 1262–1298.
33. Peter Nannestad, Gunnar Lind Haase Svendsen, and Gert Tinggaard Svendsen (2008), "Bridge over troubled water? Migration and social capital", *Journal of Ethnic and Migration Studies* 34:4, pp. 607–631.
34. Hein De Haas (2010), "The internal dynamics of migration processes: A theoretical inquiry", *Journal of Ethnic and Migration Studies*, 36:10, pp. 1587–1617.
35. Elizabeth Fussell and Douglas S. Massey (2004), "The limits to cumulative causation: International migration from Mexican urban areas", *Demography*, 41, pp. 151–171.

36. Gabriel Bădescu and Paul E. Sum (2009), "Coming Home: First and Second Order Effects of Returning Migrants on Romanian Economic Development through the Changing Dynamics of Social Capital", *Romanian Journal of Communication and Public Relations*, 11:3, pp. 17–34.
37. James T. Fawcett (1989), "Networks, Linkages, and Migration Systems", *The International Migration Review*, 23:3, pp. 671–680, p. 671.
38. Anon. (n.d.), "Empowering Migrants", https://www.iom.int/migapp. Accessed 29 July 2023.
39. Anon. (2017), "UN Migration Agency Launches Mobile App for Migrants in Mesoamerica", *International Organisation for Migration* website, 6 October, https://www.iom.int/news/un-migration-agency-launches-mobile-app-migrants-mesoamerica. Accessed 28 July 2023.
40. Anon. (2022), "MigApp, trusted mobile information for migrants around the world", *Innovation in Social Care Foundation* website, https://isocial.cat/en/migapp-trusted-mobile-information-for-migrants-around-the-world/. Accessed 29 July 2023.
41. Anon. (2018), "Empowering Migrants with MigApp", *International Organisation for Migration* website, 9 February, https://www.iom.int/news/empowering-migrants-migapp. Accessed 29 July 2023.
42. Boochani in Ben Doherty (2015), "Day of the Imprisoned Writer: Behrouz Boochani—detained on Manus Island", *The Guardian*, 14 November. https://www.theguardian.com/books/2015/nov/15/day-of-the-imprisoned-writer-behrouz-bouchani-detained-on-manus-island. Accessed 21 July 2023.
43. Omid Tofighian (2022), "Writing in Languages of Freedom: Introduction", in Behrouz Boochani (2022), *Freedom, Only Freedom: The Prison Writings of Behrouz Boochani*, (trans. Moones Mansoubi and Omid Tofighian), London: Bloomsbury, pp. 82–86, p. xviii.
44. Boochani, *No Friend*, p. 8.
45. Boochani, *No Friend*, p. 82.
46. Boochani, *No Friend*, p. xv.

47. Janet Phillips (2015), " Asylum seekers and refugees: what are the facts?", *Australian Parliament* website, 2 March, https://www.aph.gov.au/about_parliament/parliamentary_departments/parliamentary_library/pubs/rp/rp1415/asylumfacts. Accessed 29 July 2023.
48. Amnesty International (2016), *Island of Despair: Australia's "Processing" of Refugees on Nauru*, London: Amnesty International Ltd., https://www.amnesty.org/en/documents/asa12/4934/2016/en/. Accessed 26 December 2024, pp. 10–11.
49. Anon. (n.d.), "Australia's borders are closed to unauthorised maritime arrivals", *Operation Sovereign Borders*, https://osb.homeaffairs.gov.au/. Accessed 29 July 2023.
50. Boochani, *Freedom*, p. 1.
51. Anon., "Australia's borders".
52. Phillips, "Asylum Seekers".
53. Phillips, "Asylum Seekers".
54. Mansoubi, "Translating Manus", p. 12.
55. Renata Summo-O'Connell (2018), "Art, the Uncontainable Other in Contemporary Detention Camps", *Seismopolite: Journal of Art and Politics*, 20. https://www.seismopolite.com/art-the-uncontainable-other-in-contemporary-detention-camps. Accessed 21 July 2023.
56. Boochani and his collaborators eventually construct a "Manus Island Theory", which articulates a more developed critique of the Australian immigration system and its offshore detention regime. See for example Behrouz Boochani (2017), "A Kyriarchal System: New Colonial Experiments/New Decolonial Resistance", *9th Annual Maroon Conference Magazine*; Behrouz Boochani (2017), "A Letter from Manus Island" (trans. Omid Tofighian), *The Saturday Paper*, 9 December; Behrouz Boochani (2018) "Manus prison poetics/our voice: revisiting 'A Letter From Manus Island', a reply to Anne Surma", *Continuum*, 32:4, pp. 527–531; Behrouz Boochani (2018), "Manus Prison Theory" (trans. Omid Tofighian), *The Saturday Paper*, 11 August; Tofighian, "Writing in Languages", pp. xiii–xxi.
57. Early on, he invokes Giorgio Agamben's "state of exception", "a political process where the language of war is used to justify an increase in government powers" (Behrouz Boochani (2016), "Australia, exceptional in its brutality" (trans. Moones

Mansoubi), *Overland*, 25 April. See also Suvendrini Perera and Joseph Pugliese (2015), "Offshore detention 'black sites' open door to torture", *The Conversation*, 26 August); meanwhile, Anne McNevin explores the concept of "epistemic violence", the way in which "language becomes part of a structure of violence"; Anne Surma compares Boochani's "political poetics" to Jacques Ranciere's "structure[s] of rationality"; and Mansoubi writes of the Australian system using Adorno's ideas of "in group-out group differentiation". Anne McNevin (2022), "Epistemic Violence and the Man Who Loves Ducks", in Boochani, *Freedom*, pp. 82–86; Anne Surma (2022), "Boochani's 'Political Poetics' Subverting and Reimagining the Fiction of Real Politics", in Boochani (2022), *Freedom*, pp. 223–226; Moones Mansoubi (2022), "Translating Manus and Nauru: Refugee Writing", in Boochani, *Freedom*, pp. 11–19.

58. Boochani, "Manus prison poetics", p. 528.
59. Boochani, "Manus prison poetics", p. 528.
60. Tofighian, "Writing in Languages", p. xvii.
61. Boochani, *No Friend*, p. 122.
62. Boochani, *No Friend*, pp. 110–112.
63. Boochani, *No Friend*, p. 165.
64. "Probably the life most worth living is a solitary one", Boochani, *No Friend*, pp. 132–133.
65. Boochani, *No Friend*, pp. 125.
66. Behrouz Boochani (2015), "Love Song of a Caged Bird", *Writing Through Fences*, https://www.writingthroughfences.org/writers/behrouz-boochani/. Accessed 26 December 2024.
67. Boochani, "Manus prison poetics", p. 528.
68. Boochani, *No Friend*, p. 123.
69. Boochani, *No Friend*, p. 124.
70. Boochani, *No Friend*, pp. 137–140.
71. Boochani, *No Friend*, p. 284.
72. Boochani, *No Friend*, pp. 284–285.
73. Boochani, *No Friend*, p. 299.
74. McNevin, "Epistemic Violence", p. 82.
75. Behrouz Boochani (2016), "Life on Manus: island of the damned" (trans. Moones Mansoubi), *The Saturday Paper*, 4 March.
76. Boochani, *No Friend*, p. 295.

77. Boochani, *No Friend*, p. 301.
78. Boochani, *No Friend*, p. 320.
79. Rita Sakr (2020), "Decolonial Imaginaries of Sanctuary in Behrouz Boochani's Work", *Journal of Migration and Culture*, 11:2, pp. 231–249, p. 231.
80. Arianna Grasso (2022), "On Documentation, Language and Social Media", in Boochani, *Freedom*, p. 243.
81. Peter Wohlleben (2016), *The Hidden Life of Trees: What They Feel, How They Communicate: Discoveries from a Secret World* (trans. Jane Billinghurst), London: William Collins, p. 10.
82. E. Kathryn Barto et al. (2012), "Fungal superhighways: do common mycorrhizal networks enhance below ground communication?", *Trends in plant science*, 17:11, pp. 633–637, p. 633.
83. Barto et al., "Fungal superhighways", pp. 633–635.
84. Barto et al., "Fungal superhighways", p. 635; Kevin J. Beiler et al. (2010), "Architecture of the wood-wide web: Rhizopogon spp. genets link multiple Douglas-fir cohorts", *New Phytologist*, 185:2, pp. 543–553, p. 543; Monika A. Gorzelak et al. (2015), "Inter-plant communication through mycorrhizal networks mediates complex adaptive behaviour in plant communities", *Annals of Botany: Plants*, 7, plv050, p. 3; Suzanne W. Simard (2018), "Mycorrhizal networks facilitate tree communication, learning, and memory", *Memory and learning in plants*, pp. 191–213, p. 192.
85. Simard, "Mycorrhizal networks", p. 200.
86. Kazimierz Trebacz, Halina Dziubinska, and Elzbieta Krol (2006), "Electrical signals in long-distance communication in plants", *Communication in plants: Neuronal aspects of plant life*, pp. 277–290; Miguel Angel Garcia-Servin, Magdalena Mendoza-Sanchez, and Luis Miguel Contreras-Medina (2021), "Electrical signals as an option of communication with plants: A review", *Theoretical and Experimental Plant Physiology*, 33:2, pp. 125–139.
87. Simard, "Mycorrhizal networks", pp. 194–197.
88. Simard, "Mycorrhizal networks", pp. 202–203.
89. Gorzelak et al., "Inter-plant communication", p. 1.
90. See for example Beiler et al., "Architecture"; Simard, "Mycorrhizal networks".
91. Gorzelak et al., "Inter-plant communication", p. 9.
92. Simard, "Mycorrhizal networks", p. 207.

93. Gorzelak et al., "Inter-plant communication", p. 5.
94. Gorzelak et al., "Inter-plant communication", p. 9.
95. Gorzelak et al., "Inter-plant communication", p. 10.
96. Barto et al., "Fungal superhighways".
97. Simard, "Mycorrhizal networks", p. 198.
98. Beiler et al., "Architecture". Some of these ideas have been questioned, especially when they have emerged from popular rather than scientific literature, and especially in terms of their tendency to anthropomorphise chemical processes. See for example Justine Karst, Melanie D. Jones, and Jason D. Hoeksema (2023), "Positive citation bias and overinterpreted results lead to misinformation on common mycorrhizal networks in forests", *Nature Ecology and Evolution*, 7, pp. 501–511; Nils Henriksson et al. (2023), "Re-examining the evidence for the mother tree hypothesis–resource sharing among trees via ectomycorrhizal networks", *New Phytologist*, 239, pp. 19–28; David G. Robinson et al. (2024), "Mother trees, altruistic fungi, and the perils of plant personification", *Trends in Plant Science*, 29:1, pp. 20–31; Aisling Irwin (2024), "The 'Mother Tree' idea is everywhere—but how much of it is real?", *Nature*, 26 March, https://www.nature.com/articles/d41586-024-00893-0. Accessed 23 September 2024.
99. Boochani, *No Friend*, pp. 218–219.
100. Boochani, *No Friend*, p. 220.
101. Boochani, *No Friend*, p. 111.
102. Behrouz Boochani (2020), "'White Australia' Policy Lives On in Immigrant Detention", *New York Times*, 20 September.
103. Boochani, *No Friend*, p. 112.
104. Boochani, *No Friend*, p. 125.
105. Boochani, "'White Australia'".
106. Helena Zeweri (2022), "Carceral Coloniality as a History of the Present", in Boochani, *Freedom*, pp. 245–248, p. 247.
107. Arnold Zable, former president of Pen International's Melbourne office, quoted in Ben Doherty (2015), "Day of the imprisoned writer: Behrouz Boochani–Detained on Manus Island", *The Guardian*, 14 November.
108. Boochani, *No Friend*, p. 107.
109. Boochani, *No Friend*, p. 92.

110. Ben Doherty (2022), "Collaborating with Behrouz Boochani: Journalists Against a System", in Boochani, *Freedom*, pp. 20–23, p. 22.
111. Boochani, *No Friend*, p. 128.
112. Tofighian, "Writing in Languages", pp.xiii-xiv.
113. Omid Tofighian (2018), "Translator's Tale: A Window to the Mountains", in Boochani, *No Friend*, pp. 375–398, p. 380.
114. Tofighian, "Translator's Tale", p. 384.
115. Tofighian reports the first such conference appearance at "Refugee Alternatives: Improving Policy, Practice, and Public Support", UNSW, 2017.
116. Erik Jensen (2022), "Words that Escaped from Prison", in Boochani, *Freedom*, pp. 127–129, p. 127.
117. Maria Rae, Rosa Holman and Amy Nethery (2018), "Self-represented Witnessing: The Use of Social Media by Asylum Seekers in Australia's Offshore Immigration Detention Centres", *Media, Culture & Society*, 40:4, pp. 479–495, p. 487.
118. Rae et al., "Self-represented", p. 487.
119. Rae et al., "Self-represented", p. 482, citing Ito.
120. Grasso, "On Documentation", pp. 242–243.
121. Rae et al., "Self-represented", p. 492.
122. Behrouz Boochani, interviewed in Enza Capobianco (2017), "Behind the documentary that made it out of Manus", *ACMI*, 19 May. https://www.acmi.net.au/stories-and-ideas/manus-island-documentary-chauka/. Accessed 23 September 2024.
123. Michelle Nayahamui Rooney (2018), "Chauka, where are you? Notes on an ethnographic poem", *Griffith Review*, 59: Commonwealth Now, pp. 181–190, p. 183.
124. Behrouz Boochani (2018), *Facebook* post, 17 January, https://www.facebook.com/BehrouzBoochaniJournalist/posts/325405521278163. Accessed 21 July 2023.
125. Rooney, "Chauka", p. 183.
126. Omid Tofighian (2019), "Chauka Calls—a Photo Essay", *Alphaville: Journal of Film and Screen Media*, 18, pp. 205–217, p. 206.
127. Behrouz Boochani (2016), "What It's Like In Solitary Confinement On Manus Island", *The Huffington Post*, 22 August.
128. Ben Doherty and Nick Evershed (2014), "Manus Island asylum seekers put in solitary confinement at a rate of three a week", *The Guardian*, 12 December.

129. Pynt in Doherty and Evershed, "Manus Island".
130. Boochani, "A Kyriarchal System".
131. Tofighian, "Chauka Calls", p. 208.
132. Janet Galbraith (2019), "A reflection on Chauka, Please Tell Us the Time", *Alphaville: Journal of Film and Screen Media*, pp. 193–198, pp. 194–195.
133. Sakr, "Decolonial Imaginaries", pp. 243–244.
134. Rooney, "Chauka", p. 187.
135. Tofighian, "No Friend", p. 362.
136. Timothy Morton (2016), *Dark Ecology: For a Logic of Coexistence*, New York: Columbia University Press, pp. 141–154.
137. Morton, *Dark Ecology*, p. 67.
138. Timothy Morton (2012), "Art in the Age of Asymmetry: Hegel, Objects, Aesthetics", *Evental Aesthetics*, 1, p. 132.
139. Manus Regional Processing Centre was officially shut down on 31 October 2017 following a judgement by the PNG authorities that its operation was constitutionally illegal. Yet unable or unwilling to return to their places of origin, and still refused entry to Australia, the remaining 600 refugees were effectively cast adrift. For three weeks they refused to leave, initiating a 23-day stand-off with authorities. Still the only access to the outside world was through mobile technology, though by now the mycorrhizal network was well-established and humanitarian agencies, the world's media, and even sympathetic jurisdictions elsewhere (including Canada, New Zealand, and Obama-era USA) were sensitive to their plight. Eventually, on 23 July 2020—coincidentally his 37th birthday—Boochani was granted refugee status in New Zealand.

Bibliography

Ahuja, Subhash Chander, Uma Ahuja, and Siddharth Ahuja (2014), "Coconut-History, Uses, and Folklore", *Asian Agri-History*, 18:3, pp. 221–248.

Amnesty International (2016), *Island of Despair: Australia's "Processing" of Refugees on Nauru*, London: Amnesty International Ltd., https://www.amnesty.org/en/documents/asa12/4934/2016/en/. Accessed 26 December 2024, pp. 10–11.

Anon. (2012), "Manus project fills farming extension service gap", *Our Pacific Ways*, 5 October, https://ourpacificways.wordpress.com/2012/10/05/manus-project-fills-farming-extension-service-gap/. Accessed 20 September 2024.

Anon. (2017), "UN Migration Agency Launches Mobile App for Migrants in Mesoamerica", *International Organisation for Migration* website, 6 October, https://www.iom.int/news/un-migration-agency-launches-mobile-app-migrants-mesoamerica. Accessed 28 July 2023.

Anon. (2018), "Empowering Migrants with MigApp", *International Organisation for Migration* website, 9 February, https://www.iom.int/news/empowering-migrants-migapp. Accessed 29 July 2023.

Anon. (2022), "MigApp, trusted mobile information for migrants around the world", *Innovation in Social Care Foundation* website, https://isocial.cat/en/migapp-trusted-mobile-information-for-migrants-around-the-world/. Accessed 29 July 2023.

Anon. (2024), "Papua New Guinea: Coconut (*cocos nucifera*)", *World Sensorium / Conservancy: Plantings*, https://worldsensorium.com/papua-new-guinea/. Accessed 23 September 2024.

Anon. (n.d.), "Australia's borders are closed to unauthorised maritime arrivals", *Operation Sovereign Borders*, https://osb.homeaffairs.gov.au/. Accessed 29 July 2023.

Anon. (n.d.), "Empowering Migrants", https://www.iom.int/migapp. Accessed 29 July 2023.

Atapattu, Anjana J. and Shashi S. Udumann (2024), "Leveraging Agroforestry Principles for Nature-Based Climate-Smart Solutions for Coconut Cultivation", in Walter Leal Filho, Gustavo J. Nagy, and Desalegn Yayeh Ayal (eds) (2024), *Handbook of Nature-Based Solutions to Mitigation and Adaptation to Climate Change*, Cham: Springer International, pp. 1–28.

Bartha-Mitchell, Kathrin (2024), *Cosmological Readings of Contemporary Australian Literature: Unsettling the Anthropocene*, Abingdon: Routledge.

Barto, E. Kathryn, Jeffrey D. Weidenhamer, Don Cipollini, and Matthias C. Rillig (2012), "Fungal superhighways: do common mycorrhizal networks enhance below ground communication?", *Trends in plant science*, 17:11, pp. 633–637.

Beiler, Kevin J., Daniel M. Durall, Suzanne W. Simard, Sheri A. Maxwell and Annette M. Kretzer (2010), "Architecture of the wood-wide web: Rhizopogon spp. genets link multiple Douglas-fir cohorts", *New Phytologist*, 185:2, pp. 543–553.

Betzold, Carola (2015), "Coconut Power: Biofuel from Coconuts in the Solomon Islands", in Godfrey Baldacchino (ed.) (2015) *Entrepreneurship in Small Island States and Territories*, New York: Routledge, pp. 164–179.

Boochani, Behrouz (2015), "Love Song of a Caged Bird", *Writing Through Fences*, https://www.writingthroughfences.org/writers/behrouz-boochani/. Accessed 26 December 2024.

Boochani, Behrouz (2016), "Life on Manus: island of the damned" (trans. Moones Mansoubi), *The Saturday Paper*, 4 March.

Boochani, Behrouz (2016), "Australia, exceptional in its brutality" (trans. Moones Mansoubi), *Overland*, 25 April.

Boochani, Behrouz (2016), "What It's Like In Solitary Confinement On Manus Island", *The Huffington Post*, 22 August.

Boochani, Behrouz (2017), "A Kyriarchal System: New Colonial Experiments / New Decolonial Resistance", *9th Annual Maroon Conference Magazine*.

Boochani, Behrouz (2017), "A Letter from Manus Island" (trans. Omid Tofighian), *The Saturday Paper*, 9 December.

Boochani, Behrouz (2018) "Manus prison poetics/our voice: revisiting 'A Letter From Manus Island', a reply to Anne Surma", *Continuum*, 32:4, pp. 527–531.

Boochani, Behrouz (2018), *Facebook* post, 17 January, https://www.facebook.com/BehrouzBoochaniJournalist/posts/325405521278163. Accessed 21 July 2023.

Boochani, Behrouz (2018), "Manus Prison Theory" (trans. Omid Tofighian), *The Saturday Paper*, 11 August.

Boochani, Behrouz (2018), *No Friend but the Mountains: The True Story of an Illegally Imprisoned Refugee* (trans. Omid Tofighian), London: Picador.

Boochani, Behrouz (2020), "'White Australia' Policy Lives On in Immigrant Detention", *New York Times*, 20 September.

Boochani, Behrouz (2022), *Freedom, Only Freedom: The Prison Writings of Behrouz Boochani*, (trans. Moones Mansoubi and Omid Tofighian), London: Bloomsbury.

Brooksbank, John, Simon Towle, and Paul Lynch (2017), "Sustainable energy for Manus Province", paper delivered at *PNG Impact Conference*, University of Papua New Guinea, 12–13 December 2017. https://researchonline.jcu.edu.au/53140/1/Sustainable_energy_for_Manus_Province.pdf. Accessed 20 September 2024.

Bădescu, Gabriel and Paul E. Sum (2009), "Coming Home: First and Second Order Effects of Returning Migrants on Romanian Economic Development through the Changing Dynamics of Social Capital", *Romanian Journal of Communication and Public Relations*, 11:3, pp. 17–34.

Capobianco, Enza (2017), "Behind the documentary that made it out of Manus", *ACMI*, 19 May. https://www.acmi.net.au/stories-and-ideas/manus-island-documentary-chauka/. Accessed 23 September 2024.

Coca, Nithin (2020), "Coconut Oil Could Be the Next Disastrous Biofuel", *Rainforest Journalism Fund*, 23 February, https://rainforestjournalismfund.org/stories/coconut-oil-could-be-next-disastrous-biofuel. Accessed 23 September 2024.

Coca, Nithin (2020), "Palm Vs. Coconut Oil: What's the More Sustainable Choice?", *Rainforest Journalism Fund*, 8 April, https://rainforestjournalismfund.org/stories/palm-vs-coconut-oil-whats-more-sustainable-choice. Accessed 23 September 2024.

De Haas, Hein (2010), "The internal dynamics of migration processes: A theoretical inquiry", *Journal of Ethnic and Migration Studies*, 36:10, pp. 1587–1617.

Demerath, Ellen Wrchota (1997), "The impact of economic modernization on adult nutritional status in Manus, Papua New Guinea", Ph.D. thesis, University of Pennsylvania.

Doherty, Ben (2015), "Day of the Imprisoned Writer: Behrouz Boochani – detained on Manus Island", *The Guardian*, 14 November. https://www.theguardian.com/books/2015/nov/15/day-of-the-imprisoned-writer-behrouz-bouchani-detained-on-manus-island. Accessed 21 July 2023.

Doherty, Ben and Nick Evershed (2014), "Manus Island asylum seekers put in solitary confinement at a rate of three a week", *The Guardian*, 12 December.

Fawcett, James T. (1989), "Networks, Linkages, and Migration Systems", *The International Migration Review*, 23:3, pp. 671–680.

Foale, Mike (2003), *Coconut Odyssey: The Bounteous Possibilities of the Tree of Life*, Canberra: Australian Centre for International Agricultural Research.

Foale, Mike and Hugh Harries (2011), "Farm and Forestry Production Marketing Profile for Coconut (*Cocos nucifera*)", in C.R. Elevitch (ed.) (2011), *Specialty Crops for Pacific Island* Agroforestry, Holualoa: Permanent Agriculture Resources.

Fussell, Elizabeth and Douglas S. Massey (2004), "The limits to cumulative causation: International migration from Mexican urban areas", *Demography*, 41, pp. 151–171.

Galbraith, Janet (2019), "A reflection on Chauka, Please Tell Us the Time", *Alphaville: Journal of Film and Screen Media*, pp. 193–198.

Garcia-Servin, Miguel Angel, Magdalena Mendoza-Sanchez, and Luis Miguel Contreras-Medina (2021), "Electrical signals as an option of communication with plants: A review", *Theoretical and Experimental Plant Physiology*, 33:2, pp. 125–139.

Gómez-Falcón, N. E., L.A. Sáenz-Carbonell, A. Andrade-Torres, L.A. Lara-Pérez, M. Narváez, and C. Oropeza (2023), "Arbuscular mycorrhizal fungi increase the survival and growth of micropropagated coconut (*Cocos nucifera* L.) plantlets", *In Vitro Cellular & Developmental Biology—Plant*, 59, pp. 401–412.

Gorzelak, Monika A., Amanda K. Asay, Brian J. Pickles, and Suzanne W. Simard (2015), "Inter-plant communication through mycorrhizal networks mediates complex adaptive behaviour in plant communities", *Annals of Botany: Plants*, 7, plv050.

Henriksson, Nils, John Marshall, Mona N. Högberg, Peter Högberg, Andrea Polle, Oskar Franklin, and Torgny Näsholm (2023), "Re-examining the evidence for the mother tree hypothesis–resource sharing among trees via ectomycorrhizal networks", *New Phytologist*, 239, pp. 19–28.

Irwin, Aisling (2024), "The 'Mother Tree' idea is everywhere—but how much of it is real?", *Nature*, 26 March, https://www.nature.com/articles/d41586-024-00893-0. Accessed 23 September 2024.

Karst, Justine, Melanie D. Jones, and Jason D. Hoeksema (2023), "Positive citation bias and overinterpreted results lead to misinformation on common mycorrhizal networks in forests", *Nature Ecology and Evolution*, 7, pp. 501–511.

Kaur, Karandeep, Navnidhi Chhikara, Poorva Sharma, M.K. Garg, and Anil Panghal (2019), "Coconut Meal: Nutraceutical importance and food industry application", *Foods and Raw Materials*, 7:2, pp. 419–427.

KIK (2019), "Overview of the Coconut Industry", *Kokonas Indastri Koporesen*, 22 February, https://www.kik.com.pg/kokonas-tok-column-12/#:~:text=From%20the%20total%20households%20engaging,full%20time%20or%20part%20time. Accessed 23 September 2024.

Lee, Everett S. (1966), "A Theory of Migration", *Demography*, 3:1, pp. 47–57.

Light, Ivan, Parminder Bhachu, and Stavros Karageorgis (1989), "Migration Networks and Immigrant Entrepreneurship", *UCLA: Institute for Social Science Research*, 5: California Immigrants in World Perspective, n.p.

Massey, Douglas S., Joaquin Arango, Graeme Hugo, Ali Kouaouci, Adela Pellegrino, and J. Edward Taylor (1993), "Theories of International Migration: A Review and Appraisal", *Population and Development Review*, 19:3, pp. 431–66.

Nannestad, Peter, Gunnar Lind Haase Svendsen, and Gert Tinggaard Svendsen (2008), "Bridge over troubled water? Migration and social capital", *Journal of Ethnic and Migration Studies* 34:4, pp. 607–631.

Nirukshan, Gertrude Scynthya, Sanathanie Ranasinghe, and Steven Sleutel (2022), "The effect of biochar on mycorrhizal fungi mediated nutrient uptake by coconut (*Cocos nucifera* L.) seedlings grown on a sandy Regosol", *Biochar*, 4:68.

Nurzannah, S. E., I. Purnamasari, D.R. Siagian, and K.E.L. Ramija (2022), "Potential of *Trichoderma* and mycorrhizae as biological agents for controlling *Ganoderma boninense* in oil palm", *IOP Conference Series: Earth and Environmental Science*, 974:012097.

Palloni, Alberto, Douglas S. Massey, Miguel Ceballos, Kristin Espinosa, and Michael Spittel (2001), "Social capital and international migration: A test using information on family networks", *American Journal of Sociology*, 106:5, pp. 1262–1298.

Perera, Suvendrini and Joseph Pugliese (2015), "Offshore detention 'black sites' open door to torture", *The Conversation*, 26 August

Phillips, Janet (2015), " Asylum seekers and refugees: what are the facts?", *Australian Parliament* website, 2 March, https://www.aph.gov.au/about_parliament/parliamentary_departments/parliamentary_library/pubs/rp/rp1415/asylumfacts. Accessed 29 July 2023.

Rae, Maria, Rosa Holman and Amy Nethery (2018), "Self-represented Witnessing: The Use of Social Media by Asylum Seekers in Australia's Offshore Immigration Detention Centres", *Media, Culture & Society*, 40:4, pp. 479–495.

Robinson, David G., Christian Ammer, Andrea Polle, Jürgen Bauhus, Roni Aloni, Peter Annighöfer, Tobias I. Baskin, Michael R. Blatt, Andreas Bolte, Harald Bugmann, Jerry D. Cohen, Peter J. Davies, Andreas Draguhn, Henrik Hartmann, Hubert Hasenauer, Peter K. Hepler, Ulrich Kohnle, Friederike Lang, Magnus Löf, Christian Messier, Sergi Munné-Bosch, Angus Murphy, Klaus J. Puettmann, Iván Quiroz Marchant, Peter H. Raven, David Robinson, Dale Sanders, Dominik Seidel, Claus Schwechheimer, Peter Spathelf, Martin Steer, Lincoln Taiz, Sven Wagner, Nils Henriksson, and Torgny Näsholm (2024), "Mother trees, altruistic fungi, and the perils of plant personification", *Trends in Plant Science*, 29:1, pp. 20–31.

Rooney, Michelle Nayahamui (2018), "Chauka, where are you? Notes on an ethnographic poem", Griffith Review, 59: Commonwealth Now, pp. 181–190. https://www.griffithreview.com/articles/chauka-where-are-you-notes-ethnographic-poem. Accessed 31 July 2023.

Sakr, Rita (2020), "Decolonial Imaginaries of Sanctuary in Behrouz Boochani's Work", *Journal of Migration and Culture*, 11:2, pp. 231–249.

Searles, P. J. (1928), "The Most Valuable Tree in the World", *The Scientific Monthly*, 27:3, pp. 271–280.

Senarathne, S. H. S. and I. M. P. S. Ilangamudali (2018), "Effect of arbuscular mycorrhizal fungi based biofertilizer on coconut seedlings growth in nursery", *Cord*, 34:2, pp. 30–41.

Simard, Suzanne W. (2018), "Mycorrhizal networks facilitate tree communication, learning, and memory", *Memory and learning in plants*, pp. 191–213.

Summo-O'Connell, Renata (2018), "Art, the Uncontainable Other in Contemporary Detention Camps", *Seismopolite: Journal of Art and Politics*, 20, n.p. https://www.seismopolite.com/art-the-uncontainable-other-in-contemporary-detention-camps. Accessed 21 July 2023.

Summo-O'Connell, "Art, the Uncontainable Other", n.p.

Talleraas, Cathrine (2020), "Migration", in Antonio De Lauri (ed.) (2020), *Humanitarianism: Keywords*, Leiden and Boston: Brill, pp. 133–135.

Tofighian, Omid (2019), "Chauka Calls—a Photo Essay", *Alphaville: Journal of Film and Screen Media*, 18, pp. 205–217.

Trebacz, Kazimierz, Halina Dziubinska, and Elzbieta Krol (2006), "Electrical signals in long-distance communication in plants", *Communication in plants: Neuronal aspects of plant life*, pp. 277–290.

United Nations (n.d.), "Migration", *United Nations: Global Issues*, https://www.un.org/en/global-issues/migration#:~:text=Some%20people%20move%20in%20search,disasters%2C%20or%20other%20environmental%20factors. Accessed 28 July 2023.

Vieira, Fabrícia, Hortência E. P. Santana, Meirielly Jesus, Joana Santos, Preciosa Pires, Manuela Vaz-Velho, Daniel Pereira Silva, and Denise Santos Ruzene (2024), "Coconut Waste: Discovering Sustainable Approaches to Advance a Circular Economy", *Sustainability*, 16:7, 3066.

Wohlleben, Peter (2016), *The Hidden Life of Trees: What They Feel, How They Communicate: Discoveries from a Secret World* (trans. Jane Billinghurst), London: William Collins.

CHAPTER 9

Disturbance

It's not customary to truck in full-size maples to build the set of a Broadway show; ordinarily, timber or even Styrofoam will suffice. But when theatre designer Beowulf Boritt conceived of his set for *Come From Away*, nothing but real trees would do. "The individuality of a tree is like the individuality of a person", he suggests; "You can't create it as beautifully as it grows in nature".[1] So a range of beech and maple trunks were selected from upstate New York, trees that are native to a broad stretch of North America and that would also have been found on the island of Newfoundland, where this particular show is set. "Dealing with them is difficult", Boritt admits, "since they're all 24 feet long and a few tons"[2]—significantly heavier than your average set-piece—but "once the trees were felled, they were taken to an upstate scenery shop, trimmed and flame-proofed, their bottoms sealed to prevent leakage".[3] Then, the trees were positioned either side of the minimalist stage of the Schoenfeld Theatre, to frame a backdrop which through clever lighting effects subtly transitions between multiple scenes. In its most evocative state, the set presents a scene that is fixed into the cultural consciousness of Americans: the clear blue sky above Manhattan on 11 September 2001. Angled against the skyscape of the backdrop are two more "raggedly broken" trees: "unspoken stand-ins for the Twin Towers".[4] "It started out as a reference to these three broken I-beams in the rubble of Ground Zero", explains Boritt; "I wanted it to look like a hurricane came through and

K. Savage and D. Symonds, *Ecologies of Community in Performance*,
https://doi.org/10.1007/978-3-031-94343-0_9

ripped the tops off those two trees".[5] For *Come From Away* is a powerful Broadway musical set on 9/11, possibly the most devastating day in US memory. Perhaps the devastation of that day can only be understood by evoking the brute power of a hurricane.

Maples and beech are generally resilient trees. With deep roots, maples are adaptable to various soil conditions, are reasonably tolerant to temperature extremes, common diseases, and moderate drought, and since they grow fast they quickly develop mature resilience and can survive for several decades.[6] Meanwhile, beech trees are resistant to rot, insect damage, and resilient to drought, and with a dense, close-grained texture strengthening their wood, they boast long and healthy lifespans.[7] While these qualities may not be precisely the reasons Boritt chose beech and maple for his set, there is at least a poetic connection to their use in *Come From Away*: "That is what the trees are about for me: a sense of resilience that feels appropriate for this story. The country and world were damaged on that day, but hopefully we have deep enough roots to keep life going".[8] Even so, neither species is infallible, and with current rates of climate change they are susceptible to the rapid temperature rises that are underway. "By 2200, in about seven maple generations, the climate could change so much that it resembles a past extremely hot climate that the world hasn't seen for roughly 1.5 million maple generations, or 50 million years", warns arboriculturalist Jake Grossman; "Maples will probably be able to survive somewhere on Earth in this new, hot climate, but they absolutely will not be able to evolve to be adapted to it in seven generations".[9] It's a similar story for beech: "Projected climate change will foster a progressive decrease of beech growth",[10] write scientists; indeed, this is already underway in Europe, where declining numbers have been attributed to "increasing temperatures, the impact of extreme climatic events and long-term changes of environmental conditions".[11]

Whether devastation is brought about by terrorist attacks, the power of natural forces like hurricanes, or the destruction of our habitat through climate change, we can use environmental thinking to try to understand the impact. In ecological terms, these phenomena are known as *disturbance events*. In this chapter we will explore the ecological phenomena of *disturbance*. We will learn from general models of social-ecological disturbance, and we will apply these to the *disturbance event* of 9/11. We'll focus on the impact on the tiny Newfoundland community of Gander, which became home to around 6,500 grounded refugees when US airspace was shut down in the wake of the attacks. Our point of reference will be the Broadway musical *Come From Away*, and the way in which it relives this encounter. In particular, we'll consider how the show's development incorporated co-production with the Gander community, and shared experience with those who remember the event. In this, we will turn to theories of ***ecological memory*** and how it enables ***ecosystem resilience***. We'll note how *Come From Away* therefore creates a poignant response to a particularly impactful social-ecological disturbance.

Disturbance Events

When hijacked planes crashed into the towers of the World Trade Center on 11 September 2001, the remote town of Gander in Newfoundland experienced one of the most extraordinary events in its history. Thirty-eight planes were rerouted from their transatlantic flights and ordered to touch down at Gander Airport. As global news networks covered events in New York City, few people would have realized what repercussions were being faced by this tiny community 1100 miles away. That was all to change as the story of the diverted airlines found its own coverage.

No one would disagree that the terrorist attack of 9/11 marked "an event that is massively destructive and rare".[12] Its immediate impact was the collapse of two of New York's most iconic buildings; the deaths of almost 3000 individuals and the hospitalization of around 7000 more; the loss of a whole neighbourhood of Manhattan; the disruption of thousands of New Yorkers displaced from their homes or places of work; and the unprecedented closure of the US stock market for four days. Elsewhere, simultaneous attacks on Washington DC hijacked two more planes, flying one into the Pentagon and causing the deaths of 184 people, and crashing another into a field in Pennsylvania, killing 40 on board. Beyond this, the impact rippled further afield as "secondary disturbances", "indirect and deferred in time",[13] hit other communities. New York's economy was hit by the loss of 143,000 jobs a month, and by losses in wages of $2.8 billion in the first three months alone[14]; dozens more residents and emergency workers subsequently died from illnesses contracted during the recovery effort, and by 2018, over 10,000 people had been diagnosed with 9/11-related cancer.[15] Meanwhile, across America, a heightened anxiety about the threat of terrorism pervaded people's lives for at least a decade after the attacks.[16]

A ***disturbance event***, according to one of the most commonly cited sources, is "any relatively discrete event in time that disrupts ecosystem, community, or population structure and changes resources, substrate availability, or the physical environment".[17] Disturbances vary in scale, effect, and impact, and can be as discrete as a tree falling, or as chronic as glacial change. Certain disturbances like hurricanes, earthquakes, or wild fires may occur frequently enough for them to be part of the cyclical balance of an ecosystem; others, such as climate change, sea-level rises, or glacial flow may be so long-term that ecologists can plan for them and in some cases influence them. The catch-all phrase ***disturbance event*** is therefore applied to a broad range of dissimilar events with specific and unique contexts. Regardless, the disturbance will be characterized by a perceptible change in a system, relative to its original reference state. "Disturbances differ in type, spatial scale, frequency, and their effects, making disturbance sometimes a difficult concept to define", warns one account; "but it is clear that disturbance is a major factor controlling ecosystem dynamics".[18]

Natural ecosystems,[19] according to ecologists, will respond to disturbance through ecological processes—a series of successional changes like those we tracked in chapter five, in which different species emerge into the disturbed ecosystem, shifting over time to create a new reference state. By contrast, sites dominated by humans—social-ecological-technological systems (SETS)—are more likely to respond in more active ways: disaster management regimes kick in, led by injections of money, aid, and reconstruction from governments, charities, and NGOs. The aim might be to return the ecosystem to its pre-disturbance equilibrium, rather than allowing successional change. Even so, the ecosystem and its community will still have been changed, not least through "new awareness" and "unique opportunities". This is known in disturbance ecology as "social-ecological succession".[20]

"In urban systems", write McPhearson and Tidball, "disturbance can dramatically affect the built and social environment in addition to the biophysical environment". They note the potential impact on "technological systems", "built infrastructure", "travel pathways", "social networks", and even "perceptions of the system".[21] Nancy Grimm and her colleagues go further, considering the extent to which disturbance can be understood in urban environments by using a variation of the modelling customarily applied to natural ecosystems. They use a disturbance model originally conceptualized in 2011,[22] and through a series of urban disturbance case studies—including a look at the Great Fire of Baltimore in 1904—they reveal how the general principles can be applied to social-ecological-technological systems.

A General Model of Disturbance

The general disturbance model features three aspects: disturbance *drivers*; disturbance *mechanisms*; and disturbance *characteristics/impact*.[23] The *disturbance drivers* will be "either *pulses* or *presses* of climatic and atmospheric force, kinetic movement of physical materials, biotic action such as consumption, chemical transformation such as fire, and social disruptions such as demographic shifts, institutional ruptures, or redirection of finances", write Grimm et al.[24]; in other words, the drivers are the environmental circumstances that precipitate something else. In respect of the Great Fire of Baltimore, for example, they recognize that its drivers—the environmental circumstances—were a combination of "lax zoning and warehousing codes", "a lack of national standards for fire hydrant fittings", "high winds", and "freezing temperatures"[25]: none of these established the fire itself, but they all contributed to its disturbance. The *disturbance mechanisms*—the constituents of disturbance that actually cause or initiate the disturbance—can be physical (wind; flood) or biological (disease), as they often are in nature, but in a SETS context they can also be "human management activities" or "social disruptions": farming or settlement, migration, criminal activity, or even decision-making by authorities.[26] Referring to the Great Fire of Baltimore's disturbance mechanism as "combustion", Grimm et al. don't state explicitly what the precise cause was (someone discarding a match? the spark from a machine? an explosion of volatile material?), though as one contemporaneous reporter wrote, "it is believed that the cigar ashes from a passerby fell through the grates in the sidewalk, igniting blankets in the basement".[27] *Disturbance characteristics* are defined by the "onset, duration, and release" of the disturbance, recognizing that these "may arise suddenly or gradually, be long-lived, and have a rapid or a protracted release". The impacts will be "a qualitatively or quantitatively measurable alteration of the structure of the system", but also in urban systems a social impact is common, resulting in "financial losses, loss of social capital, disruptions in networks of productive social interactions, and changes in institutional capacity". Like many urban disturbances, the characteristics and impact of the Great Fire of Baltimore make for grim reading: it raged for two full days; its physical impact destroyed 57 hectares of the city and more than 1500 buildings; and, although surprisingly few deaths were recorded, its social impact caused damages of $3.8 billion in today's money and led to the loss of 35,000 jobs.[28]

These aspects—drivers, mechanisms, and characteristics—account for the discrete disturbance event; and the event itself will typically generate immediate impacts, as we have seen. However, in natural as well as social-ecological systems, the disturbance event is just part of a *disturbance process*: it will fundamentally change the make-up of the system, but it may then initiate a chain of subsequent effects that lead to the succession of a new system. Thus another set of drivers comes into play. The response to the Great Fire of Baltimore, for example, involved drivers including the civic pride of residents to restore their city; pressure from the city's elite for new safety regulations; and goading by a national press which blamed Baltimore for having sub-standard fire hydrants. The recuperation of the system then ensued, led by politicians, authorities, and the people: rebuilding the city; establishing new building and zoning regulations; and creating new standards for water hydrants. Within two years, the city was converted into a new, successional social-ecological-technological system, and lessons about city planning, fire prevention, and insurance cover had been taken on board not just by its authorities, but also further afield.[29] "We need to take advantage of any disturbance and consider it as an opportunity to learn and build resilience for larger future disturbances", write Juli Pausas and Alexandro Leverkus, pointing to the lessons learned from the Great Fire.[30] In this respect, as Erika Svendsen acknowledges, "disturbance, understood as system change, typically radiates beyond the point of impact".[31]

9/11 as a Disturbance Event

Such was the complex disturbance event of 9/11: we might see its *driver* as the extremist resentment of fundamentalist terrorists to the capitalist dominance of America; its *mechanism* being commercial airlines used as missiles; and its *character/impact* to have been immediate devastation to the city of New York and its population. Yet as a catalyst, the events of 9/11 also triggered a wider *disturbance process* reflecting a nation consumed by fear, and a political leadership faced by an unprecedented emergency. If the driver of 9/11 itself had been extremist anti-American sentiment, the drivers of the 9/11 responses were national pride, fear, and an ideological commitment to "good" against "evil"; the mechanism used for the retaliatory disturbance was the US military and a coalition of forces from other democracies; and the character of the retaliation was a decades-long "war against terror".

Within months, President George Bush had invoked the notion of an "axis of evil", referring to states such as North Korea, Iran, and Iraq who were openly hostile to the US and other western democracies.[32] As the repercussions of the atrocity played out, coalition forces would unite to drive the Taliban from power in Afghanistan, freeze the financial assets of perceived terrorist organizations across the globe, invade Iraq in pursuit of weapons of mass destruction, and hunt down hostile leaders like Saddam Hussein and Osama Bin Laden. By way of mourning the tragedy and commemorating the deaths of victims from 93 nations, the National 9/11 Memorial and Museum was set up, and a memorial called "Reflecting Absence" now occupies the footprints of where the World Trade Center towers previously stood. A national "Patriot Day" is commemorated on 11 September every year across America and in military bases around the world. Attitudes towards Muslim communities have hardened, with Republican voters especially associating Islam with violence, and with American Muslims in turn reporting greater discrimination.[33] Meanwhile, airlines and immigration bodies reported "an immediate and substantial impact on international travel worldwide".[34] Further effects rippled around the globe, influencing markets, international relations, foreign policies, and cultural sensibilities.[35]

The complexity of the disturbance process initiated by the events of 9/11 is in part because of the unprecedented scale and significance of this particular terrorist atrocity; to no small degree, however, it has been due to the ultra-complex make-up of New York as a globalized urban ecosystem. "Urban systems are highly open", observe Grimm et al., "and massive throughput of materials and consumption of energy are typical".

> Within urban systems, commuting links locations and social groups, while traffic, wires, and pipes convey resources and wastes into, through, and out of the urban system. Furthermore, information flows take the form of money, news, regulations, cultural markers, and so on [...], which are also highly open to external systems.[36]

This is true of any ecosystem, to be sure—and we have seen already not only how ecosystems function according to the logic of energy flows but also how both natural and social ecosystems have communication and transmission networks that reach further afield. As a global nexus for trade, a prominent hub for transport and communication, a major population centre, and a bucket-list cultural destination, New York City interacts with a particularly complex and extensive network of multiple secondary systems.

Perhaps the most charged secondary system among New York's network interactions on the day of 9/11 was the global airline industry, which not only had three major New York airports serving flight paths that criss-cross the western hemisphere, but which had also been appropriated as the disturbance mechanism for the terrorist atrocity itself: two planes were literally flown into the tallest structures in the city. It was as a secondary system impacted by the primary event that the airline industry was forced to ground its planes, thereby establishing dispersed sites of tertiary disturbance around the globe.

At 8:46 a.m. EST, the moment when American Airlines Flight 11 crashed into the North Tower, there were 4,546 civilian aircraft over US airspace.[37] All of them had to be landed, along with around 400 planes heading across the Atlantic. Flights less than halfway through their journey could be rerouted back to their points of origin, but those with too little fuel to backtrack were required to find an alternative runway without entering US airspace. 252 aircraft landed in 15 airports across Canada.[38] As well as the international airports in Montreal, Quebec, Toronto, Ottawa, Vancouver, and Halifax, smaller regional airports were assigned as many aircraft as they could accommodate: places such as Winnipeg in Manitoba, Calgary in Alberta, Moncton in New Brunswick, and in Newfoundland—the first available landmass for planes crossing the Atlantic—St. John's, Stephenville, and Gander.[39] In each of these localized ecosystems, a very different disturbance played out.

As it happened, Gander Airport was well set up to take in planes. It had been operating as a refuelling point for transatlantic flights since being established as a military base in the 1930s, when it had been the largest airport in the world. Although in recent years the capacity of larger planes has generally made such refuelling unnecessary, causing Gander's traffic to decline, the airport with its unusually long runway was still identified as an alternative landing site for the NASA space shuttle. Flight traffic control on 9/11 estimated that the airport could take up to 50 planes; in the end, 38 civilian and 4 military aircraft landed, stranding 6,595 passengers in a town whose own population numbered fewer than 10,000. This, not surprisingly, brought its own disturbance to the town.

As a disturbance, we can view the Gander event in the same way we observed the Great Fire of Baltimore: the driver of an international emergency precipitated the disturbance; the mechanism of 38 jumbo jets and 4 military planes delivered a significant influx of people; and the immediate impact on the Gander community was to have to accommodate, feed, and support almost twice the usual population. As with those other manifestations of disturbance, a wider process reveals further drivers, mechanisms, and characteristics/impact: the drivers of humanity and need; the mechanisms of support and friendship; and the characteristics and impact of a community being thrown together, friendships being forged, and people from diverse backgrounds and origins bonding in extraordinary circumstances.

Gander: The Ecosystem's Reference State

Newfoundland is characterized by its remoteness: as *Come From Away* tells us, it is "Somewhere in the middle of nowhere". It is also enormous, covering 407,000 km^2, which is approximately twice the size of Great Britain; yet over 50% of its population of around half a million live in what are considered "rural and small town areas",[40] whose economy, livelihood, and prospects are strained by infrastructural challenges. Cell phone coverage is inadequate,[41] transport infrastructure is thin, and like other remote communities, Newfoundland suffers from "significant challenges of economic and cultural erosion", as Freeman puts it.[42] "Out-migration has become a fact of life", reports Jenny Higgins; "adults and families with young children [...] move to urban centres in search of jobs or for better access to education, health care and other services. Those who remain are aging and there are often not enough year-round residents to support the maintenance of schools, churches, medical clinics, shopping centres, and other facilities residents of larger towns and cities often take for granted".[43]

Gander Airport was already primed to respond to emergencies, as it had done in 1985 when an American charter plane crashed on take-off killing around 250 troops and crew on board. It had also planned a response at the turn of the century for the anticipated "Millennium Bug", which might have required the airport as an emergency destination if aviation systems went down.[44] Like other locations in the province, Gander Airport was designated an Emergency Operations Centre, and was response-ready for action on 9/11 even before planes were diverted from US air space.[45]

The town had also developed a strong sense of community. "Community has always been an important part of life in rural Newfoundland", reports Jennifer Butler, noting that "geographically people identify themselves as 'belonging' somewhere rather than 'being from' or 'living' somewhere".[46] Along with that comes an additional sense of being "friendly, hospitable, and helpful".[47] To be sure, this communitarian spirit—"interdependence and neighbourliness"[48]—was something that had been promoted as part of the governance strategy of the province when in response to a report titled *Community Matters* (1995), Newfoundland and Labrador authorities devolved the responsibility for regional affairs to the regions themselves.[49] Thus the island relies on what Butler calls the "politics of community": manoeuvres by central administrations to encourage "self-governing, sustainable communities".[50]

As Butler reports, "this shift to the individual and community level has fostered the growth of community groups and self-help programs",[51] and Gander boasts a considerable number of organizations including the Rotary Club, the Lions Club, the Royal Canadian Legion, and the Women's Institute. As we have seen, it's groups such as these that Alexis de Tocqueville admired in the 1830s, and that Robert Putnam considers to be fundamental to a well-bonded society. "People who trust others are all-round good citizens", Putnam writes, "and those more engaged in community life are both more trusting and more trustworthy".[52] Such features stood out as Gander residents welcomed 6,500 newcomers in the wake of the attacks.

The 9/11 Disturbance at Gander

Townsfolk were already in action as the planes started to land from 14:27 Newfoundland Time. "All of the churches in town were placed on notice", reports Defede, "as well as the fraternal organizations such as the Lions Club, the Knights of Columbus, and the Royal Canadian Legion hall".[53] A state of emergency was declared, and arrangements made to clear the passengers through security, equip them with essentials, and find them accommodation. The Red Cross's experienced disaster coordinator Des Dillon was put in charge of procedures, and "within a couple of hours", notes Defede, he "had two hundred volunteers at the airport". Even so, it was several hours before some passengers were allowed to disembark. Security concerns were high, and not only was this exacerbating the logistics of handling 38 unexpected planes, but the ongoing strike by government employees—including the bus drivers—presented an additional obstacle. "Amazingly", remarks Defede, "as soon as the drivers realized what was happening, they laid down their picket signs, setting their own interests aside, and volunteered *en masse* to work around the clock carrying the passengers wherever they needed to go".[54]

It is this sort of magnanimous gesture that the passengers recollect: Laine Zizka, interviewing passengers later, reports "memories of being greeted by smiling volunteers and tables of donated food".[55] "At times, it seemed like all 10,000 men, women and children in town were volunteering in some capacity", remembers Kevin Tuerff[56]; "the first thing that happened was somebody came to me with like a little box of food and a toothbrush, and it was a complete stranger, just walked up to me and said, 'here'", recalls Craig Dykers.[57] In its own way, this local story of 9/11 has become as iconic as the tragedy itself—in part because of the way it draws positivity and humanity out of the horrendous events. To "offer hospitality, the welcoming of strangers"[58] is a demonstration of the small-town generosity that John McKnight and Peter Block champion in their book *Abundant Community*.

Come From Away

In 2010, the Canadian Music Theatre Project at Sheridan College in Ontario commissioned musical theatre writers Irene Sankoff and David Hein to write *Come From Away*, a musical based on the events experienced by the Gander residents and the airline refugees. The show has been an international success, subsequently developed at Goodspeed Musicals and the National Alliance for Musical Theatre, before transferring to regional runs in San Diego, Seattle, Washington DC, and Toronto, and then hitting Broadway and other global cities from 2017.

Drawn from interviews, the musical is in some ways a verbatim piece that gains its character from depicting and using the reports of real people. We meet Claude Elliott, the mayor of Gander; Oz Fudge, the police constable; Bonnie Harris, director of the Gander SPCA; and Beulah Davis, a composite character reflecting the female staff of Gander Academy, including her namesake Beulah Cooper. We also meet plane passengers Hannah O'Rourke from Brooklyn, who lost her firefighter son Kevin on the day of 9/11; Nick Marson and Diane Kirschke (called Diane Gray in the show), passengers who fall in love through the ordeal; and Kevin Tuerff and Kevin Jung, a couple at the start of the story who end up separating. Prominent among the narratives is the tale of Captain Beverley Bass, pilot of American Airlines flight 49, who sings one of the show's hit songs, "Me and the Sky".

Come From Away documents the immediate situation with humour: "We've got almost seven thousand people who might be spending the night", yells the mayor; "Jaysus, that's near as many as we got living here in town!".[59] Despite that challenge, we see the incomers housed and fed; we see medicines, toiletries, and sanitary products being gathered; and we see phone lines and computers being set up so that passengers can communicate with their loved ones. More immediate challenges are caused by language barriers, cultural insecurities, and out-and-out suspicion: "We've got passengers from all over the world coming off these busses, all speaking different languages", says one character; "We can't even say 'hello'".[60] Thus a host of contextual factors is brought into play that demonstrate the complexities of handling this disturbance in terms of both logistics and cultural assumptions. When the character Ali is discovered speaking Arabic to his family on the phone, some of the other *come-from-aways* misunderstand: "Why doesn't he speak English?" asks one; "You telling your Muslim friends where to bomb next?" demands another; "Go back where you came from!"[61] Elsewhere, cultural differences are reconciled when the bus driver realizes he can communicate with a family from Africa by referencing the bible:

> Now, obviously I can't read it, but their bible—it'll have the same number system ours does. So I ask to see it and I'm searching for something… and then—in Philippians 4:6—I give them their bible and I'm pointing and saying, look! Philippians 4:6—Be anxious for nothing. Be anxious for nothing.[62]

This is just one example of a story of crisis management surprizing us, for rather than dwelling on the negative impacts of this disturbance—though it doesn't pull its punches in terms of 9/11's human tragedy—*Come From Away* reflects an overwhelmingly positive perspective.

It is also another way in which our disturbance analogy finds parallels in ecology, since although disturbance may bring a negative impact to an ecosystem, it can also create opportunities, or *positive feedback*. For example, windstorms may cause trees to fall, whose fallen hulks can create food sources for insects and other fauna.[63] Elsewhere, the "opportunity" of forest fires can be accountable for the healthy regeneration of understory flora taking advantage of *gap phase* replacement: "The new post-fire environment (with fertilized soils, higher radiation, and reduced competition) favors species that were not dominant or perhaps not even present in the pre-fire environment",[64] explains one study, noting that the gradual replacement of one species of flora with another through successional regrowth is a factor in the ecosystem's long-term resilience. In a similar way, Gander's experience of the 9/11 disturbance event—ostensibly one that precipitated the negative impact of crisis—was responsible for significant positive feedback in terms of community bonding, intercultural awareness, and ultimately, resilience. These dynamics have been captured in the *ecological memory* of the disturbance.

Ecological Memory and the Resilience of a Community

Numerous accounts chronicle the six days that unfolded in Gander in the wake of 9/11. In 2002 journalist Jim Defede published an account in *The Day The World Came to Town*[65]; a radio play by Caroline and David Stafford, *The Day the Planes Came*, was broadcast by the BBC in 2008[66]; a TV movie, *Diverted*, aired in 2009,[67] and a documentary, *Operation Yellow Ribbon*, was screened on NBC in 2010.[68] A further documentary, *You Are Here*, from 2018, depicted not only the September 2001 experiences at Gander, but also the subsequent coverage in various media[69]; while *9/11: Cleared for Chaos*, aired on the Discovery Channel on the 2019 anniversary.[70] In 2018, airline passenger Kevin Tuerff brought out his chronicle, *Channel of Peace*,[71] and in 2021 another passenger, Elias Kanaris did the same with *Leading From the Stop*.[72] Also in 2021 Mac Moss released his account, *Flown Into the Arms of Angels*,[73] which widens the narrative to look at the whole of Newfoundland and Labrador. But the biggest splash of all has been caused by *Come From Away* since its debut on Broadway in 2017.[74]

Overwhelmingly, these narratives present the same reading of the story: not only are the anecdotes, events, and people often the same, showing us a local community rising to the challenge and coping admirably, but the accounts report the events and the response of the townsfolk as being somehow life-changing, positively affirming, and community-building. As the mayor of Gander says in his closing remarks to the show, "tonight we honour what was lost, but we also commemorate what was found". It's a sentiment echoed in the "real-life" accounts: "On that tragic day", writes Tuerff, "the people of Gander restored my faith in humanity".[75] Defede concurs: "They affirmed the basic goodness of man at a time when it was easy to doubt such humanity still existed".[76]

Throughout, *Come From Away* reiterates subtle phrases that emphasize themes of togetherness, community, resilience, or hope: "Out of the darkness, suddenly brightness, everything changes, darkness and trees" ("Lead us Out of the Darkness"); "So here we are where the world has come together" ("Dover Fault"); "Somewhere in the middle of nowhere in the middle of clear, blue air, you found your heart but left a part of you behind" ("Somewhere in the Middle of Nowhere"). In this respect, the longer-term disturbance process of Gander in the wake of 9/11 has memorialized the events of disaster management into a feel-good, uplifting celebration of community resilience, "integrating the crisis into the way [the] community narrates itself",[77] as Zizka puts it. This positive spin is exactly the perspective of *Come From Away*'s creative team: "It is not a 9/11 story", Hein has remarked; "it's a 9/12 story; it's about the way the community responded to it".[78] Crucially, in the typical view of these narratives, resilience is recognized "as being grown and sustained through community and interaction", Zizka suggests elsewhere.[79] "The resilience stories individuals tell, deconstruct, and retell function to construct the living narrative of a community—the identity mythos by which it defines and shapes itself".[80] Thus as much as New York City's experience of 9/11 was a defining cultural moment of devastation, Gander's experience was a defining moment of civic pride: "For America, the anniversary [of 9/11] is a day of mourning", writes Tuerff; "In Gander, it is a reminder of how they helped the world".[81]

Such a singular memorialization of the story recalls for us the phenomenon of *ecological memory*: the way in which an ecosystem becomes fixed into what Garry Peterson calls a *landscape pattern*. Peterson refers to ecological memory as "the degree to which an ecological process is shaped by its history".[82] According to other environmentalists, it takes into account "the persistent effects, or *legacies*, of past events",[83] "the accumulated wisdom and experiences of maturity",[84] and reflects "the ability of the past to influence the present trajectory of ecosystems".[85] As an environmental concept, it has been fundamentally associated with disturbance. One aspect of ecological memory is the tendency for ecosystems to return to their residual state following a disturbance, something which Jill Johnstone et al. attribute to the character that defines the ecosystem's reference state (*information legacy*), the species that are native to that system (*material legacy*), and the affinity for those legacy elements to dominate in any succession.[86] But an additional aspect "remembered" by the ecosystem is the disturbance itself, or at least the disturbance influence on the ecosystem, which may variously weaken, strengthen, or prime the system to respond with the knowledge of that memory in its ongoing behaviour. For example, Peterson notes that wildfires will typically burn only patches of a system, leaving what he calls a "patch mosaic" on the landscape. Since the burned and unburned patches will grow back to different states of maturity, they will have a differing likelihood of catching in subsequent fire events, meaning that over time and through multiple events, the landscape influences the spread of fire into certain patches and not others, and therefore, the landscape pattern "tends to be maintained rather than destroyed". He compares this to a "feedback loop" in which "memory causes landscape pattern to influence fire spread and vice versa". It's not just fire that triggers such a feedback loop—he also notes that "animal behaviour […] has the potential to encode memory into landscape pattern",[87] and elsewhere, other scholars have studied similar memory correlations in soil respiration,[88] the dynamics of freeze/thaw cycles,[89] and instances of coral bleaching.[90] Ecologically, such memories are all around us: they're in the valley systems formed by glaciers; they're in the meandering of rivers; they're found in the sort of ecological features that become places of pilgrimage or devotion—unique places imbued with a special meaning; and they are especially found when subsequent voices—poets or pilgrims—capture the memory for posterity and crystallize it pointedly into a moment.

At one point towards the end of *Come From Away* the middle-aged couple Diane and Nick visit the Dover Fault, a local feature near Gander which reportedly marks the point at which the continents formerly joined. They have met on one of the diverted planes, and fallen in love through the events of the disturbance, but with the prospect of the planes soon taking off and transporting them home, they fear that their whirlwind romance may just have been a brief fling. Here, at the Dover Fault, they comment on the feature's history, and it is through their implicit invocation of ecological memory that the scene gains a particular resonance:

> Five hundred forty million years ago, the continents of the world crashed together right here. And two hundred million years ago, they separated again, moving apart from each other. [...] But a little piece of them was left behind.[91]

Nick takes a photo of Diane on a cheap disposable camera, capturing a memory, fixing the moment in time, preserving a legacy of this meaningful encounter; and to reinforce the point, he sings: "Stop the world. Take a picture. Try to capture to ensure this moment lasts".[92] It's one of many poignant sequences in the show, in which multiple layers of signification coalesce, like the strata of the Dover Fault itself, or the palimpsests of its presence over time.

Such is also the way we memorialize the impact of disturbance events like 9/11, seeding our own banks of memory. In their book *Staging Loss*, Andrew Westerside and Michael Pinchbeck refer to that event's memorialization, recalling the commemorative news footage of the disaster seen some years later:

> Here, as we looked at events unfold on screen just as we did in 2001, the moment is lived again; the past, in its collision with the present, saw us commemorating not only the seismic shock of 9/11, but also the news coverage of what took place—the commemoration of an archive.[93]

In memorializing Gander's community story through the magnification of a multi-award-winning Broadway show, *Come From Away* heightens the resonance of the communal narrative to an even greater extent. Having had productions in several different countries, the show has by now become fixed in the cultural repertoire, witnessed by upwards of a million theatregoers, stored in the archive in which Broadway documents its present for the future. The world has, metaphorically, stopped—at least to notice and affirm the legacy of this show, this narrative, and these characters, and to share the unique humanity captured in the commemorative and communal retelling of a moment. As Helen Newall suggests, "these experiences are momentary hierophanies (manifestations of the sacred)",[94] "moments of *epoché* or sacred differentiation of time and space"[95] that "articulate the kind of communal and reflective space that performance provides".[96] And as Erika Svendsen has put it,

> These living memorials are created by and create narratives that are a part of our cultural fabric, serving social ecological functions and meeting community needs particularly in periods of transition from loss to grief to memory.[97]

In the case of *Come From Away*, it is the significance of community that features most prominently. "It's about community, and it's about kindness", states Petrina Bromley, who played Bonnie Harris in the Broadway production.[98] "For once", acknowledged *The Guardian*, "the hype is warranted: it turns out to be the most unlikely of commercial hits, a celebration of the power of community".[99]

Shared Experience and Co-production

Perhaps it's no coincidence that this show not only narrates a story about the shared experience of a community, but that it also recounts that story communally. All theatre can be seen as a collaborative act of co-production, witnessed in a shared moment of time by a gathering attentive to its tale; all theatre, arguably, aspires to those "momentary hierophanies" of "shared experience". But *Come From Away* is a show whose ensemble production and spirit of togetherness is particularly poignant—especially in some of the locations (notably, New York City and Gander) in which it has been performed. "What makes this show unique", suggests musical arranger Ian Eisendrath, is that "you have these twelve people coming together to form an ensemble: very representative of what happened in Newfoundland on that day [...]. And they are singing together and we are feeling that power of twelve really strong performers becoming one and sometimes representing fifteen thousand"; "There's this great metaphor underneath", comments David Hein; "which is that we're all the same underneath it all so there's something about casting it and having people play multiple parts that underlines that".[100]

More than this, the show's musical language makes use of the local sounds of Newfoundland; it sings, if you like, in the folk voices of the Gander community. From the very beginning, the instrumental and vocal palette stands out for its "incredible original authentic folk music",[101] heard in "Welcome to the Rock", epitomized by the use of instruments such as the Newfoundland Ugly Stick, the Celtic Bodhrán, and the Uillean Pipes, and dramatized in the staging of the traditional Newfoundland "Screech In" ceremony, in which *come-from-aways* must kiss a fish in order to become an honorary member of the community.[102] And by way of acknowledging the *come-from-aways* who are so much a part of this insider–outsider tale, other moments deftly spin the cultural sound of diversity through the score: the song "Prayer", for example, interweaves three songs of worship from different cultures, the Christian hymn "Make Me a Channel of Your Peace", the Jewish rite "Oseh Shalom", and the Islamic prayer "Salah". "Making the prayers of the world work together in harmony was a huge amount of work", writes Hein; "however, like a metaphor for religions finding common ground, the effort was well worth it".[103]

With such rich plurality woven through various dimensions of the text (its verbatim co-creation; its workshopped rehearsal and development; its tight ensemble performance; its musical and dramatic language; its overt theatricality), the act of experiencing this production is also made to feel like a moment of connection, involvement, and even co-creation. To audiences for whom the original events were particularly significant, the impact of such connection is especially poignant. "It resonates here in a way that I don't think it's resonated anywhere else", remarked producer Sue Frost about the Broadway run.[104] The production released a limited number of tickets for 9/11 families in a gesture that endeared the show to those most affected by the tragedy. Director Christopher Ashley also commented: "The audience gets more emotional, quicker [in New York]. I feel like the audience brings […] their own story to the theater, and they're watching our show to build onto their own story".[105]

Meanwhile, for the Ganderites, the opportunity to have their story presented onstage as if they were embodying it themselves has been profound: "To see yourself portrayed with that level of accuracy in terms of content, word-for-word replication, and even pauses in […] delivery where I actually used them is absolutely surreal", remarks Brian Mosher; "I still fill up every time I see *Come From Away*".[106] The relationship between the performers and the real people they portrayed has been sustained. "I've never done a show where I've felt so connected to the audience afterwards", reports performer Jim Walton, who joined the Broadway cast to play Nick; "I feel something deeply universal that I have in common with these people and with all people".[107] Throughout the musical's development from Sheridan to Broadway, the Ganderites have been involved, attending productions in several of the cities and being invited on stage at the Broadway opening. Cast members have struck up direct relationships with their real-life counterparts, and many have travelled to Gander to meet the residents in person.[108] Early in the development of the musical, by way of a thankyou, *Come From Away* was performed in Gander for the local communities. Two sold-out shows in the ice hockey rink, which itself had been used as a giant walk-in freezer during the 9/11 emergency, raised money for local charities, and brought the real and fictionalized communities face-to-face.

Come From Away: The Global Megamusical

Part of *Come From Away*'s "universal" appeal, undoubtedly, has been its literal and metaphorical themes of togetherness, community, and solidarity: "Because we come from everywhere", goes the tagline, "we all come from away". Implicit in that tagline is a recognition that "we all" have the capacity to see ourselves as part of a global community, to travel, to communicate, and to enjoy the privileges of the connected world while still recognizing the local communities we each inhabit. If the events of 9/11 disrupted that narrative for a period, it was only for Transatlantic travel, international trade, and global possibility to bounce back, at least for those with the mobility privileges of the global North. And it must be acknowledged that such privileges have also been magnified—further privileged—by the success of *Come From Away*, through its exploitation of the global North's cultural ability to circulate "our" stories. There's no clearer vehicle—no clearer *disturbance mechanism*—for that privilege than the global infrastructure of the megamusical circuit, onto which *Come From Away* has been super-produced in replica productions in New York, Toronto, Dublin, London, Sydney, and Melbourne, and in a national US tour.

What *Come From Away* does not really address is a more nuanced (ordinary) landscape of this Newfoundland scenario once those two extraordinary disturbances (9/11 in New York; 9/11 in Gander) have been stripped away; nor does it admit the other ordinary voices into this wider landscape—already, perhaps, marginalized voices who are further marginalized by this tale. It's fair to say that we do see some glimpses of the real-life issues facing Newfoundlanders: local school bus drivers are on strike early in the show, for instance, though this is presented more as an inconvenience than a commentary on the long-term economic pressures in Newfoundland. Yet *Come From Away* does not focus on other, longer-term, and real-life insecurities: such economic and social realities remain very much offstage, side-lined in deference to what Freeman calls "the more palatable story of its generous inhabitants".[109] We don't see the arguably more formative and chronic *disturbance events* of social division (regional inequality, class and race privilege, environmental destruction) that are in part enabled by the mechanisms of things like the musical theatre industry, the airline industry, or the social-ecological-technological systems operations of spaces like New York City, all manifestations of the global capitalism that has marked our anthropocentric dominance over the ecosphere. It's the *charismatic* disturbance of 9/11 that gets attention. All art is ecological, to be sure; but not all art is *ecological.*

Instead *Come From Away* becomes, in effect, a simplistic tale of disaster and redemption, strangers and Samaritans: two events and two protagonists. And while it might be because of that simplicity that the story speaks to us in such a compelling way, capitalizing on broad brushstrokes we can recognize rather than particularizing the narrative with specific details, it is nevertheless a story that is redolent of a new Millennium, content to leave behind anything and anyone that can't keep up with its breathless chase towards development.

In fact, those privileges have paid Gander back with dividends, for this small-town 9/11 story has inspired a boom in tourism, "an influx of 'come from aways'", as Barry Freeman puts it. "Provincial tourism spiked about 30 per cent between 1996 and 2002",[110] suggests one report, with visitor numbers expanding from less than 400,000 in 1998 to over 500,000 in 2015; jobs in the tourism industry ballooning from 13,000 to 18,000; and tourist expenditures now topping $1b per annum.[111] Thus the successional change of an ecosystem. To be sure, these statistics are not due exclusively to the influence of *Come From Away*, but rather a confluence of high-profile Newfoundland stories including the 2001 film *The Shipping News* and the 2002 TV series *Random Passage* (neither of which covers 9/11). But they are evidence that our use of contemporary storytelling mobilities to tell broad brushstroke narratives serves as a springboard, not only for cultural capital, but also for socioeconomic and investment capital, something which is just as urgent for rural and remote locations as it is for cultural centres: "There is some consensus in the province that investments in arts and culture on and beyond the island are critical to its future", writes Barry Freeman.[112] For without that sort of investment, that inward interest, there's a risk that the community itself will cease to grow.

To everyone's surprise, on the stage of the Schoenfeld Theatre, tiny buds began to appear from the trunks of the maples. It was a shock to the production team, since even though these were genuine trees, none of them had roots or access to sunlight. "The tree was dead, chopped down with two others for the set", remarks Hoffman.[113] Romantically, poetically, or wishfully, the souvenir book of the show attributes the miraculous growth to "the warmth of the stage lights and the human affection in the theater".[114] Beowulf Boritt's explanation is more rationale: "Trees hibernate in winter and they store a lot of moisture in them", he conjectures; "Obviously, there's enough of a spectrum of light in the lighting grid that they're getting some kind of nutrient and there's one little tree that really doesn't know it has been chopped down because it's still growing".[115] That at least sounds more plausible than the trees being spurred into life by the affection of the audience. Even so, "It's such a perfect metaphor for a show about hope", remarks Boritt.[116] And as the souvenir book concludes, "On-stage—and off—it seems that, with *Come From Away*, hope springs eternal".[117]

Notes

1. Beowulf Boritt interviewed in Zachary Stewart (2017), "A Tree Grows on Broadway: *Come From Away* and the Cutting Edge of Green Set Design", *Theatermania*, 7 May.
2. Boritt in Stewart, "A Tree Grows".
3. Barbara Hoffman (2017), "Broadway Musical Uses Real Trees for Touching 9/11 Lesson", *New York Post*, 1 June.
4. Hoffman, "Broadway Musical".
5. Irene Sankoff and David Hein (2015), *Come From Away* (libretto), included in Laurence Maslon, Irene Sankoff and David Hein (2019), *Come From Away: Welcome to the Rock*, London and New York: Omnibus Press, p. 117.
6. Anon. (n.d.), "Are Red Maples Resilient?" *Maple Living*, https://mapleliving.com/are-red-maples-resilient/. Accessed 20 February 2024; Leonardo Gabriel Fernandez (2024), "Top 10 Most Resilient Plants: Survivors Against the Odds", *Leotechevo*,

https://leotechevo.com/most-resilient-plants/#Maple_Trees. Accessed 20 February 2024.

7. James Brown (2024), "Are Beech Trees Strong? (The ULTIMATE Guide to Their Strength)", *Tree Pursuits*, https://treepursuits.com/are-beech-trees-strong/#What_Makes_Beech_Trees_Resilient. Accessed 20 February 2024; Roman Gebauer et al. (2020), "The resistance and resilience of European beech seedlings to drought stress during the period of leaf development", *Tree Physiology*, 40:9, pp. 1147–1164.
8. Boritt in Stewart, "A Tree Grows".
9. Jake Grossman interviewed in Deborah Blackwell (2020), "How Plants Adapt to Climate Change", *The Harvard Gazette*, 5 August, https://news.harvard.edu/gazette/story/2020/08/arboretum-examines-climate-changes-impact-on-maple-trees/. Accessed 20 February 2024.
10. Edurne Martinez del Castillo et al. (2022), "Climate-change-driven growth decline of European beech forests", *Communications Biology*, 5:163, https://www.nature.com/articles/s42003-022-03107-3. Accessed 20 February 2024.
11. Martinez del Castillo et al., "Climate-change-driven".
12. Edward J. Rykiel Jr. (1985), "Towards a definition of ecological disturbance", *Australian Journal of Ecology*, 10:3, pp. 361–365, p. 361.
13. Battisti et al., *An Introduction*, p. 17.
14. Lydia Polgreen (2004), "Study Confirms 9/11 Impact On New York City Economy: Economic Effect of Sept. 11 Attack", *New York Times*, 30 June, p. B6.
15. Erin Durkin (2018), "September 11: nearly 10,000 people affected by 'cesspool of cancer'", *The Guardian*, 11 September, https://www.theguardian.com/us-news/2018/sep/10/911-attack-ground-zero-manhattan-cancer. Accessed 8 February 2024.
16. Hannah Hartig and Carroll Doherty (2021), "Two Decades Later, the Enduring Legacy of 9/11", *Pew Research Center*, 2 September, https://www.pewresearch.org/politics/2021/09/02/two-decades-later-the-enduring-legacy-of-9-11/. Accessed 8 February 2024.
17. Steward T.A. Pickett and P.S. White (1985), *The Ecology of Natural Disturbance and Patch Dynamics*, San Diego and London: Academic Press, p. 7.

18. Donald B. Zobel, Joseph A. Antos, and Dylan G. Fischer (2021), "Secondary disturbance following a deposit of volcanic tephra: a 30-year record from old-growth forest understory", *Canadian Journal of Forest Research*, 51:10, pp. 1541–1549, p. 1541.
19. Natural ecosystems are the focus of most literature on disturbance, and the presence or impact of humans on these ecosystems is typically seen in itself as a form of disturbance. This has been called a "naive ecological habit". Nancy B. Grimm et al. (2017), "Does the ecological concept of disturbance have utility in urban social–ecological–technological systems?", *Ecosystem Health and Sustainability* 3: 1, e01255, pp. 1–18, p. 1.
20. Timon McPhearson and Keith G. Tidball (2013), "Disturbances in urban social-ecological systems: Niche opportunities for environmental education", in Marianne E. Krasny and Justin Dillon (eds) (2014), *Trading zones in environmental education: Creating transdisciplinary dialogue*, New York: Peter Lang, pp. 193–230, pp. 193–194.
21. McPhearson and Tidball, "Disturbances", pp. 197–198.
22. Debra P.C. Peters et al. (2011), "Cross-system comparisons elucidate disturbance complexities and generalities", *Ecosphere* 2, pp. 1–26, p. 9.
23. Grimm et al., "Does the ecological", p. 4. The Peters model posits a subtly different set of components: "(1) environmental drivers, (2) initial system properties, and (3) physical and biological mechanisms of effect". Peters et al., "Cross-system comparisons", p. 1.
24. Grimm et al., "Does the ecological", p. 4. Even in Peters et al. there is a recognition of "anthropogenic" drivers ("people, economics, cultural"), see Peters et al., "Cross-system comparisons", p. 10.
25. Grimm et al., "Does the ecological", p. 7.
26. Grimm et al., "Does the ecological", p. 5.
27. Russell K. George (2004), *Mayor and City Council of Baltimore v. Baltimore and Philadelphia Steamboat Company*, 65 A. 353, 104 Md. 485 (Dec. 19, 1906).
28. Grimm et al., "Does the ecological", pp. 5–8.
29. See Lucas Pancini (2018), "Why Are the Cat Bonds Taking Over the Reinsurance Market?", *Revista Brasileira de Risco e Seguro*, 14: 23, pp. 19–32.

30. Juli G. Pausas and Alexandro B. Leverkus (2023), "Disturbance ecology in human societies", *People and Nature*, 5, pp. 1082–1093, p. 1090.
31. Erika S. Svendsen (2005), *Living memorials project: year 1 social and site assessment*, 333, US Department of Agriculture, p. 8.
32. Office of the Press Secretary, (2002), "President Delivers State of the Union Address", 29 January, https://georgewbush-whitehouse.archives.gov/news/releases/2002/01/20020129-11.html. Accessed 8 February 2024.
33. Hartig and Doherty (2021), "Two Decades Later", n.p.
34. Derekh Cornwell and Bryan Roberts (2010), "The 9/11 Terrorist Attack and Overseas Travel to the United States: Initial Impacts and Longer-Run Recovery", *Office of Immigration Statistics*, https://www.dhs.gov/sites/default/files/publications/9_11%20and%20Travel%20to%20US%202010.pdf. Accessed 8 February 2024.
35. See for example Vincent Richman, Michael R. Santos, and John T. Barkoulas (2005), "Short-and long-term effects of the 9/11 event: the international evidence", *International Journal of Theoretical and Applied Finance* 8:07, pp. 947–958.
36. Grimm et al., "Does the ecological", p. 4.
37. Jim Defede (2002), *The Day The World Came To Town: 9/11 in Gander, Newfoundland*, New York: Harper Collins, p. 6.
38. Defede, *The Day*, p. 57.
39. Defede, *The Day*, pp. 7, 23.
40. Rob Greenwood (2009), "Rural and Regional Development in Newfoundland and Labrador: Emerging Challenges and Opportunities", *The Harris Centre, Memorial University*, https://www.mun.ca/harriscentre/policy/conferencesworkshops/Credit_Unions_Gander_May_2009.pdf. Accessed 30 June 2021.
41. David Holt (2019), "Reliable cellular signals still an issue in rural Newfoundland and Labrador", *Saltwire Network*, 19 September, https://www.saltwire.com/newfoundland-labrador/news/reliable-cellular-signals-still-an-issue-in-rural-newfoundland-and-labrador-354311/. Accessed 30 June 2021.
42. Barry Freeman (2017), "The Need of a Good Story: Understanding *Come From Away*'s Warm Reception", *Canadian Theatre Review*, 171, pp. 52–58, p. 56.

43. Jenny Higgins (2008), "Rural Depopulation", *Heritage: Newfoundland and Labrador*, https://www.heritage.nf.ca/articles/society/depopulation.php. Accessed 30 June 2021.
44. Joseph T. Scanlon (2003), "Helping the other victims of September 11: Gander uses multiple EOCs to deal with 38 diverted flights", *Australian Journal of Emergency Management*, 18:2, pp. 108–121, p. 111.
45. Scanlon, "Helping", p. 110.
46. Jennifer L. Butler (2008), "Rebuilding rural Newfoundland: Neoliberal discourse and Narratives of Rural Development", Masters Thesis, Windsor: University of Windsor, p. 5.
47. Dona Lee Davis (2003), "In the Beginning: Region, Crisis, and Occupational Choice Among Newfoundland's Youth", in Reginald Byron (ed.) (2003), *Retrenchment and Regeneration in Rural Newfoundland*, Toronto: University of Toronto Press, pp. 177–98, p. 184.
48. J.D. House (2003), "Does Community Really Matter in Newfoundland and Labrador? The Need for Supportive Capacity in the New Regional Economic Development", in Byron, *Retrenchment*, pp. 226–267.
49. Task Force on Community Economic Development in Newfoundland and Labrador (1995), *Community Matters: The New Regional Economic Development. Report of the Task Force on Community Economic Development in Newfoundland and Labrador*, St John's: Atlantic Canada Opportunities Agency.
50. Butler, "Rebuilding", p. 8.
51. Butler, "Rebuilding", p. 4.
52. Robert D. Putnam (2000), *Bowling Alone: The Collapse and Revival of American Community*, New York: Simon & Schuster, p. 137.
53. Defede, *The Day*, pp. 54–56.
54. Defede, *The Day*, p. 57.
55. Laine Zizka (2021), "You Are Here: Narrative Construction of Identity and Community Resilience in Newfoundland during and after 9/11", Masters thesis, Texas Christian University, p. 40.
56. Kevin Tuerff (2017), *Channel of Peace: Stranded in Gander on 9/11*, Print on Demand: River Grove Books, p. 27.

57. Angela Antle (dir.) (2018), *Gander's Ripple Effect: How a Small Town's Kindness Opened on Broadway*, CBC/Radio-Canada, *YouTube*, https://www.youtube.com/watch?v=LTNDRvUqVQA. Accessed 04 April 2020.
58. John McKnight and Peter Block (2010), *The Abundant Community: Awakening the Power of Families and Neighborhoods*, San Francisco: Berrett-Koehler Publishers, Inc., p. 67.
59. Maslon, Sankoff and Hein, *Come From Away*, p. 49.
60. Maslon, Sankoff and Hein, *Come From Away*, p. 76.
61. Maslon, Sankoff and Hein, *Come From Away*, pp. 130–131.
62. Maslon, Sankoff and Hein, *Come From Away*, p. 79.
63. Brian R. Sturtevant and Marie-Josée Fortin (2021), "Understanding and modeling forest disturbance interactions at the landscape level", *Frontiers in Ecology and Evolution*, 9: 653647, p. 7.
64. Francisco Lloret and Paul H. Zedler (2009), "The Effects of Forest Fire on Vegetation", in Artemi Cerdà and Peter R. Robichaud (2009), *Fire effects on soils and restoration strategies*, Boca Raton: CRC Press, pp. 257–298, p. 265.
65. Defede, *The Day*.
66. Caroline Stafford and David Stafford (2008), *The Day The Planes Came* (radio play), BBC Radio 4, 24 June, dir. Marc Beeby.
67. Tony Marchant (2009), "*Diverted* (TV drama), CBC, 8 March, dir. Alex Chapple.
68. Tom Brokaw and Shirley Brooks-Jones (2010), *Operation Yellow Ribbon* (TV documentary), NBC, 27 February, dir. Tom Brokaw.
69. Moze Mossanen (2018), *You Are Here: A Come From Away Story* (TV documentary), dir. Moze Mossanen.
70. Gary Lang (2019), *9/11: Cleared for Chaos* (TV documentary), 11 September, dir. Gary Lang.
71. Tuerff, *Channel of Peace*.
72. Elias Kanaris (2021), *Leading From the Stop: Positive Influence and Heartfelt Resilience in Times of Adversity*, New York: Aviva Publishing.
73. Mac Moss (2021), *Flown Into the Arms of Angels: Newfoundland and Labrador's Unsung Heroes Of 9/11*, St John's: Flanker Press.
74. Irene Sankoff and David Hein (2017), *Come From Away* (Broadway musical), Schoenfeld Theatre, New York, 12 March, dir. Christopher Ashley.

75. Tuerff, *Channel of Peace*, p. 59.
76. Defede, *The Day*, p. 7.
77. Zizka, "You Are Here", pp. 3–4.
78. Alan Seales (moderator) (2017), "Come From Away: Broadway Cast and Creatives", Talks at Google, 7 April, *YouTube*, https://www.youtube.com/watch?v=P0XtepC6_Ck&t=1s. Accessed 04 April 2020.
79. Laine Zizka, and Amorette Hinderaker (2023), "You Are Here: Narrative Construction of Identity and Community Resilience in Newfoundland during and after 9/11", *Communication Studies*, 74:4, pp. 285–301, p. 286.
80. Zizka and Hinderaker, "You Are Here", p. 286.
81. Tuerff, *Channel of Peace*, p. 74.
82. Garry D. Peterson (2002), "Contagious Disturbance, Ecological Memory, and the Emergence of Landscape Pattern", *Ecosystems*, 5:4, pp. 329–338, p. 329.
83. Jill F. Johnstone et al. (2016), "Changing disturbance regimes, ecological memory, and forest resilience", *Frontiers in Ecology and the Environment*, 14:7, pp. 369–378, p. 369.
84. Crawford S. Holling (2001), "Understanding the complexity of economic, ecological, and social systems", *Ecosystems*, 4, pp. 390–405, p. 398.
85. Terry P. Hughes et al. (2019), "Ecological memory modifies the cumulative impact of recurrent climate extremes", *Nature Climate Change*, 9, pp. 40–43, p. 40.
86. Johnstone et al., "Changing disturbance", p. 369.
87. Peterson, "Contagious Disturbance", pp. 336–8.
88. Rodrigo Vargas et al. (2011), "On the multi-temporal correlation between photosynthesis and soil CO_2 efflux: reconciling lags and observations", *New Phytologist*, 191, pp. 1006–1017.
89. Juergen Kreyling, Carl Beierkuhnlein and Anke Jentsch (2010), "Effects of soil freeze-thaw cycles differ between experimental plant communities", *Basic and Applied Ecology*, 11, pp. 65–75.
90. Hughes et al., "Ecological memory".
91. Maslon, Sankoff and Hein, *Come From Away*, p. 160.
92. Maslon, Sankoff and Hein, *Come From Away*, p. 160.

93. Michael Pinchbeck and Andrew Westerside (2018), "Staging Loss: an Introduction", in Michael Pinchbeck and Andrew Westerside (eds) (2018), *Staging Loss: Performance as Commemoration*, Cham: Palgrave Macmillan, pp. 1–15, p. 6.
94. Helen Newall (2018), "Commemoration: Sacred Differentiation of Time and Space in Three World War I Projects", in Pinchbeck and Westerside, *Staging Loss*, pp. 59–76, p. 60.
95. Newall, "Commemoration", p. 72.
96. Pinchbeck and Westerside, "Staging Loss", p. 8. Newall associates these charged moments with Mihaly Czikszentmihalyi's notion of *flow*, a state in which the individual enters an almost transcendent state of presence and often creativity.
97. Svendsen, *Living Memorials Project*, p. 8.
98. Michael Paulson (2022), "As 'Come From Away' Closes, a Newfoundlander Heads Back Home", *The New York Times*, 3 October.
99. Alison Croggon (2019), "Come From Away review—a feelgood 9/11 musical? Believe the hype", *The Guardian*, 21 July.
100. Seales, "Come From Away".
101. Craine, Debra (2019), "Come From Away: The feelgood story from 9/11 that's become something to sing about", *The Times*, 9 February, https://www.thetimes.co.uk/article/come-from-away-the-feelgood-story-from-9-11-that-s-become-something-to-sing-about-h5v9djssw. Accessed 30 June 2021; See also Mitterauer, David, "'Enough of that Dance-Hall Razzmatazz': The Cultural Impact of Music in Kevin Major's *Lead Me Home* and the Role of the Musician as Racial Ambiguity", *Newfoundland and Labrador Studies*, 34:2, pp. 193–229.
102. Anon. (n.d.), "The Newfoundland Screech-In Ceremony", *Newfoundland Times*, https://newfoundlandtimes.ca/newfoundland-screech-ceremony/. Accessed 30 June 2021.
103. Maslon, Sankoff and Hein, *Come From Away*, p. 129.
104. Kelli Boyle (2017), "Tony Nominees Share How 'Come From Away' Inspires The Families Of 9/11 Victims", *Elite Daily*, 4 May, https://www.elitedaily.com/entertainment/celebrity/tony-nominees-come-from-away-911/1892803. Accessed 20 February 2024.
105. Boyle, "Tony Nominees".
106. Maslon, Sankoff and Hein, *Come From Away*, p. 191.

107. Ruthie Fierberg (moderator) (2019), "Come From Away: A Broadway 2nd-Anniversary Celebration", 92nd Street Y, 8 May, *YouTube*, https://www.youtube.com/watch?v=dVYpaUXn580. Accessed 04 April 2020.
108. Jeryl Brunner (2019), "The Stars Of 'Come From Away' Take An Epic Trip To Newfoundland Where Their Musical Takes Place", *Forbes*, 11 January, https://www.forbes.com/sites/jerylbrunner/2019/01/11/the-stars-of-come-from-away-take-an-epic-trip-to-newfoundland-where-their-musical-takes-place/?sh=782cfefd3734. Accessed 30 June 2021.
109. Freeman, "The Need", p. 56.
110. Freeman, "The Need", p. 57.
111. Clare J.A. Mitchell and Meghan Shannon (2018), "Exploring cultural heritage tourism in rural Newfoundland through the lens of the evolutionary economic geographer", *Journal of Rural Studies*, 59, pp. 21–34, p. 22.
112. Freeman, "The Need", p. 57.
113. Hoffman, "Broadway musical".
114. Maslon, Sankoff and Hein, *Come From Away*, p. 222.
115. Boritt in Stewart, "A Tree Grows".
116. Hoffman, "Broadway musical".
117. Maslon, Sankoff and Hein, *Come From Away*, p. 222.

Bibliography

Anon. (n.d.), "Are Red Maples Resilient?" *Maple Living*, https://mapleliving.com/are-red-maples-resilient/. Accessed 20 February 2024.

Anon. (n.d.), "The Newfoundland Screech-In Ceremony", *Newfoundland Times*, https://newfoundlandtimes.ca/newfoundland-screech-ceremony/. Accessed 30 June 2021.

Antle, Angela (dir.) (2018), *Gander's Ripple Effect: How a Small Town's Kindness Opened on Broadway*, CBC/Radio-Canada, *YouTube*, https://www.youtube.com/watch?v=LTNDRvUqVQA. Accessed 04 April 2020.

Battisti, Corrado, Gianluca Poeta, and Giuliano Fanelli (2016), *An Introduction to Disturbance Ecology: A Roadmap for Wildlife Management and Conservation*, Cham: Springer.

Blackwell, Deborah (2020), "How Plants Adapt to Climate Change", *The Harvard Gazette*, 5 August, https://news.harvard.edu/gazette/story/2020/08/arboretum-examines-climate-changes-impact-on-maple-trees/. Accessed 20 February 2024.

Boyle, Kelli (2017), "Tony Nominees Share How 'Come From Away' Inspires The Families Of 9/11 Victims", *Elite Daily*, 4 May, https://www.eliteddaily.com/entertainment/celebrity/tony-nominees-come-from-away-911/1892803. Accessed 20 February 2024.

Brokaw, Tom and Shirley Brooks-Jones (2010), *Operation Yellow Ribbon* (TV documentary), NBC, 27 February, dir. Tom Brokaw.

Brown, James (2024), "Are Beech Trees Strong? (The ULTIMATE Guide to Their Strength)", *Tree Pursuits*, https://treepursuits.com/are-beech-trees-strong/#What_Makes_Beech_Trees_Resilient. Accessed 20 February 2024.

Brunner, Jeryl (2019), "The Stars Of 'Come From Away' Take An Epic Trip To Newfoundland Where Their Musical Takes Place", *Forbes*, 11 January, https://www.forbes.com/sites/jerylbrunner/2019/01/11/the-stars-of-come-from-away-take-an-epic-trip-to-newfoundland-where-their-musical-takes-place/?sh=782cfefd3734. Accessed 30 June 2021.

Butler, Jennifer L. (2008), "Rebuilding rural Newfoundland: Neoliberal discourse and Narratives of Rural Development", Masters Thesis, Windsor, Ontario: University of Windsor.

Byron, Reginald (ed.) (2003), *Retrenchment and Regeneration in Rural Newfoundland*, Toronto: University of Toronto Press.

Cerdà, Artemi and Peter R. Robichaud (2009), *Fire effects on soils and restoration strategies*, Boca Raton: CRC Press.

Cornwell, Derekh and Bryan Roberts (2010), "The 9/11 Terrorist Attack and Overseas Travel to the United States: Initial Impacts and Longer-Run Recovery", *Office of Immigration Statistics, Homeland Security*, March, https://www.dhs.gov/sites/default/files/publications/9_11%20and%20Travel%20to%20US%202010.pdf. Accessed 8 February 2024.

Craine, Debra (2019), "Come From Away: The feelgood story from 9/11 that's become something to sing about", *The Times*, 9 February, https://www.thetimes.co.uk/article/come-from-away-the-feelgood-story-from-9-11-that-s-become-something-to-sing-about-h5v9djssw. Accessed 30 June 2021.

Croggon, Alison (2019), "Come From Away review—a feelgood 9/11 musical? Believe the hype", *The Guardian*, 21 July.

Defede, Jim (2002), *The Day The World Came To Town: 9/11 in Gander, Newfoundland*, New York: Harper Collins.

Durkin, Erin (2018), "September 11: nearly 10,000 people affected by 'cesspool of cancer'", *The Guardian*, 11 September, https://www.theguardian.com/us-news/2018/sep/10/911-attack-ground-zero-manhattan-cancer, Accessed 8 February 2024.

Fernandez, Leonardo Gabriel (2024), "Top 10 Most Resilient Plants: Survivors Against the Odds", *Leotechevo*, https://leotechevo.com/most-resilient-plants/#Maple_Trees. Accessed 20 February 2024.

Fierberg, Ruthie (moderator) (2019), "Come From Away: A Broadway 2nd-Anniversary Celebration", 92nd Street Y, 8 May, *YouTube*, https://www.youtube.com/watch?v=dVYpaUXn580. Accessed 04 April 2020.

Freeman, Barry (2017), "The Need of a Good Story: Understanding *Come From Away*'s Warm Reception", *Canadian Theatre Review*, 171, pp. 52–58.

Gebauer, Roman, Roman Plichta, Josef Urban, Daniel Volařík, and Martina Hájíčková (2020), "The resistance and resilience of European beech seedlings to drought stress during the period of leaf development", *Tree Physiology*, 40:9, pp. 1147–1164.

George, Russell K. (2004), *Mayor and City Council of Baltimore v. Baltimore and Philadelphia Steamboat Company*, 65 A. 353, 104 Md. 485 (Dec. 19, 1906).

Greenwood, Rob (2009), " Rural and Regional Development in Newfoundland and Labrador: Emerging Challenges and Opportunities", *The Harris Centre, Memorial University*, https://www.mun.ca/harriscentre/policy/conferencesworkshops/Credit_Unions_Gander_May_2009.pdf. Accessed 30 June 2021.

Grimm, Nancy B., Steward T. A. Pickett, Rebecca L. Hale, and Mary L. Cadenasso (2017), "Does the ecological concept of disturbance have utility in urban social–ecological–technological systems?", *Ecosystem Health and Sustainability* 3: 1, e01255, pp. 1–18.

Hartig, Hannah and Carroll Doherty (2021), "Two Decades Later, the Enduring Legacy of 9/11", *Pew Research Center*, 2 September, n.p., https://www.pewresearch.org/politics/2021/09/02/two-decades-later-the-enduring-legacy-of-9-11/. Accessed 8 February 2024.

Higgins, Jenny (2008), "Rural Depopulation", *Heritage: Newfoundland and Labrador*, https://www.heritage.nf.ca/articles/society/depopulation.php. Accessed 30 June 2021.

Hoffman, Barbara (2017), "Broadway Musical Uses Real Trees for Touching 9/11 Lesson", *New York Post*, 1 June.

Holling, Crawford S. (2001), "Understanding the complexity of economic, ecological, and social systems", *Ecosystems*, 4, pp. 390–405.

Holt, David (2019), "Reliable cellular signals still an issue in rural Newfoundland and Labrador", *Saltwire Network*, 19 September, https://www.saltwire.com/newfoundland-labrador/news/reliable-cellular-signals-still-an-issue-in-rural-newfoundland-and-labrador-354311/. Accessed 30 June 2021.

Hughes, Terry P., James T. Kerry, Sean R. Connolly, Andrew H. Baird, C. Mark Eakin, Scott F. Heron, Andrew S. Hoey, Mia O. Hoogenboom, Mizue Jacobson, Gang Liu, Morgan S. Pratchett, William Skirving, and Gergely Torda (2019), "Ecological memory modifies the cumulative impact of recurrent climate extremes", *Nature Climate Change*, 9, pp. 40–43.

Johnstone, Jill F., Craig D. Allen, Jerry F. Franklin, Lee E. Frelich, Brian J. Harvey, Philip E. Higuera, Michelle C. Mack, Ross K. Meentemeyer, Margaret R. Metz, George L.W. Perry, Tania Schoennagel, Monica G. Turner (2016), "Changing disturbance regimes, ecological memory, and forest resilience", *Frontiers in Ecology and the Environment*, 14:7, pp. 369–378.

Kanaris, Elias (2021), *Leading From the Stop: Positive Influence and Heartfelt Resilience in Times of Adversity*, New York: Aviva Publishing.

Krasny, Marianne E. and Justin Dillon (eds) (2014), *Trading zones in environmental education: Creating transdisciplinary dialogue*, New York: Peter Lang.

Kreyling, Juergen, Carl Beierkuhnlein and Anke Jentsch (2010), "Effects of soil freeze-thaw cycles differ between experimental plant communities", *Basic and Applied Ecology*, 11, pp. 65–75.

Lang, Gary (2019), *9/11: Cleared for Chaos* (TV documentary), 11 September, dir. Gary Lang.

Marchant, Tony (2009), "*Diverted* (TV drama), CBC, 8 March, dir. Alex Chapple.

Martinez del Castillo, Edurne, Christian S. Zang, Allan Buras, Andrew Hacket-Pain, Jan Esper, Roberto Serrano-Notivoli, Claudia Hartl, Robert Weigel, Stefan Klesse, Victor Resco de Dios, Tobias Scharnweber, Isabel Dorado-Liñán, Marieke van der Maaten-Theunissen, Ernst van der Maaten, Alistair Jump, Sjepan Mikac, Bat-Enerel Banzragch, Wolfgang Beck, Liam Cavin, Hugues Claessens, Vojtěch Čada, Katarina Čufar, Choimaa Dulamsuren, Jozica Gričar, Eustaquio Gil-Pelegrín, Pavel Janda, Marko Kazimirovic, Juergen Kreyling, Nicolas Latte, Christoph Leuschner, Luis Alberto Longares, Annette Menzel, Maks Merela, Renzo Motta, Lena Muffler, Paola Nola, Any Mary Petritan, Ion Catalin Petritan, Peter Prislan, Álvaro Rubio-Cuadrado, Miloš Rydval, Branko Stajić, Miroslav Svoboda, Elvin Toromani, Volodymyr Trotsiuk, Martin Wilmking, Tzvetan Zlatanov, and Martin de Luis (2022), "Climate-change-driven growth decline of European beech forests", *Communications Biology*, 5:163, https://www.nature.com/articles/s42003-022-03107-3. Accessed 20 February 2024.

Maslon, Laurence, Irene Sankoff and David Hein (2019), *Come From Away: Welcome to the Rock*, London and New York: Omnibus Press.

McKnight, John and Peter Block (2010), *The Abundant Community: Awakening the Power of Families and Neighborhoods*, San Francisco: Berrett-Koehler Publishers, Inc.

Mitchell, Clare J.A., and Meghan Shannon (2018), "Exploring cultural heritage tourism in rural Newfoundland through the lens of the evolutionary economic geographer", *Journal of Rural Studies*, 59, pp. 21–34.

Mitterauer, David, "'Enough of that Dance-Hall Razzmatazz': The Cultural Impact of Music in Kevin Major's *Lead Me Home* and the Role of the

Musician as Racial Ambiguity", *Newfoundland and Labrador Studies*, 34(2), pp. 193–229.

Moss, Mac (2021), *Flown Into the Arms of Angels: Newfoundland and Labrador's Unsung Heroes Of 9/11*, St John's: Flanker Press.

Mossanen, Moze (2018), *You Are Here: A Come From Away Story* (TV documentary), dir. Moze Mossanen.

Office of the Press Secretary (2002), "President Delivers State of the Union Address", 29 January, https://georgewbush-whitehouse.archives.gov/news/releases/2002/01/20020129-11.html. Accessed 8 February 2024.

Paulson, Michael (2022), "As 'Come From Away' Closes, a Newfoundlander Heads Back Home", *The New York Times*, 3 October.

Pausas, Juli G. and Alexandro B. Leverkus (2023), "Disturbance ecology in human societies", *People and Nature* 5, pp. 1082–1093.

Peters, Debra P.C., Ariel E. Lugo, F. Stuart Chapin III, Steward T. A. Pickett, Michael Duniway, Adrian V. Rocha, Frederick J. Swanson, Christine Laney, and Julia Jones (2011), "Cross-system comparisons elucidate disturbance complexities and generalities", *Ecosphere* 2, pp. 1–26.

Peterson, Garry D. (2002), "Contagious Disturbance, Ecological Memory, and the Emergence of Landscape Pattern", *Ecosystems*, 5:4, pp. 329–338.

Pickett, Steward T. A. and White, P.S. (1985), *The Ecology of Natural Disturbance and Patch Dynamics*, San Diego and London: Academic Press.

Pinchbeck, Michael, and Andrew Westerside (eds) (2018), *Staging Loss: Performance as Commemoration*, Cham: Palgrave Macmillan.

Polgreen, Lydia (2004), "Study Confirms 9/11 Impact On New York City Economy: Economic Effect of Sept. 11 Attack", *New York Times*, 30 June, p. B6.

Putnam, Robert D. (2000), *Bowling Alone: The Collapse and Revival of American Community*, New York: Simon & Schuster.

Richman, Vincent, Michael R. Santos, and John T. Barkoulas (2005), "Short-and long-term effects of the 9/11 event: the international evidence", *International Journal of Theoretical and Applied Finance* 8:07, pp. 947–958.

Rykiel Jr, Edward J. (1985), "Towards a definition of ecological disturbance", *Australian Journal of Ecology*, 10:3, pp. 361–365.

Sankoff, Irene and David Hein (2017), *Come From Away* (Broadway musical), Schoenfeld Theatre, New York, 12 March, dir. Christopher Ashley.

Scanlon, Joseph T. (2003), "Helping the other victims of September 11: Gander uses multiple EOCs to deal with 38 diverted flights", *Australian Journal of Emergency Management*, 18:2, pp. 108–121.

Seales, Alan (moderator) (2017), "Come From Away: Broadway Cast and Creatives", Talks at Google, 7 April, *YouTube*, https://www.youtube.com/watch?v=P0XtepC6_Ck&t=1s. Accessed 04 April 2020.

Stafford, Caroline and David Stafford (2008), *The Day The Planes Came* (radio play), BBC Radio 4, 24 June, dir. Marc Beeby.

Stewart, Zachary (2017), "A Tree Grows on Broadway: *Come From Away* and the Cutting Edge of Green Set Design", *Theatermania*, 7 May.

Sturtevant, Brian R., and Marie-Josée Fortin (2021), "Understanding and modeling forest disturbance interactions at the landscape level", *Frontiers in Ecology and Evolution*, 9: 653647.

Svendsen, Erika S. (2005), *Living memorials project: year 1 social and site assessment*, Vol. 333, US Department of Agriculture, Forest Service, Northeastern Research Station, Northeastern Area State and Private Forestry.

Task Force on Community Economic Development in Newfoundland and Labrador (1995), *Community Matters: The New Regional Economic Development. Report of the Task Force on Community Economic Development in Newfoundland and Labrador*, St John's: Atlantic Canada Opportunities Agency.

Tuerff, Kevin (2017), *Channel of Peace: Stranded in Gander on 9/11*, Print on Demand: River Grove Books.

Vargas, Rodrigo, Dennis D. Baldocchi, Michael Bahn, Paul J. Hanson, Kevin P. Hosman, Liisa Kulmala, Jukka Pumpanen, and Bai Yang (2011), "On the multi-temporal correlation between photosynthesis and soil CO_2 efflux: reconciling lags and observations", *New Phytologist*, 191, pp. 1006–1017.

Zizka, Laine (2021), "You Are Here: Narrative Construction of Identity and Community Resilience in Newfoundland during and after 9/11", Masters thesis, Texas Christian University.

Zizka, Laine, and Amorette Hinderaker (2023), "You Are Here: Narrative Construction of Identity and Community Resilience in Newfoundland during and after 9/11", *Communication Studies*, 74:4, pp. 285–301.

Zobel, Donald B., Joseph A. Antos, and Dylan G. Fischer (2021), "Secondary disturbance following a deposit of volcanic tephra: a 30-year record from old-growth forest understory", *Canadian Journal of Forest Research*, 51:10, pp. 1541–1549.

CHAPTER 10

Rewilding

Isabella Tree's rewilding journey began with a realization: farming the family estate was no longer profitable; not only that, but the land itself had degraded beyond recognition. Gone were the wildflower meadows and natural hedgerow habitats; gone were the birds and most of the insects. "We are living in a desert compared to our gloriously wild past",[1] she realized; and she was clear about what had caused it: "Changes in land use and, in particular, intensive farming, have altered the landscape beyond anything our great-grandparents would recognize".[2] Taking a radical step, Tree and her husband began rewilding part of their estate. Over three years they removed agricultural chemicals from the soil, prepared the ground, and introduced around 150 fallow deer. "For the first time", she recalls, "we were doing something *with* the land, rather than battling against it".[3] Natural grazing controlled the foliage but enabled sufficient growth to encourage biodiversity; and the effects were striking: "Suddenly, that first summer, we were walking knee-deep in wildflowers, kicking up grasshoppers, bees and butterflies". More than that, "the sound of insects was astonishing", she recalls; "something we hadn't even realised we'd been missing".[4] The couple rolled out the scheme to the rest of the estate, bringing in carefully chosen wild herbivores to help the land find its natural equilibrium.

K. Savage and D. Symonds, *Ecologies of Community in Performance*,
https://doi.org/10.1007/978-3-031-94343-0_10

Tree and her husband Charlie Burrell were inspired by the work of Dutch conservationist Frans Vera, whose concept of "Nature Development" (Natuurontwikkeling) had responded to a similar crisis in the Netherlands.[5] The Dutch government had become aware that around a million acres of land had been lost to over-farming. This led to a widespread "disappearance of species" and a "breakdown of natural processes", exacerbated by "the influences of pollutants, eutrophication and desiccation on the reduced and fragmented areas".[6] The Netherlands was named an "ecological disaster area", with existing conservation efforts judged inadequate. A more "offensive" strategy was needed, which would enable them to "restore independently functioning ecosystems", "give space to natural processes", and "increase biodiversity".[7]

Vera realized that the fragmentation of the land into individual farms and fields had effectively created a series of disconnected islands, "like a Persian carpet cut into millions of small rectangular pieces, each fashioned by the plough and the spade over thousands of years of cultivation".[8] He devised a scheme to link up the fragments in order to create a larger and better functioning landscape. He relocated wild Heck cattle and Konik horses to substitute for some of the now-extinct grazing herbivores which would have originally roamed the region. Over time, with equilibrium between the grazers and the vegetation, the ecosystem stabilized.

It was an approach Tree and Burrell mimicked, and as she documents in her account, *Wilding*, its effect has been startling. By 2009 there were "fifteen 'UK Biodiversity Action Plan Priority' species" on the estate[9]; by 2012, there were 34 different kinds of butterfly[10]; by 2016, 441 species of moth, sixty-two species of bee, and thirty species of wasp.[11] More recently, by way of a deliberate reintroduction, the estate has also welcomed its first beavers.[12] As she reflects, "the right number of the right species of grazing animals introduced into even a relatively small, isolated area can have an exponential impact on biodiversity".[13]

But it is when Tree discusses the knock-on effects of this rewilding throughout the ecosystem and beyond that the value of rewilding becomes apparent. She discusses the purifying impact of earthworms solving "the great concerns of our time"[14]; she writes about the benefits to human mental health that are gained by grazing animals on pasture rather than feeding them grain[15]; and her discussion of the beavers is unequivocal: "the filtration system created by the beavers dramatically reduces pollution levels", she writes; "levels of nitrates and phosphates in the water [...] are also reduced to virtually nothing", she goes on;

"soil run-off is captured, too". Indeed, the beaver, she suggests, is "the ultimate flood-control engineer",[16] preventing disastrous downstream consequences for thousands of households.

In short, the positive benefits of rewilding can reach far beyond its immediate ecosystem, and can have an exponential impact on multiple other factors. Not only does biodiversity in the vicinity gain, but so can the wider environment, human health, public safety, and economic cost.

Rewilding: Crisis, Cause, and Solution

Rewilding is a concept that in this early part of the twenty-first century has hit the zeitgeist. Despite their differences, environmentalists have outlined an understanding of what it involves: it is a response to "major human disturbance"; it is a quest to "restore natural processes"; and its aim is to result in a "self-sustaining and resilient ecosystem".[17] Rewilding initiatives start from a recognition of crisis; they identify a cause; they propose a rewilding solution; and through its implementation, they bring about change.

In most cases, the crisis is that an ecosystem has become so degraded that it is unsustainable. Such concerns are familiar, and are raised by activists all the time. For Dave Foreman, who coined the term "rewilding" back in 1992, the crisis had created a "world of wounds"[18]; for Frans Vera, as we have seen, the Netherlands had been turned into an "ecological disaster area"[19]; and for Isabella Tree, the crisis was no different: "we are living in a desert compared to our gloriously wild past".[20]

The cause is also easily identified: it's the long-term abuse of land through rapid industrialization; or the burning of fossil fuels; or deforestation; or agriculture—collectively, activities that we can view as "man's unnatural treatment of nature and its sad results".[21] In Foreman's words, "the land is scarred by plow, combine, exotic crops, pesticides, and fertilizers"[22]; for Vera and the Netherlands, the cause was "pollutants, eutrophication and desiccation"[23]; and according to Tree, it was "changes in land use and, in particular, intensive farming".[24]

The solution becomes obvious: we need to reverse the negative effects of man-made crisis; to reduce global warming and slow down rising sea-levels; and to restore pre-industrial levels of biodiversity. The UN General Assembly has designated the current decade (2021–2030) the "decade on ecosystem restoration", and rewilding—replacing straightforward conservation—has emerged as the new model for the principles, policies, and practices of making that change. Its aim is "to prevent, halt and reverse the degradation of ecosystems on every continent and in every ocean".[25] With origins in Aldo Leopold's land ethic and Arno Naess's deep ecology movement, rewilding requires a move away from anthropocentric principles and towards a more sustainable, *ecocentric* vision of the world[26]: "a paradigm shift in the relationship between humans and nature",[27] as scholars have put it. "It will only succeed", suggests the UN, "if everyone plays a part".[28]

Between 2019 and 2021 a Rewilding Task Force set up by the CEM (International Union for Conservation of Nature Commission on Ecosystem Management) put together a survey drawing on 59 rewilding "pioneers" from various initiatives across the world. Building on its responses, the task force established ten principles for rewilding:

1. Rewilding utilizes wildlife to restore trophic interactions.
2. Rewilding employs landscape-scale planning that considers core areas, connectivity, and coexistence.
3. Rewilding focuses on the recovery of ecological processes, interactions, and conditions based on reference ecosystems.
4. Rewilding recognizes that ecosystems are dynamic and constantly changing.
5. Rewilding should anticipate the effects of climate change and where possible act as a tool to mitigate impacts.
6. Rewilding requires local engagement and support.
7. Rewilding is informed by science, traditional ecological knowledge, and other local knowledge.
8. Rewilding is adaptive and dependent on monitoring and feedback.
9. Rewilding recognizes the intrinsic value of all species and ecosystems.
10. Rewilding requires a paradigm shift in the coexistence of humans and nature.[29]

Inevitably, there are challenges, especially if rewilding involves reintroducing species. Just in terms of outlay, it can demand significant investment: land is taken out of commercial use, and relocating species can involve hefty transportation costs. In terms of ethical and welfare concerns, there can also be points to note: "More than 70% of reintroductions fail", remarks one study, asking "what happens to the organisms involved in failed introductions?".[30] More than this, there can be secondary effects on existing habitats and their native species,[31] and any interference with a habitat can result in unforeseen consequences.[32] As Graham Lawton observes, even passive rewilding ("letting natural processes re-establish by withdrawing human activities"[33]) is not without its problems; "Just leaving land to go its own way may actually lead to a further reduction in biodiversity, not an increase".[34] To top everything off, there is a hearts-and-minds consideration to any rewilding, since human communities will undoubtedly be affected. Isabella Tree remarks on the public concern that the Knepp project provoked as fields were left to grass over and take on a "wilder" appearance. That sort of aesthetic gripe may not seem terribly significant when we consider the wider environmental stakes, but like it or not, the impacts of rewilding can be substantial.

The Landscape of the Arts in the UK

Throughout this book, we've been exploring the dynamics of ecological thinking, asking what can be gained from applying those perspectives to the arts. We've done this with a focus on communities and how they can be read as performative. We've seen how ecological thinking can unlock new ways of understanding the arts. We've been building up to this final chapter, because we're going to propose that the tenets of *rewilding* can be brought into play. We'll revisit our recognition that conservation solutions come from a perspective of a crisis and its cause, and then we will track through recent studies of the UK arts landscape to demonstrate a comparable crisis in that environment. We'll turn to a "State of the Arts" report (2024) to establish how far the ecosystem has declined, and to Robert Hewison and Jen Harvie to identify the cause of that decline. By means of first steps towards a solution, we'll zoom in on some discrete examples of regeneration: the case of St Ives, which we've already encountered; of Feversham Primary Academy in South Yorkshire; and of the ACE-funded "Transported" in the East Midlands. We'll propose our own principles for rewilding the arts in the UK, and we'll turn to *panarchy* theory, models of *adaptive governance*, and various *scaling* strategies to conceptualize how rewilding might work. But we'll also suggest that stand-alone examples of arts regeneration only go so far, and that even the principles of rewilding only offer a theoretical blueprint. To fully achieve the fundamental change that we and others have called for, we'll suggest, a radically new way of thinking needs to be conceived.

We're going to take as a case study the landscape of the arts in the UK. This is the domain with which we're most familiar, and it's a field which has informed our work and our research. Of course, the UK is a very specific region, localized to a particular set of politics and pressures; and we're viewing these from a discrete atomic point—the mid-2020s. There is also inevitably a preoccupation with experiences and concerns that our conditions privilege—these won't be enjoyed by all, and as an example our case study is therefore contingent, and needs to be viewed in context. However, many of these circumstances are closely aligned to those of other countries and other arts settings, especially in the global North, and the conclusions we will draw from observing this case are ones that can be applied further afield. Our point is that all of these environments work as ecosystems, operating through their network relations, energy flows, and holistic dynamics. My ecosystem may differ from yours, just as my arts scene differs; but the health of our respective ecosystems rests on principles that are commonly shared with the environment.

When we talk of a crisis in the UK arts landscape, then, we are following the logic of our metaphor: our ecosystem is on the brink of collapse, its health has been terminally weakened, and it has suffered from multiple seasons of mismanagement and decline. The obvious follow-on is that we need to revive the ecosystem. We think we have a way forward; it's based on an established ecological approach. This approach is called *rewilding*, and our proposed solution is that we should work together in *rewilding the arts*.

Recognizing a Crisis

In common with the environment, UK arts is in crisis; the cause is man-made; we need to reverse the negative effects; and to do that we need to implement change. This is not a trite comparison: the environmental crisis is undoubtedly existential; but so too, we will claim, is the crisis in the arts. Just as environmental desertification leads to catastrophic consequences, so too does cultural desertification. The crisis in the arts, we suggest, has been precipitated by long-term and systematic failures from the top: ideologies and policies that have over time and through successive governments depleted our cultural landscape, destroyed our awareness, and then ignorantly or cynically used powerful rhetoric to deny the damage. Just as climate deniers are now being held to account, we need to hold to account the sort of culture deniers that have presided over waves of recent management decline.

Don't just take our word for it: industry voices have been crying out that the arts is in crisis. In June 2024 *Arts Professional* reported a "huge decline" in the UK Arts sector's health[35]; in July the BBC covered news of "the perilous situation facing arts sponsorship in the UK"[36]; in October *Art Review* warned that "the UK's arts education crisis is about to get even worse"[37]; and in December industry bodies warned that "hundreds of theatres face closure [...] after years of cutbacks and underfunding".[38] Meanwhile, "The State of the Arts" report from the Campaign for the Arts and the University of Warwick revealed "a crisis in the UK's art sector". In a press release unveiling the report, they announced that the UK arts sector was "on the brink of collapse".[39] Among its findings, "The State of the Arts" reports a drop in cultural events in the UK between 2018 and 2023, with a significant number of venues, producers, and promoters [having] scaled back their "operations or shut down"[40]; lower UK spending on the arts than in almost any other country in Europe, and a reduction of 6% between 2010 and 2022 compared to "increases of 22% in Germany, 25% in France and 70% in Finland"[41]; precarious and low paid working conditions, especially of freelance workers, with median earnings of arts workers falling "below relative poverty and potentially absolute poverty thresholds"[42]; and significant engagement challenges for the disabled and ethnic minorities.[43]

Granted, the past few years have not been easy for a number of reasons: the global pandemic with its enforced lockdown; the UK's clumsy exit from the European Union; a brutal war on the continent that has pushed up the price of energy; another in the Middle East that has precipitated protests and campaigns; escalating inflation that has driven up the cost of living. Sometimes reports on the crisis in the arts mention these as factors contributing to the decline. But other commentators point to more systemic issues born out of the deep-rooted political debasement of UK arts. That has been more long-term, going back at least to 2010, and arguably further. A whole generation has grown up in a landscape that has been left to decay.

Identifying the Cause

Robert Hewison is one of many commentators who have attempted to diagnose the cause. In his 2014 book *Cultural Capital: The Rise and Fall of Creative Britain* he argues that the neoliberal approach of successive British governments towards the arts has drained them of their resources, diluted the way they are valued in society, removed creative experiences from people's lives, and undermined the cultural make-up of the country.

One of the issues, Hewison suggests, is that the government motivation for funding the arts was not straightforwardly about benefiting culture itself. Instead, it was about instrumentalizing the arts to serve political purposes such as improving social inclusion, community cohesion, and building a better society. Looking back to 1997, he criticizes New Labour, despite that government ostensibly injecting money and energy into supporting culture. He argues that rebranding the arts as the "creative industries" was a demonstration of "managerialism, instrumentalism, [and] centralization", "an act of cultural capitalism on a grand scale",[44] and he even refers to the arts being "sold into bonded labour".[45] Under the subsequent administration, David Cameron's Conservative Coalition with the Liberal Democrats from 2010 to 2015, the situation worsened. In what Hewison calls "an even purer experiment in neoliberalism", the arts by now weren't even being put to political ends: "the Coalition abandoned any expectations of arts-driven social regeneration", he laments, "and withdrew its support for publicly funded culture in the hope that the market would provide".[46]

Hewison's bleak reading of the period is similar to Jen Harvie's critique in her 2013 book *Fair Play: Art, Performance, and Neoliberalism*. For her, neoliberalism is tainted by an "auto-regenerative propensity to cultivate destruction" (note the ecological metaphor)[47]; and in "increasingly casting art practice as economic practice and the artist as entrepreneur",[48] she suggests, the neoliberal push towards a creative economy forces art into a particular, cynical role: "it insists that art prioritizes self-interest and individualism", she writes; "it requires art to acquiesce to creative destruction as an apparently inevitable by-product of innovation"; and "it obliges art relentlessly to pursue productivity, permanent growth, and profit".[49] The end result, Harvie assesses, is to "deprioritize and potentially endanger" the arts, to damage "collaborative sociality", and even to jeopardize humanity. For as she concludes, "a commitment to exploiting—or 'sweating'—assets may, after all, include exploiting human assets".[50]

Those words were written in 2013, and Hewison's in 2014; a decade on and the heavy weight of that free market ideology has pressurized the system even more. When the Centre for Contemporary Arts in Brighton closed in 2023, its director pointed the finger of blame at "government funding cuts over the past 12 years". He cited "sustained, politically motivated attacks on the creative sector" and "the wanton destruction of our most valuable institutions and creative industries".[51] Meanwhile, when Birmingham City Council withdrew funding entirely from the city's cultural institutions in early 2024, it was the authorities at a local level who were singled out for criticism: "Our city's council, regional authorities and key agencies have simply not placed enough value on culture over the years, despite its proven social and economic benefits".[52] All of this had been done while government soundbites trumpeted a rhetoric of boosterism and growth.[53] As the country limped towards the elections of 2024, with its public services desperately impoverished and its economy having ground to a halt, the clarion call of growth was sounded by every political party.

One of the most significant problems in instrumentalizing the arts is that its many wider benefits begin to suffer—qualities that make the arts appeal to audiences, and that incentivize our engagement: aesthetic appeal, emotional connection, experience, involvement, empathy, or affect. If value is reduced to a measure of profit and loss and the ecosystem gets engineered towards productivity, the results are—as with so many environments in the wild—potentially disastrous (damaging collaborative sociality; jeopardizing humanity itself; leading to creative destruction; or in that other scenario, scarring the land; leaving a world of wounds). Let's not forget: maximizing the yield (of the topsoil) leads to a catastrophic depletion (of nutrients and biodiversity). Eventually the landscape becomes barren. Soil becomes dirt. A crisis is on our hands.

Indeed, perhaps the quest for growth has been one of the root causes of this crisis and others. That is certainly the argument that celebrity environmentalist David Attenborough makes in his 2020 book *A Life on Our Planet*,[54] and it's also the logic of other environmentalists: "the era of ecosystem management via incremental increases in efficiency is over", states Crawford Holling—in other words, you can't sweat the system indefinitely. "We are now in an era of transformation", he argues, proposing a need for resilience and flexibility to inform ecosystem management, and for solutions to be guided by an ability "to cope, innovate, and adapt".[55] These writers, of course, are referring to the environment, but in other contexts the warnings have begun to emerge: "There are now growing doubts over whether the promise of growth can be perpetuated",[56] writes Patrick Deneen, while Jason Hickel's book *Less is More* champions the ideas of "eco-socialism": "post-growth" and "degrowth".[57] "People need to understand that 'growth' is not the same as social progress", Hickel argues.[58] From the perspective of our ecological thinking, such arguments are just as applicable in the landscape of the arts.

What's Wrong with Growth?

Politicians who talk about economic growth have often made reference to another metaphor, conjuring up the image of a pie in one of a number of clumsy culinary metaphors that were particularly popular with the 2010–2024 Conservative government in the UK.[59] "We need to grow the economic pie so that everyone gets a bigger slice", proposed Prime Minister Liz Truss at the Tory party conference in 2022.[60] The assumption seemed to be that the pie (or economy) could expand, enabling more of its largesse to go round. But as a metaphor, this one seemed flawed. In our experience, pies don't grow. If they do, it's due to their pastry crusts rising as the layers get filled with hot air. Pies also get sliced up and eaten, at which point they're reduced to crumbs, which can hardly be the story politicians are seeking to sell. In short, the idea of a pie seems an ill-thought through and cumbersome metaphor for growth.

On the other hand, there are plenty of ecological examples of growth that might be instructive in terms of conceptual metaphors. These reveal growth to be nuanced, not least because in nature it can often rely on processes of metamorphosis as early-stage organisms transform into their adult state (butterflies, frogs, ants, etc.) and literally change form. In other cases, growth may not be metamorphosis precisely, but might still involve a transformation as, for example, a seed germinates and pushes roots down into the soil in the first stage of developing into a plant or tree. Here, organic growth is hardly a one-directional process: the root first grows down(ish), and then proceeds in multiple directions in its belowground medium before shoots break the surface of the soil and grow ostensibly up(ish) to reach out for sunlight. By the time a tree is fully grown, it will have literally branched out in all directions, making versatile and efficient use of its immediate environment and, by virtue of processes such as seed dispersal and relationships with other organisms such as mycorrhizae, also stretching its influence in multiple other dimensions.

More than this, it is an inevitable and important part of a deciduous tree's health to build into its growth cycle a process of *abscission* as the winter months approach, shedding its leaves and reducing its leaf biomass.[61] Growth in this case also involves shrinkage, though significantly the leaf fall transports nutrients to ground level and subsequently into the belowground ecology in another example of a tree's rich engagement with its surroundings.

Beyond the individual tree, the ecosystem of a forest will also adapt to seasonal variation, predator threat, and climate change as processes of decay, collapse, and regeneration complexify assumptions of growth. This may improve levels of ecosystem functionality, ecosystem health, species richness, and biodiversity, though on the whole it seems to be nature's aim to sustain a kind of equilibrium rather than expansion (growth) per se. As such, forest management practices have developed to assist ecosystem sustainability over the years, and some of these, such as coppicing, pollarding, and thinning, also involve a considered reduction strategy.

Meanwhile, in terms of our use of the land as a resource, the growth paradigm of industrial agrilogistics has led to problematic monocultures that are both ecologically damaging to the environment and socially damaging to communities: among these are often cited the expansion of avocado farms in South America, palm oil plantations in the Pacific, and wheatfields in the US. Growth (expansion) in these instances has evidently been, by some measures, counter-productive, and perhaps for more sustainable practices of land consumption, a reversion to degrowth or one of its alternatives such as *regenerative farming* might be more tenable.

In short, any simplistic assumption of growth—and particularly the way in which it is projected by neoliberal politicians—seems an inadequate point of reference for generating (or even conceptualizing) a healthy, balanced ecosystem. And as many commentators have shown, the dynamics of economics are themselves nuanced and multi-layered, requiring a far more complex conceptual understanding than can be gleaned from the metaphor of a pie.

Underneath the topsoil of the arts in the UK, the nutrients have indeed been depleted, thanks in no small part to the depletion of creative disciplines within education. Commentary is consistent on this point: "Music is literally disappearing from our schools", remarked Lord Black of Brentwood to the House of Lords, as "a direct result of government policy".[62] Other observers agree: "arts education has been progressively marginalised and devalued in English state schools",[63] write Ashton and Ashton. In its most recent manifestation, this erosion began to happen from about 2010, when the Coalition government slashed funding for local authorities, abolished the "Building Schools for the Future" programme, instituted the (arts-less) EBacc system and the associated Progress 8, and drastically reduced arts budgets. At the same time, the Secretary of State for Education Michael Gove was making the claim that "every child should receive a strong, knowledge based cultural education and should have the opportunity to learn and play a musical instrument and to sing".[64] Perhaps this was in recognition that "a high percentage of 11–15 year olds in England" only engage with the arts at school rather than outside.[65] Yet rather than achieving such aspirations, the reforms precipitated a significant decline. They "caused pupils to move away from arts subjects such as dance, music and art, and towards more traditional academic subjects".[66] As another report observed, "the combined effects of funding cuts, university admission advice, school responses to audit and markets, and the ongoing school subject hierarchy which places arts near the bottom" created "a 'perfect storm'".[67] Figures since 2010 show a drop of 47% in pupils taking arts GCSEs, while A-levels in arts disciplines also slumped by 29%[68]; the number of arts teachers fell by 27%, and the number of hours spent on arts teaching in schools fell by 23%.[69] These decreases have had an impact right through the education system. Despite the then-government's 2022 claim that "Music and arts education is the second highest centrally funded part of the curriculum behind sport",[70] and despite its stated initiatives to restore arts opportunities to young people in the UK,[71] the Office for Students ruled in 2021 that at higher education level, "subjects in music, dance, drama and performing arts; art and design; media studies; and archaeology" were "not among its strategic priorities", and that their funding would therefore "be subject to a reduction of 50 per cent".[72]

The damage within the sector is not just about levels of engagement; the reduction in creative and artistic experiences at a formative age can lead to more substantive loss—terminal ecosystem decline. "Engaging with the arts improves quality of life and supports our health and well-being", reports "The State of the Arts",[73] citing World Health Organisation findings that show "a robust impact of the arts on both mental and physical health".[74] Children lacking early exposure to structured creativity, self-expression, or imaginative play risk a loss of development in confidence and other skills such as problem-solving, focus, or the ability to work with others[75]; they may also be at risk of underdeveloped neurological, emotional, or behavioural qualities later in life.[76] Ultimately, to deprive a society of its creative skills risks damaging people's lives.

At the time of writing it is not yet clear how the policies of the New Labour government might evolve, though in its manifesto, *Change*, Labour reiterated its support for the arts, with a specific pledge to form a National Music Education Network[77]; and on taking up its position in May of 2024 the new administration almost immediately launched a curriculum and assessment review of the British education system, aiming to bring "improved access to music, art, sport and drama, as well as vocational subjects".[78] Even so, in statements from both before and after the election, messaging has conformed to the neoliberal chase for growth.[79] More detail came in its plan for the Arts, Culture, and Creative Industries, which was titled *Creating Growth*. While acknowledging that "arts and culture have intrinsic value", the main assertion, made in the foreword by Thangam Debonnaire, the then Shadow Secretary State for Culture, Media, and Sport but previously a professional cellist, is that the arts "have enormous economic value to the UK, and huge growth potential".[80] Debonnaire points out that "Their story is already one of growth against the odds, with their value (GVA) increasing by 5% in real terms between 2019 and 2022, compared with 2% growth in the overall economy".[81] The document goes on to assert that growth will be achieved by boosting skills; supporting creative SMEs; introducing at least one creative subject into the EBacc; collaborating on a public–private finance model; fixing post-Brexit challenges experienced by touring musicians; tackling ticket touts at venues; continuing to support the BBC; instituting a review of the Arts Council; and forming that National Music Education Network.[82] Arts leaders generally rallied enthusiastically to Labour promises and to the leadership of Debonnaire.[83] But the growth-led message is still there: "Creative output and imagination are the raw materials for the creative industries, which are worth £125bn to our economy".[84] Alongside the repeated language of growth economics, we're bound to notice in this document the continued casual use of ecological metaphor, as it pledges to get "more investment into the ecosystem",[85] and build "pathways from grassroots art and culture to creative careers".[86] "Arts, culture and the creative industries are a complex ecosystem", it cries; "failing to strengthen this ecosystem is a failure to support the ability of British creative industries to grow".[87] Quite how it intends to strengthen the ecosystem is not fully detailed, though plenty of voices have chipped in to advocate for solutions like increased local authority funding, more art provision in schools, and an end to privatization.[88] To be fair, the new government followed through on some of its pledges, with Chancellor Rachel Reeves in her October 2024 budget allocating £3m to improve school pupils' awareness of creative career paths through the Creative Careers Programme; £25m for local authorities in the North East to invest in a new film studio; and other financial assistance for gaming start-ups, grassroots music venues, and clusters of creative firms situated outside London. There were also initiatives to boost grant funding to museums, and to increase tax relief for creative businesses, though the flip side of these hand-outs was a much-lamented increase in National Insurance contributions from employers. This was a mixed bag, then, and the real policy approaches to arts and culture have yet to be revealed as they emerge over the coming years. As it happens, Thangam Debonnaire lost her Bristol Central seat in the 2024 General Election, ironically to Carla Denyer of the Green Party. New minister Lisa Nandy took over the call for growth, reiterating the "national missions of this Government": "driving economic growth into our towns and cities; drawing on the wealth of talent that exists everywhere; and flying the flag for British culture and values on the world stage".[89] In February 2025, Nandy announced a wave of new public funding for culture and the arts[90]; though quite what the trade-off is expected to be between investment and return is yet to be fully revealed. It's still early days for Sir Keir Starmer's New Labour government.

Towards a Solution

Meanwhile, in an act of literal rewilding, the government granted a licence to the National Trust to release beavers into the wild as part of the Purbeck Beaver Project in Dorset—the first official release of beavers into the wild for several centuries.[91] Just as environmentalists have been calling for actions such as this—a rewilding of nature—to be put into place, we call for a comparable rewilding of the arts to follow suit. There are positive signs that such an approach could be workable and effective: isolated examples of "rewilding" suggest the effects of restoring a vibrant cultural and artistic landscape can be transformative.

We've seen this already in our St Ives case study in chapter five, where after some problematic starts, the town retrenched and began to review its regeneration as an ecosystem. Led by new Tate St Ives initiatives and careful allocation of the Town Deal spend, the various arts projects at St Ives Theatre, the Guildhall, the Palais de Danse, and the Leach Pottery are complemented by other schemes tackling traffic congestion, developing a low carbon transport hub, offering more affordable homes, providing enterprise opportunities for small businesses, and creating a community orchard and garden. Offering public consultation and debate has meant that the local communities feel a part of the redevelopments, while recruiting two youth engagement project officers has meant that young communities engage and have a say.

The case of Feversham Primary Academy in inner-city Bradford is another example of a significant turnaround. Once languishing at the bottom of the league tables and in OFSTED's[92] special measures, the arts in this school have been foregrounded for the last ten years, with the result that creative literacy, social behaviour, and academic results have all improved exponentially. The school's headteacher, Naveed Idrees, "had always championed music and arts in education", reports music coordinator Jimmy Rotheram, "as a way of developing children's confidence, social skills, concentration and their 'spiritual' side".[93] When Idrees came to the failing school in 2013, he introduced compulsory music classes into the curriculum, with additional sessions available outside school time. Soon, further opportunities in drama and art were offered.[94] The results have been spectacular: "Learning behaviour is now excellent, with improved listening and focus", Rotheram notes; "High self-esteem means less bullying, improved attainment and higher attendance. Children and staff love coming to school!". The school's success has been rewarded: Idrees won the national Headteacher of the Year award in 2019. "The arts are the bedrock of academic success", he notes[95]; and Jimmy Rotheram agrees: "Thanks to our creative curriculum and our work as a team, we have gone from special measures to being in the top 10% of schools in the country".[96] In 2022 Feversham Primary Academy achieved an "outstanding" report from OFSTED, bringing this story full circle: "The creative and well-thought-out curriculum, with music and drama at its core makes a significant contribution to developing pupils' very positive attitudes to learning and outcomes", the report states; "Pupils' superb spiritual, moral, social and cultural development is nurtured from the outset".[97]

Just these examples are enough to show us that the arts can indeed have an exponential impact and can lead to transformative change. Even so, these examples stand out as quite unusual, relying on the vision and leadership of cultural environmentalists and their commitment to tending the soil of their patch. But if the ecosystem of the arts as a whole is to recover, this sort of vision will need to be adopted elsewhere. As *The Guardian* has emphasized, "tackling the creativity crisis in state schools is a matter of urgency, both for pupils and for the country as a whole".[98] It's an urgency that sounds remarkably similar to calls by the environmental lobby in the 1990s.

Thankfully, there are other cultural environmentalists. "Transported", an Arts Council-funded organization, is one, working quite literally with seed producers in Lincolnshire. Focused on the boroughs of Boston and South Holland in the south east of the county, it has since 2012 been supported by ACE's "Creative People and Places Fund", a scheme deliberately set up "to ensure people living in the places which have the least engagement with creativity and culture have more opportunities to take part".[99] Transported has taken arts projects into unconventional spaces to engage with participants, communities, and audiences unfamiliar with the arts—spaces such as the warehouses of Spalding-based seed producer Elsoms, and the wagons of local haulier FreshLinc. "It's not about doing art *to* people", remarks Chief Executive of the Arts Council Darren Henley; "it's about doing art *with* people and *for* people".[100] The 2014 project "Elsoms Creates", for example, staged a dance piece by Assault Events, inspired by the stories and lives of employees at Elsoms Seeds.[101] Meanwhile, the "Haulage" project, running throughout 2014–2017 in several iterations, saw FreshLinc's drivers select artist Alisha Miller and local design company Electric Egg to decorate their forty-foot articulated lorries with newly commissioned artworks.[102] "Our projects have ended up being of positive value to the businesses and contributed to their business objectives", notes Transported Director Nick Jones, pointing to benefits to the participating businesses in productivity, staff retention, morale, recruitment, and ambition[103]; "it makes them feel good about themselves, their contribution, and proud of the place that they work for",[104] Jones claims. The businesses agree: "It's been good for staff morale", acknowledges Robin Hancox, managing director of FreshLinc;

"a lot of people have benefited from it directly, some indirectly, and to me that is what the whole project has been about—not just with our staff but with the local community generally".[105]

If the arts really can unlock such significant benefits, as has been shown by just these examples, it is clearly worth us sowing further seeds in a comprehensive rewilding of the landscape. But the piecemeal activities of individual visionaries or localized projects will not be sufficient[106]; recall the "Persian carpet" that Frans Vera noted in his Netherlands approach to rewilding. Finding a way to rewild the arts holistically across the nation will take a concerted effort from stakeholders at every level.

We've spent a considerable part of this chapter addressing the arts crisis in our particular ecosystem, the UK. We've targeted successive government policies based on a growth agenda as damaging to the system, we've recognized a depletion of nutrients in the undersoil leading to inhabitants being starved of skills and engagement, but we've also identified a handful of localized instances of microsystem health. The discussion so far in this chapter is rooted in the UK experience, for obvious reasons—it is the system we know best; yet we'll argue that the solutions can be implemented in other systems and can bring benefit to other communities. It's worth noting that studies elsewhere have also noted similar declines: a report in 2011 by Nick Rabkin and Erik Hedburg speculated that the early-2000s decline in American arts participation, especially among low income, black, and Hispanic communities, was linked to reductions in arts education throughout the 80s, 90s, and 00s[107]; meanwhile a report in the *New York Times* signalled that a drop in government funding was seen as the cause for arts decline in the Netherlands[108]; and Soraya Roberts in Toronto saw Canadian decline as "part of a larger failure in a country that doesn't take art seriously".[109] In short, the story is familiar across multiple neoliberal democracies, suggesting that there is a fundamental incompatibility, at least in the global North, between contemporary economic policies and a healthy arts culture.[110] Fundamentally, the trend seems to result from a lack of support from governments, who prioritize other assumptions about growth.

Plant a New Truffula

There are two recurring themes that have emerged from our research throughout this project. The first is a reiterated call for wholesale change from many of the environmental voices. That's not just piecemeal or superficial change, mind you; it's a call for a fundamentally different way of thinking. We saw it in Bateson's call for "better thinking", Guattari's "transversal thinking", and Morton's call to "wake up and smell the total coffee" in chapter two; we saw it in Cowen's "root and branch change" in chapter four; and we saw it in Thunberg's "whole new way of thinking" in chapter six. These are environmentalists and philosophers talking about our attitudes to the ecosystem of planet Earth, our natural ecosphere; but their calls are just as appropriate for the ecosystem of the arts, our cultural ecosphere. The second theme we have encountered is the push for a whole ecosystem approach—we've seen what happens when we don't adopt such a holistic mindset: we saw the risks of over-farming the land, as with the Thneed in chapter one; of imbalancing a biome, as with St Ives in chapter five; of focusing on a single charismatic species, as with the pandas in chapter six; and of prioritizing growth, which we've explored in this chapter. And we've seen that the way of resolving problems is through holistic solutions, collective efforts, community action, and diverse voices, which we've encountered throughout our book.

The first thing to acknowledge is that the solution is not *just* about "sowing seeds". That's just one more of those casual metaphors we critiqued in our introduction. Rewilding the arts, in other words, is not just about getting every child to "play a musical instrument and to sing". Look at Isabella Tree's project at Knepp, whose objective was a holistic rewilding of the entire ecosystem: before they could introduce beavers to the system they had to implement a wholesale and long-term set of processes on the estate, removing chemicals from the soil, preparing the ground, bringing in fallow deer; and that was just the beginning. In her book Tree documents the entire approach, revealing also how they had to gather expert environmental advice, source species from different regions, gain the support of local communities, leverage funding from organizations, and obtain permissions from authorities. In other words, the complexities of rewilding involve an ecosystem approach: multiple dimensions and considerations that interweave and inform one another in a sophisticated adaptive process.

Of course, it's possible to start from the position of a single species, especially if that species is an ecosystem engineer: it's possible just to introduce beavers and hope that something happens. That would be a near-to passive, single species rewilding strategy, perhaps the equivalent of giving a child a violin. Such approaches have been successful, and after all, beavers are able to colonize areas independently and re-engineer the ecosystem themselves.[111] Even so, single species rewilding schemes are not always successful: the beavers may not take to a particular ecosystem, just as children don't always take to the violin. Even at Knepp, the first attempt at introducing a pair of beavers failed: "within weeks, both had escaped in different directions"[112]; and even in St Ives, the teleporting in of a major international gallery upended the system. Typically, more managed processes are required, recognizing a holistic approach to the ecosystem(s) affected and bearing in mind further considerations. Reintroductions have to think about the trickle-down effects; win over hearts and minds; negotiate licences with authorities; and in most cases consider the boundaries of the area which is to be repopulated. There are guidebooks on how to do this,[113] and following a number of reintroductions in the UK, Europe, and America, there are nowadays plenty of precedents from which to learn.

Principles for Rewilding the Arts

We've seen already how the CEM survey resulted in a set of principles for rewilding. You might assume that our conclusion to this chapter could take those ten principles of rewilding and adapt them to the arts: Rewilding utilizes wildlife...; Rewilding employs landscape-scale planning...; Rewilding focuses on the recovery of ecological processes..., etc. That would be neat and tidy. No doubt we could come up with some adequate corollaries. Let's try it for size—though maybe that is just a first step:

1. Success requires long-term planning. Just as we have begun to think in decades and centuries with environmental targets, we should future-proof our commitment to the arts and our support for the wider societal and individual benefits they can bring. This should reflect a concerted strategy, clear objectives, and appropriate policies, and should project beyond the lifespan of single parliaments with commitments for future governments.
2. Success requires transformation. Just as we have introduced new paradigms and structures in pursuit of environmentalism, we should innovative in radical and creative ways to transform our future landscape of the arts.

3. Success requires reverse engineering. Just as we have recognized the damage of climate change and taken steps to reverse it, we should recognize the decline in our cultural literacy and engagement and take steps to reverse that.
4. Success requires a SMART approach. Just as we have agreed targets for net zero, global warming, and sea-level rises, we should establish strategic, measurable, attainable, realistic, and time-sensitive objectives for rewilding the arts.
5. Success requires adequate resourcing. Just as we have prioritized funding for environmental initiatives such as industrial decarbonization, we should ensure sufficient and equitable funding for rewilding the arts.
6. Success requires us to embed appropriate skills and know-how into the heart of communities, and from an early age. Just as we have included sustainability and climate change in the national curriculum, we must ensure that creativity is foregrounded in the education of our younger generations at all stages.
7. Success requires collaboration. Just as we have recognized the need for collaborative approaches to the environment, we must ensure that government agencies, local authorities, arts organizations, and community groups work meaningfully together to promote a sustainable ecosystem of the arts.
8. Success requires advocacy. Just as our awareness and support for environmental initiatives has benefited from interventions, we need advocates and profile from a diverse array of voices to raise awareness of the arts.
9. Success requires creative thinking. Just as we have benefited from pioneers like Isabella Tree, Akira Miyawaki, and Vanessa Nakate, we must seek out innovative thinkers to help unlock a mini-revolution in the arts.
10. Success requires openness to change. Just as we have been open and receptive to adapting our approaches to the environment, we must fundamentally change the way we think, in ways that we have not yet even conceptualized.

To be clear: these rewilding *principles* are just that; a basis to work from, but one whose implementation will be complex and which will take considerable time. In progressing from its principles, the Rewilding Task

Force continues to establish “a conceptual and methodological framework”. It builds on “other ecosystem management concepts, including Cultural Practices, Nature-based Solutions, Ecosystem-based Adaptation, Ecosystem Governance, Ecosystem Resilience and Protected Areas Management”.[114] In order to build a similar framework for the arts, we perhaps also need to turn to approaches like these.

As it happens, we can take a lead from another rewilding scheme, the Cornwall Beaver Project, which has established a small enclosure with the aim of “reducing the impacts of farming, climate change and biodiversity loss on agricultural landscapes and riverscapes”.[115] In reflecting on various ecosystem management practices to inform their project, landowner Chris Jones and academic Mike Jones conclude that a thorough approach to rewilding requires the “application of systems thinking to research, practice and policy”.[116] They consider panarchy theory, adaptive governance, and several kinds of scaling to expand successful practice. These approaches broadly reflect a need to acknowledge human communities, societies, and technologies as part of integrated ecosystems; a recognition of the complex, multi-layered dynamics that those systems evolve; and the participation of multiple stakeholders in co-produced decision-making. Together, they offer a useful conceptual overview of how rewilding might best work: preparing a base ecosystem, monitoring and managing the implementation of a scheme, acknowledging further impacts and enabling wider systems to adapt, allowing for flexibility and change within the processes, expanding the potential of schemes if they are successful, and influencing policy where appropriate. It’s worth sketching out briefly how this overview of ecosystem management might work.

Panarchy Theory, Adaptive Governance, and Scaling Strategies

Ecologist Crawford Holling introduced the term “panarchy” in the early 2000s, later collaborating with Lance Gunderson to publish *Panarchy: Understanding Transformations in Human and Natural Systems*.[117] The term is used to reflect non-hierarchical integrations among ecosystems, accounting for the ways in which natural systems, human systems, human-nature systems, and social-ecological systems are interlinked. Holling turns to Schumpeter’s understanding of *innovation cycles* to explain how the dynamics work. The panarchy comprises a number of layers, each an *adaptive cycle* shaped like a figure-eight (see Fig. 10.1). The dynamics flow "between long periods of slow accumulation and transformation of resources (from exploitation to conservation, or Γ to K)", and "shorter periods that create opportunities for innovation (from release to reorganization, or Ω to α)".[118]

Each cycle therefore oscillates between the predictability of the first phase (conservation) and the uncertainty of the next (release), following which there is a period of reorganization and exploitation as the new constitution of the ecosystem builds.

Adaptive or innovation cycles in isolation are dynamic enough, but when they come to work together, as they do in Holling's vision of a panarchy, the results can be even more dynamic. Here, he conceptualizes cycles stacked on top of one another such that the emergent prospects of one can influence a larger, slower cycle above it, while the legacy remnants from one above can cascade into the smaller, faster cycle underneath. He calls these dynamics "revolt" and "remember": "the 'revolt' connection [...] can cause a critical change in one cycle to cascade up to a vulnerable

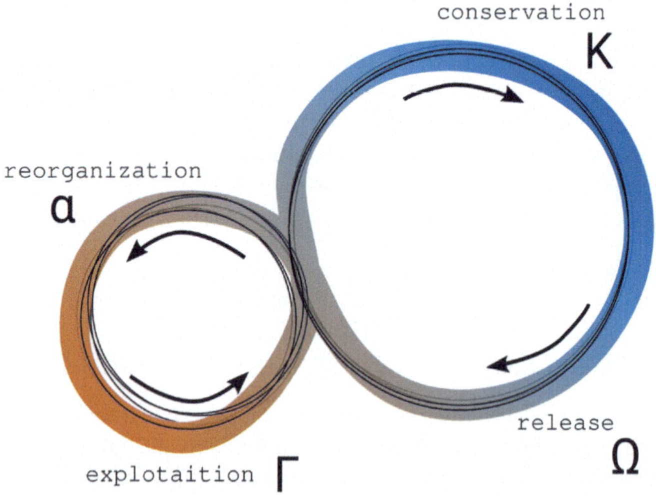

Fig. 10.1 Schematic representation of the adaptive cycle metaphor. Inspired and redrawn from Gunderson and Holling, 2002. Image credit: Hernán De Angelis, CC BY-SA 4.0, https://creativecommons.org/licenses/by-sa/4.0, via Wikimedia Commons

stage in a larger and slower one", he clarifies; meanwhile, "the 'remember' connection [...] facilitates renewal by drawing on the potential that has been accumulated and stored in a larger, slower cycle". The conceptualization of a panarchy visualizes how the adaptive cycle of one level can mutually influence the cycles of other larger/slower/above or smaller/faster/below cycles such that the whole is an ever-changing mutation.

Ecologists already recognize the processes within adaptive systems to be transferable to other domains; after all, this thinking derived from Schumpeter's economic theory in the first place. Holling offers examples of how the structure of a panarchy might be understood in various human system contexts. Whereas in a natural context, for example, the "revolt" might lead from "stands, to forests, to biomes"[119] as the dynamics move "up" the chain, in institutions the different cycles might concern "operational rules, collective choice rules, and constitutional rules"; in developing nations, he suggests, it could be "markets, infrastructure, and governance", and for knowledge systems, "local knowledge, management practice, and world view".[120]

Reaching beyond the idea that these systems are stacked (as-if) vertically, and also beyond the larger/slower/above or smaller/faster/below conceptualization, subsequent ecologists have envisaged "polycentric systems", which are "characterized by multiple and overlapping jurisdictions at different levels".[121] Focusing on these, Fikret Berkes considers the dynamics of environmental governance. He notes that "governance issues in the Anthropocene are often multilevel, cutting across the jurisdictional scale and linking decision-makers horizontally and vertically". He gives as an example the considerations that might guide authorities on biodiversity, proposing three levels that scale up in the manner of Holling: "the local perspective", he suggests, "may be largely about livelihoods, the national perspective about tourism development, and the international perspective about global biodiversity conservation".[122] However, he then notes that the dynamic influences between systems may interact in more complex ways than simply (as it were) up and down a system ladder, or back and forth between smaller and larger systems. Instead, they may be triggered at multiple levels at different times and at varying rates of change.

Thus the conceptual "scaling up" that Holling evokes in panarchy theory ("from centimeters and days to hundreds of kilometers and millennia")[123] can also be imagined in different scaling dimensions, as Michele-Ann Moore and colleagues have proposed. "Scaling out", for example, involves the "replication of successful innovations in[to] different communities (or 'niches'), with the hopes of spreading those same results to more people"[124]; this might involve initiatives to "'seed' like-minded organisations" (note again the ecological language) or to "spread principles further afield".[125] On the other hand, "scaling deep" recognizes that cultural or systemic change requires fundamental development "into the beliefs, ideas, and narratives of dominant social structures".[126] In particular, they propose "big cultural ideas" and "transformative learning" as ways of achieving such depth.[127]

Perhaps it is inevitable in social-ecological systems that such multidimensional adaptive dynamics will lead to complex manifestations of community. "To the extent that a diversity of actors is involved in the process (rather than only experts)", writes Berkes, "there are various kinds of collaborative or cooperative approaches toward adaptive governance", such as "knowledge co-production and use of problem-solving collaborative networks".[128] He notes that "These types of systems are well suited for the governance of dynamic natural resources due to their adaptability, and for their self-organizing and learning capacity",[129] and in a related discussion, James R.A. Butler et al. discuss the use of "adaptive co-management" approaches to rewilding schemes, which combine the "iterative co-learning, knowledge generation and problem-solving of adaptive management with the stakeholder collaboration, power-sharing and alternative institutions of co-management".[130] It is just these sorts of dynamics we have been encountering throughout this book, and in glimpsing some of the theories behind how they function in respect of ecosystem rewilding, we can perhaps begin to conceptualize more thoroughly a set of approaches for rewilding the arts.

Rewilding the Arts

The "State of the Arts" report considers the health of the UK arts scene using five metrics; its conclusion is that the arts is currently failing in respect of all of those. "A society in which the arts are healthy", it states, "is one in which there is robust and sustained funding, from both public and private sources, so arts initiatives can grow and flourish"; "one in which there is an extensive and diverse cultural offering: places to go and things to do"; "one in which people from all communities and backgrounds are engaging with what's on offer and participating in the arts"; "one in which high-quality arts education is valued and available to all, fostering creativity and developing the artists and audiences of the future"; and "one in which people can sustain a decent living from their work creating or supporting art".[131] Crucially, "these 'health metrics' are interconnected", the report asserts; "The arts are deeply interwoven, not only with each other but with the broader fabric of society".[132] In other words, this is exactly the complex ecosystem we have been considering, characterized by integrated layers and the adaptive cycles we have introduced. Rather than a gimmicky single species solution such as getting each child to "play a musical instrument and to sing", what the health of this system needs is an ecosystem approach: "An arts employment can follow from an arts education, which can stem from an arts engagement, which can hinge on local arts provision, which can depend on the availability of arts funding".[133] This is an approach that recognizes the important relationships across trophic levels, between flora and fauna, above- and belowground, through numerous spatial terrains, micro- and macroscopically, short- and long-term, in relation to other ecosystems, and between natural and man-made habitats, and it recognizes that before you can sow the seeds, you have to implement a wholesale set of processes—removing chemicals from the soil, preparing the ground, gathering expert advice, and gaining local support.

All of this accounts, then, for the *ecosystem approach* that our research has identified we need, and all of these ideas—the rewilding principles; the "State of the Arts" recommendations—would be reasonable steps towards rewilding if the underlying soil system was healthy. But it's not. The arts, you see, are integrated not only with each other but also with the broader landscape of society, with other sectors, and throughout our communities. The problem is that *that* landscape is suffering from desertification, deforestation, and depletion. And if we implement changes to the arts while the landscape of the wider ecosystem is still dirt, our seeds will simply die, or be blown away, or be eaten by birds. In order to fully rewild, we need to tend the wider system.

That is where our other finding comes in, because to tend the wider system will require fundamental change, will require us *to radically rethink*, to disembark from our dominant paradigm. To tackle the crisis of the arts we also need to combat the crisis in education, the failure of politics, and the ideology of economic growth; to think beyond arbitrary borders, national responsibilities, and towards an ecosphere approach; to bring in integrated policies; to dismantle our post-industrial and neoliberal constructs; to look beyond disciplinary differences; and to embrace unknown possibilities. We need disturbance in our ecology of mind; we need successional change from our current paradigms; we need a trophic cascade throughout our culture.

Shifting Baseline Syndrome

Finally, then, we should recognize one other ecological principle that underscores all of this: the idea of the *shifting baseline* that causes us over time to forget.

The 2012 Universal Pictures remake of *The Lorax* extends Dr Seuss's narrative and updates the animations using CGI.[134] The young boy of the original is called Ted, we learn, and is a resident of Thneedville, a city built on the ashes of the Once-ler's empire and walled into an artificial biome to prevent the people from either seeing or suffering from the pollution of the wasted land beyond it. The city's landscaping inside the biome is all synthetic, and corrupt capitalist Aloysius O'Hare has built his own empire from air conditioning the environment and selling bottled air to the inhabitants. The citizens themselves seem happy: they have the trappings of a comfortable existence; they have all the technological mod-cons; they have the affluent lifestyles of a population benefiting from industrial growth.

But they have forgotten.

"What are those?", asks Ted, when his girlfriend Audrey shows him a picture of a real tree. He sets out to get her one. "You would rather have some dirty, messy lump of wood that just sticks out of the ground?", scoffs his Mom. She has the latest techno-tree, a state-of-the-art "Oak-amatic"; she has been conditioned to the new normal. Only Ted's elderly grandmother remembers the former times. When Audrey reveals her wish to see a real tree, Ted is encouraged by his grandmother to break out of the city to find the Once-ler and listen to his story.

Thus the new version of *The Lorax* explicitly sets up Dr Seuss's fable to demonstrate that there has been a shift in the population's expectations. This is what ecologists recognize as "the so-called 'shifting baseline syndrome'":

> each human generation accepts as a baseline assumption the ecosystem and landscape configurations that occurred at the beginning of their careers, or in their childhood, and therefore uses these as baselines to evaluate ongoing change.[135]

Recall that Isabella Tree was responding to a shifting baseline herself: "Changes in land use and, in particular, intensive farming, have altered the landscape beyond anything our great-grandparents would recognize",[136] she states; "Each generation [...] redefines what is 'natural'. Each time the baseline drops it is considered the new normal".[137]

Not that this is a problem specifically for the arts; but instead, for the whole of our way of thinking. Over successive generations, the sweep of agrilogistics, of progress, and ultimately of neoliberalism has redefined the "natural" way of the world to create the "new normal". When politicians peddle expectations of growth, and even when they assertively call for *Change*, they are working within parameters that have already denuded the land.

This is not just about pushing back the baseline to a point at which the arts were "better"; we're not suggesting that.[138] It's about recognizing that our "normality" simply doesn't have to be the way it is. This, if you like, is a conceptual baseline. The change in the way we need to think is fundamental; it's not just *Change* within the current way of the world. What did Greta Thunberg say? "If solutions within this system are so impossible to find then maybe we should change the system itself?".[139] And what about Arne Naess, demanding "*the necessity of a substantial change in economic, social, and ideological structures*".[140] For him, as we have heard, it is "a shift from science to wisdom".[141] We could do worse than listening to the wisdom of other cultures as we seek our ecosystem health: "wellbeing within community and [...] community wellbeing are intrinsically linked".[142]

Once we have changed the way we think, then we can rewild our arts together.

Then, it's about decontaminating the soil: we've started to do this environmentally since *Silent Spring*, by recognizing the damage and acting to resolve it; we should also recognize the damage that has been done to our culture and act to resolve that too. After that, it's about replenishing the soil with nutrients: for the spirit of play and curiosity to become normalized modes of being in the world, we need to balance the nutrients of creativity and imagination, and art for art's sake; we need to nurture the arts of imagination and instinct; of aspiration and inspiration; wonder and fascination; experiment, procrastination, and getting it wrong. Next, it's about ensuring that the belowground lifeforms are active: that our seedbanks are full; that our mycorrhizal communicators reach out; that our cooperating tangle can evolve. But it's not only our belowground that needs preparing, for the dynamics of our ecosystem work at all levels. It's also about the ecosystem engineers and keystone species that influence the web; about the seeds that nature engineers to be blown by the wind or carried across the oceans; about the nutrients in the understory released as matter decays; about processes that interweave, revolt, and remember. And of course it's about the human interactions with this web; about the communities that coexist; about the technologies that hybridize and enhance; about governance and management, preserving, conserving, and allowing to adapt.

Only then can we plant seeds and expect our Truffulas to grow.

Notes

1. Isabella Tree (2018), *Wilding: The Return of Nature to a British Farm*, London: Picador, p. 10.
2. Tree, *Wilding*, p. 3.
3. Tree, *Wilding*, p. 43.
4. Knepp (2022–2023), "Restoration of Repton Park", *Knepp website*, https://knepp.co.uk/knepp-estate/history/restoration-of-repton-park/. Accessed 24 August 2023.
5. Jieles Van Baalen (1995), "Natuurontwikkeling en natuurbeleid", *De Levende Natuur*, 96:5, pp. 130–133.
6. "Het verdwijnen van soorten en het uitvallen van natuurlijke processen nog eens versterkt door de invloeden van verontreinigende stoffen, vermesting en verdroging op de verkleinde en gefragmenteerde gebieden". Van Baalen, "Natuurontwikkeling", p. 130.
7. "Ecologisch rampgebied"; "een veel offensievere richting"; "herstellen van het zelfstandig functioneren van ecosystemen"; "geven van ruimte aan natuurlijke processen"; "in dienst van verhoging van biodiversiteit". Van Baalen, "Natuurontwikkeling", p. 130.
8. Frans Vera (2009), "Large-scale nature development—the Oostvaardersplassen", *British Wildlife*, pp. 28–36, p. 28.
9. Tree, *Wilding,* p. 169.
10. Tree, *Wilding*, p. 272.
11. Tree, *Wilding*, p. 273.
12. Knepp (2022–2023), "Beaver Pen Tour", *Knepp website*, https://knepp.co.uk/view/safaris/beaver-pen-tour/#:~:text=Our%20two%20beavers%2C%20Banksy%20and,and%20coppicing%20the%20surrounding%20woodland. Accessed 24 August 2023.

13. Tree, *Wilding*, p. 161.
14. Tree, *Wilding*, pp. 268–290, p. 289.
15. Tree, *Wilding*, pp. 246–267.
16. Tree, *Wilding*, p. 239.
17. Steve Carver et al. (2021), "Guiding principles for rewilding", *Conservation Biology*, 35:6, pp. 1882–1893, pp. 1884, 1888.
18. Dave Foreman (1992), "Around the Campfire", *Wild Earth*, 2:3, IFC. The phrase "world of wounds" is borrowed from Aldo Leopold's posthumous *Sand County Almanac* (1949). Aldo Leopold (1970), *A Sand County Almanac: with essays on conservation from Round River*, New York: Ballantine Books, p. 197.
19. Van Baalen, "Natuurontwikkeling", p. 130.
20. Tree, *Wilding*, p. 10.
21. Lynn White, Jr. (1967), "The Historical Roots of Our Ecologic Crisis", *Science*, 155:3767, pp. 1203–1207, p. 1203.
22. Foreman, "Around".
23. Van Baalen, "Natuurontwikkeling", p. 130.
24. Tree, *Wilding*, p. 3.
25. United Nations (2024), "Preventing, Halting, and Reversing Loss of Nature", *United Nations Decade on Ecosystem Restoration*, https://www.decadeonrestoration.org/#:~:text=The%20UN%20Decade%20on%20Ecosystem%20Restoration%20aims%20to,will%20only%20succeed%20if%20everyone%20plays%20a%20part. Accessed 9 December 2024.
26. Sally Hawkins (2023), "Developing a framework for rewilding based on its social ecological aims", in Sally Hawkins et al. (eds) (2023), *The Routledge Handbook of Rewilding*, Oxford and New York: Taylor & Francis, pp. 42–53.
27. Carver et al., "Guiding principles", p. 1888.
28. United Nations, "Preventing", n.p.
29. Carver et al., "Guiding principles", pp. 1888–1890. The Rewilding Task Force is now branded as the Rewilding Thematic Group. See https://iucn.org/our-union/commissions/group/iucn-cem-rewilding-thematic-group. Accessed 11 December 2024.
30. David Nogués-Bravo et al. (2016), "Rewilding is the new Pandora's box in conservation", *Current Biology*, 26:3, pp. R87–R91, p. R87.

31. Carl-Gustaf Thulin and Helena Röcklinsberg (2020), "Ethical considerations for wildlife reintroductions and rewilding", *Frontiers in Veterinary Science*, 7:163.
32. Jamie Lorimer et al. (2015), "Rewilding: science, practice, and politics", *Annual Review of Environment and Resources*, 40, pp. 39–62, p. 50.
33. Henrike Schulte to Bühne et al. (2022), "The policy consequences of defining rewilding", *Ambio* 51:1, pp. 93–102, p. 94.
34. Lawton, "The Call", p. 38.
35. Neil Puffett (2024), "Arts finances in worst state for five years", *Arts Professional*, https://www.artsprofessional.co.uk/news/exclusive-arts-finances-worst-state-five-years. Accessed 27 December 2024.
36. Samira Ahmed (2024), "Arts Sponsorship in Crisis?", *Front Row*, 23 July, https://www.bbc.co.uk/programmes/m0021b8t. Accessed 27 December 2024.
37. Brian Dillon (2024), "The UK's Arts Education Crisis Is About to Get Even Worse", *Art Review*, 4 October, https://artreview.com/the-uks-arts-education-crisis-is-about-to-get-even-worse/. Accessed 27 December 2024.
38. David Maddox and Jabed Ahmed (2024), "Hundreds of theatres and museums face closure as crisis in sectors laid bare", *The Independent*, 26 December, https://www.independent.co.uk/news/uk/home-news/uk-arts-funding-regional-theatre-museum-culture-government-b2665778.html#:~:text=Hundreds%20of%20theatres%20face%20closure%20and%20more%20than,scale%20of%20the%20risk%20facing%20Britain%E2%80%99s%20cultural%20venues. Accessed 27 December 2024.
39. University of Warwick (2024), "Shocking New Report Reveals UK's Art Sector on the Brink of Collapse", *University of Warwick website*, 25 July, https://warwick.ac.uk/newsandevents/pressreleases/?newsItem=8a17841a90dfc30d0190e8ed578e1740. Accessed 9 December 2024.
40. Heidi Ashton et al. (2024), "The State of the Arts", Campaign for the Arts & Centre for Cultural and Media Policy Studies, University of Warwick, p. 5.
41. Ashton et al., "The State", pp. 14–17.
42. Ashton et al., "The State", pp. 98–100.
43. Ashton et al., "The State", pp. 8–9.

44. Robert Hewison (2014), *Cultural Capital: The Rise and Fall of Creative Britain*, London and New York: Verso, p. 7.
45. Hewison, *Cultural Capital*, p. 91.
46. Hewison, *Cultural Capital*, pp. 7–8.
47. Jen Harvie (2013), *Fair Play: Art, Performance, and Neoliberalism*, Houndmills: Palgrave, p. 14.
48. Harvie, *Fair Play*, p. 62.
49. Harvie, *Fair Play*, p. 63.
50. Harvie, *Fair Play*, pp. 72–73.
51. Anny Shaw and Hannah McGivern (2023), "Special Report: Funding cuts and weak economy send UK's visual arts into crisis", *The Art Newspaper*, 11 July, https://www.theartnewspaper.com/2023/07/11/special-report-funding-cuts-and-weak-economy-send-uks-visual-arts-into-crisis. Accessed 12 September 2023.
52. Lyle Bignon in Josh Sandiford (2024), "Budget cuts mean city faces 'cultural deprivation'", *BBC News*, 21 February, https://www.bbc.co.uk/news/articles/c51ee94ywj2o.Accessed 12 December 2024.
53. See for example the "Sector Vision" for the UK's creative industries, published by the DCMS in June 2023. Through government support (we were told), we were going to grow creative clusters across the UK, adding £50 billion more in Gross Value Added; we were going to build a highly skilled, productive and inclusive workforce for the future, supporting one million more jobs across the UK; and we were going to maximize the positive impact of the creative industries on individuals and communities, the environment and the UK's global standing (p. 6). This—the crudely economic mechanism of the creative industries—presupposes that they operate to market principles in a simplistic linear model whose cost (investment) and benefit (growth) are measured purely in financial terms. Department for Culture, Media and Sport (2023), *Creative Industries Sector Vision: A joint plan to drive growth, build talent and develop skills*, London: Department for Culture, Media and Sport.
54. David Attenborough with Jonnie Hughes (2020), *A Life on Our Planet: My Witness Statement and a Vision for the Future*, London: Witness Books.

55. Crawford S. Holling (2001), "Understanding the complexity of economic, ecological, and social systems", *Ecosystems*, 4, pp. 390–405, p. 404.
56. Patrick J. Deneen (2018), *Why Liberalism Failed*, New Haven: Yale University Press, p. 139.
57. Jason Hickel (2024), *Less is More: How Degrowth will Save the World*, London: Penguin Random House.
58. Jason Hickel in Matthew Taylor (2024), "'These ideas are incredibly popular': what is degrowth and can it save the planet?", *The Guardian*, 27 August, https://www.theguardian.com/environment/article/2024/aug/27/what-is-degrowth-can-it-save-planet. Accessed 12 December 2024.
59. For example, many UK readers will recall Boris Johnson's "oven-ready" Brexit Deal, or indeed, his "cakeist treaty" claim that the UK could have its Brexit cake and eat it. Boris Johnson, (2020), "Prime Minister's statement on EU negotiations: 24 December 2020", *Gov.UK website*, 24 December, https://www.gov.uk/government/speeches/prime-ministers-statement-on-eu-negotiations-24-december-2020. Accessed 10 March 2025; Boris Johnson (2020), "Post-Brexit trade: UK having its cake and eating it, says Boris Johnson", *BBC News*, https://www.bbc.co.uk/news/uk-politics-55486081. Accessed 10 March 2025.
60. Elizabeth Truss (2022), "Full text: Liz Truss's Tory conference speech", *The Spectator*, 5 October, https://www.spectator.co.uk/article/full-text-liz-truss-s-tory-conference-speech/. Accessed 10 March 2025. This was perhaps the most explicit invocation of the economic pie by a major politician, which was by turns excitedly championed and vilified by the press and other politicians. Nevertheless, it was not the first use of the term, and nor has the metaphor been used exclusively in the UK or exclusively by conservative politicians: even Barack Obama used the phrase as he conjured up a homely American scene: "Working Americans are finally getting a little bigger piece of the pie". Barack Obama (2016), "Remarks by the President on the Economy", *The White House*, 1 June, https://obamawhitehouse.archives.gov/the-press-office/2016/06/01/remarks-president-economy. Accessed 10 March 2025.
61. D.F. Karnosky et al. (1996), "Changes in growth, leaf abscission, and biomass associated with seasonal tropospheric ozone exposures of *Populustremuloides* clones and seedlings", *Canadian Journal of Forest Research*, 26:1.

62. Lord Black of Brentwood in House of Lords (2018), "Schools: Music Education", 18 October, *Hansard*, 793: 603.
63. Heidi Ashton and David Ashton (2023), "Creativity and the Curriculum: educational apartheid in 21st Century England, a European outlier", *International Journal of Cultural Policy*, 29:4, pp. 484–499, p. 487.
64. Michael Gove in Darren Henley (2011), *Music Education in England*, London: Department for Culture, Media and Sport, p. 38.
65. Ashton et al., "The State", p. 76.
66. Sally Weale (2021), "'Creativity crisis' looms for English schools due to arts cuts, says Labour", *The Guardian*, 15 July, https://www.theguardian.com/education/2021/jul/15/creativity-crisis-looms-for-english-schools-due-to-arts-cuts-says-labour. Accessed 12 September 2023.
67. Pat Thomson et al. (2019), "Towards an Arts Education for Cultural Citizenship", in Stewart Riddle and Michael W. Apple (2019), *Re-Imagining Education for Democracy*, Abingdon: Routledge, pp. 174–189, p. 174.
68. Between 2009–2010 and 2022–2023, GCSE music entries were down by 41%; drama by 45%; dance by 60%; and performing/expressive arts by 72%; A-level music by 46%; drama by 45%; dance by 57%; and performing/expressive arts by 68%. Ashton et al., "The State", pp. 84–87.
69. Ashton et al., "The State", pp. 78–81.
70. Anon. (2022), "Everything You Need to Know about Music in Schools", *Department for Education*, 25 June, https://educationhub.blog.gov.uk/2022/06/25/everything-you-need-to-know-about-music-in-schools/. Accessed 8 September 2023.
71. Anon. (2022), "The power of music to change lives: a national plan for music education", *Department for Education and Department for Digital, Culture, Media and Sport*, 25 June, https://assets.publishing.service.gov.uk/government/uploads/system/uploads/attachment_data/file/1086619/The_Power_of_Music_to_Change_Lives.pdf. Accessed 23 November 2023.
72. Office for Students (OfS) (2021), *Consultation on Recurrent Funding for 2021–22*, London: OfS, 26 March, https://www.officeforstudents.org.uk/publications/consultation-on-recurrent-funding-for-2021-22/. Accessed 15 September 2023, p. 17.

73. Ashton et al., "The State", p. 57.
74. Daisy Fancourt and Saoirse Finn (2019), *What is the evidence on the role of the arts in improving health and well-being? A scoping review*, Copenhagen: World Health Organization, p. 52.
75. Frank R. Lilly (2014), "Creativity in early childhood", in Thomas P. Gullotta and Martin Bloom (eds) (2014), *Encyclopedia of Primary Prevention and Health Promotion* (second edition), New York: Springer, pp. 598–609.
76. Research in this area is extensive and compelling; an overview is provided in Carol Wagner and Danielle E. Schlough (2017), "Creativity: A childhood essential", *Exploring the benefits of creativity in education, media, and the arts*, IGI Global, pp. 1–25.
77. Anon., *Change*, p. 87.
78. Anon. (2024), "Curriculum and assessment review", *Gov.UK*, https://www.gov.uk/government/groups/curriculum-and-assessment-review. Accessed 10 March 2025.
79. Anon. (2024), *Change: Labour Party Manifesto 2024*, London: David Evans, p. 86.
80. Thangam Debonnaire (2024), "Foreword", in Anon. (2024), *Creating Growth: Labour's Plan for the Arts, Culture and Creative Industries*, London: David Evans, p. 4.
81. Anon., *Creating Growth*, pp. 34–35.
82. Anon., *Creating Growth*, pp. 6–7.
83. Chris Wiegand (2024), "'Absolute outrage': arts in state schools must be a priority for next government, say theatre leaders", *The Guardian*, 14 June, https://www.theguardian.com/stage/article/2024/jun/14/absolute-outrage-arts-in-state-schools-must-be-a-priority-for-next-government-say-theatre-leaders. Accessed 12 December 2024.
84. Debonnaire in "Foreword", p. 4.
85. Debonnaire in "Foreword", p. 4.
86. Anon., *Creating Growth*, p. 7.
87. Anon., *Creating Growth*, p. 18.
88. Charlotte Higgins (2024), "Arts funding has collapsed under 14 years of Tory rule. Here are three ways Labour can fix it", *The Guardian*, 19 March, https://www.theguardian.com/commentisfree/2024/mar/19/arts-funding-austerity-collapse-tories-labour. Accessed 12 December 2024.

89. Department for Culture, Media and Sport and The Rt Hon Lisa Nandy MP (2024) "Funding for UK's growth-driving creative industries confirmed in the Budget" (press release), Department for Culture, Media and Sport, 1 November, https://www.gov.uk/government/news/funding-for-uks-growth-driving-creative-industries-confirmed-in-the-budget. Accessed 11 December 2024.
90. Lanre Bakare (2025), "Lisa Nandy announces £270m fund for England's 'crumbling' cultural infrastructure", *The Guardian*, 20 February, https://www.theguardian.com/politics/2025/feb/20/lisa-nandy-announces-fund-uk-cultural-infrastructure. Accessed 21 February 2025; Lisa Nandy (2025), "Protests and boycotts are killing off the arts in Britain", *The Independent*, 21 February, https://www.independent.co.uk/voices/lisa-nandy-arts-festival-boycott-b2702136.html. Accessed 21 February 2025.
91. Stuart Otway (2025), "Beaver wild release: a milestone for nature recovery in England", *Gov.UK*, 28 February, https://naturalengland.blog.gov.uk/2025/02/28/beaver-wild-release-a-milestone-for-nature-recovery-in-england/. Accessed 10 March 2025.
92. OFSTED is the much-maligned Office for Standards in Education, Children's Services and Skills, the current manifestation of the UK government's inspection scheme for state-run educational establishments. Such establishments are placed in "special measures" if they are deemed to be "inadequate" in their provision. OFSTED itself, however, has been the subject of much controversy on multiple occasions for what are perceived as the failings of its inspection system.
93. Jimmy Rotheram, quoted in Anon. (2018), "Singing up a storm—Feversham Primary makes the case for music education", *Music Mark*, https://www.musicmark.org.uk/news/singing-up-a-storm-feversham-primary-makes-the-case-for-music-education/. Accessed 23 November 2023.
94. Liam Maloy (2022), "How the Arts have transformed Feversham Primary", *Researchign the Arts in Primary Schools,* 18 January, https://artsprimary.com/2022/01/18/how-the-arts-have-transformed-feversham-primary/. Accessed 23 November 2023.

95. Nina Swift (2019), "Meet Britain's best headteacher—and he's right here in Yorkshire", *The Yorkshire Post*, 15 July, https://www.yorkshirepost.co.uk/education/meet-britains-best-headteacher-and-hes-right-here-in-yorkshire-1752963. Accessed 23 November 2023.
96. Rotheram in Anon., "Singing". See also Megan Jones and Amelia Parker (2023), "Rethinking Music: Beyond Face Value", *BBC Sounds*, 14 March, https://www.bbc.co.uk/sounds/play/m001k0l6. Accessed 23 November 2023.
97. Dimitris Spiliotis, Gill McCleave, and Aleksandra Hartshorne (2022), "School Report: Feversham Primary Academy", *OFSTED*, https://files.ofsted.gov.uk/v1/file/50076452. Accessed 23 November 2023.
98. The article notes with irony that a broad range of arts opportunities are available to those who have the privilege of a private sector education, like many of the politicians who have presided over the collapse. Anon. (2023), "The Guardian view on arts education: a creativity crisis", *The Guardian*, 7 February, https://www.theguardian.com/commentisfree/2023/feb/07/the-guardian-view-on-arts-education-a-creativity-crisis. Accessed 23 November 2023.
99. Anon. (2023), "Creative People and Places Projects", *Arts Council England*, https://www.artscouncil.org.uk/creative-people-and-places-0/creative-people-and-places-projects. Accessed 8 December 2023.
100. Transported Art (2017), "Transported: Arts & Business", *Transported*, https://transportedart.com/gallery/videos/. Accessed 21 December 2023.
101. Transported Art (2014), "Transported Sneak Peek: ElsomsCreates", *YouTube*, https://www.youtube.com/watch?v=qRCej1muzIs. Accessed 21 December 2023.
102. Transported Art (2014), "Transported: Haulage 2015", *YouTube*, https://www.youtube.com/watch?v=WcVj6IgW9CE. Accessed 21 December 2023; Transported Art (2017), "Transported: Haulage 2017", *YouTube*, https://www.youtube.com/watch?v=vqxNqefiEsE. Accessed 21 December 2023.

103. Transported Art "Transported: Arts & Business"; Transported Art (2023), "Transported Live", *Transported*, https://www.transportedart.com/projects/past-projects/transported-live/. Accessed 8 December 2023.
104. Transported Art, "Transported: Arts & Business".
105. Transported Art, "Transported: Arts & Business".
106. The Creative People and Places Fund continues to operate, with 142 places currently eligible for funding. Anon. (2023), "Creative People and Places National Portfolio Programme 2025–28", *Arts Council* website, https://www.artscouncil.org.uk/creative-people-and-places-0/creative-people-and-places-national-portfolio-programme-2025-28. Accessed 21 December 2023.
107. Nick Rabkin and Eric Christopher Hedberg (2011), "Arts education in America: What the declines mean for arts participation", Based on the 2008 Survey of Public Participation in the Arts. Research Report #52. Washington, DC: National Endowment for the Arts. See also Paul DiMaggio and Toqir Mukhtar (2007), "Arts Participation as Cultural Capital in the United States, 1982–2002: Signs of Decline?", in Steven J. Tepper and Bill Ivey (2007), *Engaging Art: The Next Great Transformation of America's Cultural Life*, New York: Routledge, pp. 273–305.
108. Nina Siegal (2013), "Dutch Arts Scene is Under Siege", *New York Times*, 29 January, https://www.nytimes.com/2013/01/30/arts/30iht-dutch30.html. Accessed 30 November 2024.
109. Soraya Roberts (2024), "Toronto's Arts Institutions Are Crumbling and it's Always the Same Story", *The Local*, 25 April, https://thelocal.to/hot-docs-canadian-art-institutions/. Accessed 30 November 2024;
110. See also Thomas Floyd (2023), "Fewer People are Going to Movies, Theater, and Museums, NEA Study Shows", *Washington Post*, 18 October; Angelique Chrisafis (2011), "France and the Arts: A New Revolution", *The Guardian*, 24 March, https://www.theguardian.com/world/2011/mar/24/france-arts-revolution. Accessed 30 November 2024; Alison Croggon (2019), "The Desertification of Australian Culture", *The Monthly*, October, https://www.themonthly.com.au/issue/2019/october/1569374077/alison-croggon/desertification-australian-culture. Accessed 30 November 2024; Kate Connolly (2024), "Plan to cut Berlin arts budget will 'destroy' city's culture, directors

warn", *The Guardian*, 27 November, https://www.theguardian.com/world/2024/nov/27/plan-to-cut-berlin-arts-budget-will-destroy-citys-culture-directors-warn. Accessed 30 November 2024.

111. See Virginia Thomas (2022), "Actors and actions in the discourse, policy and practice of English rewilding", *Environmental Science & Policy*, 132, pp. 83–90.
112. Knepp (2022–2024), "Beavers", *Knepp website*, https://knepp.co.uk/rewilding/reintroductions/beavers/. Accessed 4 December 2024.
113. For example, Michael M. Pollock et al. (2023), *The Beaver Restoration Guidebook: Working with Beaver to Restore Streams, Wetlands, and Floodplains*, Portland: United States Fish and Wildlife Service.
114. International Union for Conservation of Nature and Natural Resources (2024), "IUCN CEM Rewilding Thematic Group", *IUCN website*, https://iucn.org/our-union/commissions/group/iucn-cem-rewilding-thematic-group. Accessed 11 December 2024.
115. Mike Jones and Chris Jones (2023), "The Cornwall Beaver Project: navigating the social-ecological complexity of rewilding as a nature-based solution", *Frontiers in Conservation Science*, 4:1252275, p. 3.
116. Jones and Jones, "The Cornwall Beaver Project", p. 1.
117. Crawford S. Holling and Lance H. Gunderson (eds) (2002), *Panarchy: Understanding Transformations in Human and Natural Systems*, Washington DC: Island Press.
118. Holling, "Understanding", p. 397.
119. Holling, "From complex regions", p. 3.
120. Holling, "Understanding", p. 398.
121. Fikret Berkes (2017), "Environmental governance for the Anthropocene? Social ecological systems, resilience, and collaborative learning", *Sustainability* 9:7, 1232, p. 6.
122. Berkes, "Environmental governance", p. 2.
123. Holling, "From complex regions", pp. 3–4.
124. Michele-Lee Moore, Darcy Riddell, and Dana Vocisano (2015), "Scaling Out, Scaling Up, Scaling Deep: Strategies of Non-profits in Advancing Systemic Social Innovation", *The Journal of Corporate Citizenship*, 58, pp. 67–84, p. 74.

125. Moore, Riddell, and Vocisano, "Scaling Out", p. 78.
126. Moore, Riddell, and Vocisano, "Scaling Out", p. 79.
127. Moore, Riddell, and Vocisano, "Scaling Out", pp. 79–80.
128. Berkes, "Environmental governance", p. 7.
129. Berkes, "Environmental governance", p. 6.
130. James R.A. Butler et al. (2021), "Decision-making for rewilding: an adaptive governance framework for social-ecological complexity", *Frontiers in Conservation Science*, 2, 681545, p. 2.
131. Ashton et al., "The State", p. 5.
132. Ashton et al., "The State", p. 11.
133. Ashton et al., "The State", p. 5.
134. Chris Renaud (dir.) (2012), *The Lorax*, Universal Pictures.
135. Luis Balaguer et al. (2014), "The historical reference in restoration ecology: Re-defining a cornerstone concept", *Biological Conservation*, 176, pp. 12–20, p. 14.
136. Tree, *Wilding*, p. 3.
137. Tree, *Wilding*, p. 148.
138. Still, observers of the House of Lords debates may have heard the compelling 2024 speech by Lord Melvyn Bragg, vividly demonstrating a baseline shift when he recalled that "In the last century, there were the county music services, free instrumental lessons, Saturday morning music schools, orchestras and choirs. After 2020, these services were transferred into 'hubs', a private enterprise model. The local authorities lost responsibility for them and the slide began. In 2022, the number of hubs was reduced nationally from 116 to 43, in direct contradiction to consultations saying that this would be the worst possible option for state schools. The 43 hubs had to do the same work as the 116, and on the same money". The remainder of the debate is worth consulting, and corroborates much of what we have argued. Lord Bragg (2024), "Arts—Motion to Take Note", *House of Lords Debates*, 1 February, c1280.
139. Greta Thunberg (2019), *No One is Too Small to Make a Difference*, London: Allen Lane, p. 28.
140. Arne Naess (1995), "Deepness of Questions and the Deep Ecology Movement", in George Sessions (ed.) (1995), *Deep Ecology for the Twenty-First Century*, Boston: Shambhala, pp. 204–212, p. 211.

141. Stephan Bodian (1982), "Simple in Means, Rich in Ends: An Interview with Arne Naess", in Sessions, *Deep Ecology*, pp. 26–36, p. 27.
142. Judy Taylor et al. (2012), "Towards a conceptual understanding of Aboriginal and Torres Strait Islander community and community functioning", *Community Development Journal*, 47:1, pp. 94–110, p. 99.

Bibliography

Ahmed, Samira (2024), "Arts Sponsorship in Crisis?", *Front Row*, 23 July, https://www.bbc.co.uk/programmes/m0021b8t. Accessed 27 December 2024.

Anon. (2018), "Singing up a storm—Feversham Primary makes the case for music education", *Music Mark*, https://www.musicmark.org.uk/news/singing-up-a-storm-feversham-primary-makes-the-case-for-music-education/. Accessed 23 November 2023.

Anon. (2020), *Innovative Approaches to Accelerating and Scaling up Climate Technology Implementation for Mitigation and Adaptation*, Bonn: Technology Executive Committee, United Nations Framework Convention on Climate Change, https://unfccc.int/ttclear/misc_/StaticFiles/gnwoerk_static/innovative_approaches/07a2f73969c945928ffa1ec74285f356/235654758e1343f788b1f1132bb109b8.pdf. Accessed 8 December 2023.

Anon. (2022), "Everything You Need to Know about Music in Schools", *Department for Education*, 25 June, https://educationhub.blog.gov.uk/2022/06/25/everything-you-need-to-know-about-music-in-schools/. Accessed 8 September 2023.

Anon. (2022), "The power of music to change lives: a national plan for music education", *Department for Education and Department for Digital, Culture, Media and Sport*, 25 June, https://assets.publishing.service.gov.uk/government/uploads/system/uploads/attachment_data/file/1086619/The_Power_of_Music_to_Change_Lives.pdf. Accessed 23 November 2023.

Anon. (2023), "Creative People and Places National Portfolio Programme 2025–28", *Arts Council* website, https://www.artscouncil.org.uk/creative-people-and-places-0/creative-people-and-places-national-portfolio-programme-2025-28. Accessed 21 December 2023.

Anon. (2023), "Creative People and Places Projects", *Arts Council England*, https://www.artscouncil.org.uk/creative-people-and-places-0/creative-people-and-places-projects. Accessed 8 December 2023.

Anon. (2023), "The Guardian view on arts education: a creativity crisis", *The Guardian*, 7 February, https://www.theguardian.com/commentisfree/2023/feb/07/the-guardian-view-on-arts-education-a-creativity-crisis. Accessed 23 November 2023.

Anon. (2024), *Change: Labour Party Manifesto 2024*, London: David Evans, p. 87.

Anon. (2024), *Creating Growth: Labour's Plan for the Arts, Culture and Creative Industries*, London: David Evans.

Anon. (2024), "Curriculum and assessment review", *Gov.UK*, https://www.gov.uk/government/groups/curriculum-and-assessment-review. Accessed 10 March 2025.

Ashton, Heidi, and David Ashton (2023), "Creativity and the Curriculum: educational apartheid in 21st Century England, a European outlier", *International Journal of Cultural Policy*, 29:4, pp. 484–499.

Ashton, Heidi, David Brownlee, Jack Gamble, and Melanie Stavrou (2024), "The State of the Arts", Campaign for the Arts & Centre for Cultural and Media Policy Studies, University of Warwick.

Attenborough, David, with Jonnie Hughes (2020), *A Life on Our Planet: My Witness Statement and a Vision for the Future*, London: Witness Books.

Bakare, Lanre (2025), "Lisa Nandy announces £270m fund for England's 'crumbling' cultural infrastructure", *The Guardian*, 20 February, https://www.theguardian.com/politics/2025/feb/20/lisa-nandy-announces-fund-uk-cultural-infrastructure. Accessed 21 February 2025.

Balaguer, Luis, Adrián Escudero, José F. Martín-Duque, Ignacio Mola, and James Aronson (2014), "The historical reference in restoration ecology: Re-defining a cornerstone concept", *Biological Conservation*, 176, pp. 12–20.

Bateson, Gregory (1991), *A Sacred Unity: Further Steps to an Ecology of Mind*, New York: Cornelia & Michael Bessie Books.

Berkes, Fikret (2017), "Environmental governance for the Anthropocene? Social ecological systems, resilience, and collaborative learning", *Sustainability* 9:7, 1232.

Butler, James R.A., Mariella Marzano, Nathalie Pettorelli, Sarah M. Durant, Johan T. du Toit, and Juliette C. Young (2021), "Decision-making for rewilding: an adaptive governance framework for social-ecological complexity", *Frontiers in Conservation Science*, 2, 681545.

Carver, Steve, Ian Convery, Sally Hawkins, Rene Beyers, Adam Eagle, Zoltan Kun, Erwin Van Maanen, Yue Cao, Mark Fisher, Stephen R. Edwards, Cara Nelson, George D. Gann, Steve Shurter, Karina Aguilar, Angela Andrade, William J. Ripple, John Davis, Anthony Sinclair, Marc Bekoff, Reed Noss, Dave Foreman, Hanna Pettersson, Meredith Root-Bernstein, Jens-Christian Svenning, Peter Taylor, Sophie Wynne-Jones, Alan Watson Featherstone, Camilla Fløjgaard, Mark Stanley-Price, Laetitia M. Navarro, Toby Aykroyd,

Alison Parfitt, Michael Soulé (2021), "Guiding principles for rewilding", *Conservation Biology*, 35:6, pp. 1882–1893.

Chrisafis, Angelique (2011), "France and the Arts: A New Revolution", *The Guardian*, 24 March, https://www.theguardian.com/world/2011/mar/24/france-arts-revolution. Accessed 30 November 2024.

Climate Assembly UK (2020), "UK path to net zero must be underpinned by education, choice, fairness and political consensus, urges Climate Assembly", *The Path to Net Zero*, 10 September, https://www.climateassembly.uk/news/uk-path-net-zero-must-be-underpinned-education-choice-fairness-and-political-consensus-urges-climate-assembly/. Accessed 8 December 2023.

Climate Change Committee (2008), "A Legal Duty to Act", https://www.theccc.org.uk/what-is-climate-change/a-legal-duty-to-act/#:~:text=The%20Climate%20Change%20Act%202008,to%20deliver%20on%20these%20requirements. Accessed 1 December 2023.

Connolly, Kate (2024), "Plan to cut Berlin arts budget will 'destroy' city's culture, directors warn", *The Guardian*, 27 November, https://www.theguardian.com/world/2024/nov/27/plan-to-cut-berlin-arts-budget-will-destroy-citys-culture-directors-warn. Accessed 30 November 2024.

Croggon, Alison (2019), "The Desertification of Australian Culture", *The Monthly*, October, https://www.themonthly.com.au/issue/2019/october/1569374077/alison-croggon/desertification-australian-culture. Accessed 30 November 2024.

Deneen, Patrick J. (2018), *Why Liberalism Failed*, New Haven: Yale University Press.

Department for Culture, Media and Sport (2023), *Creative Industries Sector Vision: A joint plan to drive growth, build talent and develop skills*, London: Department for Culture, Media and Sport.

Department for Culture, Media and Sport and The Rt Hon Lisa Nandy MP (2024) "Funding for UK's growth-driving creative industries confirmed in the Budget" (press release), Department for Culture, Media and Sport, 1 November, https://www.gov.uk/government/news/funding-for-uks-growth-driving-creative-industries-confirmed-in-the-budget. Accessed 11 December 2024.

Dillon, Brian (2024), "The UK's Arts Education Crisis Is About to Get Even Worse", *Art Review*, 4 October, https://artreview.com/the-uks-arts-education-crisis-is-about-to-get-even-worse/. Accessed 27 December 2024.

Environment Agency, Department for Environment, Food & Rural Affairs, and Rebecca Pow MP (2023), "Collaboration urged as climate change puts water security at risk" (press release), https://www.gov.uk/government/news/collaboration-urged-as-climate-change-puts-water-security-at-risk. Accessed 8 December 2023.

Fancourt, Daisy, and Saoirse Finn (2019), *What is the evidence on the role of the arts in improving health and well-being? A scoping review*, Copenhagen: World Health Organization.

Floyd, Thomas (2023), "Fewer People are Going to Movies, Theater, and Museums, NEA Study Shows", *Washington Post*, 18 October.

Foreman, Dave (1992), "Around the Campfire", *Wild Earth*, 2:3, IFC.

Gullotta, Thomas P., and Martin Bloom (eds) (2014), *Encyclopedia of Primary Prevention and Health Promotion* (second edition), New York: Springer.

H.M. Government (2009), *The UK Renewable Energy* Strategy, London: The Stationery Office, https://assets.publishing.service.gov.uk/government/uploads/system/uploads/attachment_data/file/228866/7686.pdf. Accessed 1 December 2023.

H.M. Government (2021), *Net Zero Strategy: Build Back Greener*, London: The Stationery Office, https://assets.publishing.service.gov.uk/media/6194dfa4d3bf7f0555071b1b/net-zero-strategy-beis.pdf. Accessed 1 December 2023.

Harvie, Jen (2013), *Fair Play: Art, Performance, and Neoliberalism*, Houndmills: Palgrave.

Hawkins, Sally, Ian Convery, Steve Carver and Rene Beyers (eds) (2023), *The Routledge Handbook of Rewilding*, Oxford and New York: Taylor & Francis, pp. 42–53.

Henley, Darren (2011), *Music Education in England*, London: Department for Culture, Media and Sport.

Hewison, Robert (2014), *Cultural Capital: The Rise and Fall of Creative Britain*, London and New York: Verso.

Hickel, Jason (2024), *Less is More: How Degrowth will Save the World*, London: Penguin Random House.

Higgins, Charlotte (2024), "Arts funding has collapsed under 14 years of Tory rule. Here are three ways Labour can fix it", *The Guardian*, 19 March, https://www.theguardian.com/commentisfree/2024/mar/19/arts-funding-austerity-collapse-tories-labour. Accessed 12 December 2024.

Holling, Crawford S. (2001), "Understanding the complexity of economic, ecological, and social systems", *Ecosystems*, 4, pp. 390–405.

Holling, Crawford S. (2004), "From complex regions to complex worlds", *Ecology and Society*, 9:1, 11.

Holling, Crawford S., and Lance H. Gunderson (eds) (2002), *Panarchy: Understanding Transformations in Human and Natural Systems*, Washington DC: Island Press.

House of Lords (2018), "Schools: Music Education", 18 October, *Hansard*, 793: 603.

International Union for Conservation of Nature and Natural Resources (2024), "IUCN CEM Rewilding Thematic Group", *IUCN website*, https://iucn.org/our-union/commissions/group/iucn-cem-rewilding-thematic-group. Accessed 11 December 2024.

Jones, Megan, and Amelia Parker (2023), "Rethinking Music: Beyond Face Value", *BBC Sounds*, 14 March, https://www.bbc.co.uk/sounds/play/m001k0l6. Accessed 23 November 2023.

Jones, Mike, and Chris Jones (2023), "The Cornwall Beaver Project: navigating the social-ecological complexity of rewilding as a nature-based solution", *Frontiers in Conservation Science*, 4:1252275.

Knepp (2022–2023), "Beaver Pen Tour", *Knepp website*, https://knepp.co.uk/view/safaris/beaver-pen-tour/#:~:text=Our%20two%20beavers%2C%20Banksy%20and,and%20coppicing%20the%20surrounding%20woodland. Accessed 24 August 2023.

Knepp (2022–2023), "Restoration of Repton Park", *Knepp website*, https://knepp.co.uk/knepp-estate/history/restoration-of-repton-park/. Accessed 24 August 2023.

Knepp (2022–2024), "Beavers", *Knepp website*, https://knepp.co.uk/rewilding/reintroductions/beavers/. Accessed 4 December 2024.

Kulakiewicz, Aaron, Robert Long, and Nerys Roberts (2021), "Inclusion of sustainability and climate change in the national curriculum" (research briefing, House of Commons Library), 25 October, https://commonslibrary.parliament.uk/research-briefings/cdp-2021-0166/, Accessed 8 December 2023.

Lawton, Graham (2018), "The Call of Rewilding", *New Scientist*, 240:3199, pp. 34–38.

Leopold, Aldo (1970), *A Sand County Almanac : with essays on conservation from Round River*, New York: Ballantine Books.

Lord Bragg (2024), "Arts—Motion to Take Note", *House of Lords Debates*, 1 February, c1280.

Lorimer, Jamie, Chris Sandom, Paul Jepson, Chris Doughty, Maan Barua, and Keith J. Kirby (2015), "Rewilding: science, practice, and politics", *Annual Review of Environment and Resources*, 40, pp. 39–62.

Maddox, David and Jabed Ahmed (2024), "Hundreds of theatres and museums face closure as crisis in sectors laid bare", *The Independent*, 26 December, https://www.independent.co.uk/news/uk/home-news/uk-arts-funding-regional-theatre-museum-culture-government-b2665778.html#:~:text=Hundreds%20of%20theatres%20face%20closure%20and%20more%20than,scale%20of%20the%20risk%20facing%20Britain%E2%80%99s%20cultural%20venues. Accessed 27 December 2024.

Maloy, Liam (2022), "How the Arts have transformed Feversham Primary", *Researchign the Arts in Primary Schools,* 18 January, https://artsprimary.com/2022/01/18/how-the-arts-have-transformed-feversham-primary/. Accessed 23 November 2023.

Moore, Michele-Lee, Darcy Riddell, and Dana Vocisano (2015), "Scaling Out, Scaling Up, Scaling Deep: Strategies of Non-profits in Advancing Systemic Social Innovation", *The Journal of Corporate Citizenship*, 58, pp. 67–84.

Nandy, Lisa (2025), "Protests and boycotts are killing off the arts in Britain", *The Independent*, 21 February, https://www.independent.co.uk/voices/lisa-nandy-arts-festival-boycott-b2702136.html. Accessed 21 February 2025.

Nogués-Bravo, David, Daniel Simberloff, Carsten Rahbek, and Nathan James Sanders (2016), "Rewilding is the new Pandora's box in conservation", *Current Biology*, 26:3, pp.R87-R91.

Office for Students (OfS) (2021), *Consultation on Recurrent Funding for 2021–22*, London: OfS, 26 March, https:// www.officeforstudents.org.uk/publications/consultation-on-recurrent-funding-for-2021-22/. Accessed 15 September 2023.

Otway, Stuart (2025), "Beaver wild release: a milestone for nature recovery in England", *Gov.UK*, 28 February, https://naturalengland.blog.gov.uk/2025/02/28/beaver-wild-release-a-milestone-for-nature-recovery-in-eng land/. Accessed 10 March 2025.

Pollock, Michael M., Gregory M. Lewallen, Kent Woodruff, Chris E. Jordan, and Janine M. Castro (2023), *The Beaver Restoration Guidebook: Working with Beaver to Restore Streams, Wetlands, and Floodplains*, Portland: United States Fish and Wildlife Service.

Puffett, Neil (2024), "Arts finances in worst state for five years", *Arts Professional*, https://www.artsprofessional.co.uk/news/exclusive-arts-fin ances-worst-state-five-years. Accessed 27 December 2024.

Rabkin, Nick, and Eric Christopher Hedberg (2011), "Arts education in America: What the declines mean for arts participation", Based on the 2008 Survey of Public Participation in the Arts. Research Report #52. Washington, DC: National Endowment for the Arts.

Renaud, Chris (dir.) (2012), *The Lorax*, Universal Pictures.

Riddle, Stewart, and Michael W. Apple (2019), *Re-Imagining Education for Democracy*, Abingdon: Routledge, pp. 174–189.

Roberts, Soraya (2024), "Toronto's Arts Institutions Are Crumbling and it's Always the Same Story", *The Local*, 25 April, https://thelocal.to/hot-docs-canadian-art-institutions/. Accessed 30 November 2024;

Sandiford, Josh (2024), "Budget cuts mean city faces 'cultural deprivation'", *BBC News*, 21 February, https://www.bbc.co.uk/news/articles/c51ee94yw j2o.Accessed 12 December 2024.

Schulte to Bühne, Henrike, Nathalie Pettorelli, and Michael Hoffmann (2022), "The policy consequences of defining rewilding", *Ambio* 51:1, pp. 93–102.

Sessions, George (ed.) (1995), *Deep Ecology for the Twenty-First Century*, Boston: Shambhala.

Shaw, Anny and Hannah McGivern (2023), "Special Report: Funding cuts and weak economy send UK's visual arts into crisis", *The Art Newspaper*, 11 July, https://www.theartnewspaper.com/2023/07/11/special-report-funding-cuts-and-weak-economy-send-uks-visual-arts-into-crisis. Accessed 12 September 2023.

Siegal, Nina (2013), "Dutch Arts Scene is Under Siege", *New York Times*, 29 January, https://www.nytimes.com/2013/01/30/arts/30iht-dutch30.html. Accessed 30 November 2024.

Spiliotis, Dimitris, Gill McCleave, and Aleksandra Hartshorne (2022), "School Report: Feversham Primary Academy", *OFSTED*, https://files.ofsted.gov.uk/v1/file/50076452. Accessed 23 November 2023.

Swift, Nina (2019), "Meet Britain's best headteacher—and he's right here in Yorkshire", *The Yorkshire Post*, 15 July, https://www.yorkshirepost.co.uk/education/meet-britains-best-headteacher-and-hes-right-here-in-yorkshire-1752963. Accessed 23 November 2023.

Taylor, Matthew (2024), "'These ideas are incredibly popular': what is degrowth and can it save the planet?", *The Guardian*, 27 August, https://www.theguardian.com/environment/article/2024/aug/27/what-is-degrowth-can-it-save-planet. Accessed 12 December 2024.

Tepper, Steven J., and Bill Ivey (2007), *Engaging Art: The Next Great Transformation of America's Cultural Life*, New York: Routledge.

Thomas, Virginia (2022), "Actors and actions in the discourse, policy and practice of English rewilding", *Environmental Science & Policy*, 132, pp. 83–90.

Thulin, Carl-Gustaf and Helena Röcklinsberg (2020), "Ethical considerations for wildlife reintroductions and rewilding", *Frontiers in Veterinary Science*, 7:163.

Thunberg, Greta (2019), *No One is Too Small to Make a Difference*, London: Allen Lane .

Transported Art (2014), "Transported Sneak Peek: ElsomsCreates", *YouTube*, https://www.youtube.com/watch?v=qRCej1muzIs. Accessed 21 December 2023.

Transported Art (2014), "Transported: Haulage 2015", *YouTube*, https://www.youtube.com/watch?v=WcVj6IgW9CE. Accessed 21 December 2023.

Transported Art (2017), "Transported: Haulage 2017", *YouTube*, https://www.youtube.com/watch?v=vqxNqefiEsE. Accessed 21 December 2023.

Transported Art (2017), "Transported: Arts & Business", *Transported*, https://transportedart.com/gallery/videos/. Accessed 21 December 2023.

Transported Art (2023), "Transported Live", *Transported*, https://www.transportedart.com/projects/past-projects/transported-live/. Accessed 8 December 2023.

Tree, Isabella (2018), *Wilding: The Return of Nature to a British Farm*, London: Picador.

United Nations (2024), "Preventing, Halting, and Reversing Loss of Nature", *United Nations Decade on Ecosystem Restoration*, https://www.decadeonrestoration.org/#:~:text=The%20UN%20Decade%20on%20Ecosystem%20Restoration%20aims%20to,will%20only%20succeed%20if%20everyone%20plays%20a%20part. Accessed 9 December 2024.

University of Warwick (2024), "Shocking New Report Reveals UK's Art Sector on the Brink of Collapse", *University of Warwick website*, 25 July, https://warwick.ac.uk/newsandevents/pressreleases/?newsItem=8a17841a90dfc30d0190e8ed578e1740. Accessed 9 December 2024.

Van Baalen, Jieles (1995), "Natuurontwikkeling en natuurbeleid", *De Levende Natuur*, 96:5, pp. 130–133.

Vera, Frans (2009), "Large-scale nature development—the Oostvaardersplassen", *British Wildlife*, pp. 28–36.

Wagner, Carol, and Danielle E. Schlough (2017), "Creativity: A childhood essential", *Exploring the benefits of creativity in education, media, and the arts*, IGI Global, pp. 1–25.

Ward, Kim J., and Jonathan Prior (2020), "The reintroduction of beavers to Scotland: rewilding, biopolitics, and the affordance of non-human autonomy", *Conservation and Society*, 18:2, pp. 103–113.

Weale, Sally (2021), "'Creativity crisis' looms for English schools due to arts cuts, says Labour", *The Guardian,* 15 July, https://www.theguardian.com/education/2021/jul/15/creativity-crisis-looms-for-english-schools-due-to-arts-cuts-says-labour. Accessed 12 September 2023.

White, Jr, Lynn (1967), "The Historical Roots of Our Ecologic Crisis", *Science*, 155:3767, pp. 1203–1207.

Wiegand, Chris (2024), "'Absolute outrage': arts in state schools must be a priority for next government, say theatre leaders", *The Guardian*, 14 June, https://www.theguardian.com/stage/article/2024/jun/14/absolute-outrage-arts-in-state-schools-must-be-a-priority-for-next-government-say-theatre-leaders. Accessed 12 December 2024.

Index

K. Savage and D. Symonds, *Ecologies of Community in Performance*, https://doi.org/10.1007/978-3-031-94343-0

S

T

The manufacturer's authorised representative in the EU is Springer Nature Customer Service Centre GmbH, Europaplatz 3, 69115 Heidelberg, Germany. If you have any concerns regarding our products, please contact ProductSafety@springernature.com

Printed and bound by CPI Group (UK) Ltd, Croydon, CR0 4YY

01/10/2026

02225954-0010